Clinical Significance of C-reactive Protein

Waliza Ansar • Shyamasree Ghosh
Editors

Clinical Significance of C-reactive Protein

Springer

Editors
Waliza Ansar
Department of Zoology
Behala College
Kolkata, West Bengal, India

Shyamasree Ghosh
School of Biological Sciences
NISER Bhubaneswar, India
Orissa, India

ISBN 978-981-15-6786-5 ISBN 978-981-15-6787-2 (eBook)
https://doi.org/10.1007/978-981-15-6787-2

This Springer imprint is published by the registered company Springer Nature Singapore Pte Ltd.
The registered company address is: 152 Beach Road, #21-01/04 Gateway East, Singapore 189721, Singapore

Dedicated to our
Beloved family,
our adorable children and to
Physicians for delivering high-quality care to
all patients……

Preface

The success and credibility of good science always lies in the willingness of researchers to expose to new ideas, to modify accepted theories and independent reproduction of scientific thoughts and experimental results in light of valuable comprehension of in-hand evidences.

The goal of this book is to encompass residents, scholars, physicians, professors, critical care practitioners, and other healthcare professionals with a comprehensive view of some diseases. Though this book is not a textbook, it tries to be handy in some clinical conditions, management, recent advances and thoughts in cancers, neurodegenerative diseases, acute pancreatitis, pediatric respiratory diseases, myocardial infarction, sepsis, tropical diseases, inflammatory bowel diseases, and in veterinary science.

Research work on the inflammatory biomarker, C-reactive protein (CRP), an evolutionary conserved plasma protein since 1970s has benefited humanity and science leading to long journey of routine clinical use of CRP measurements and its future enrooting as a rational clinical molecule in various diseases. It is a multi-author book, inscribed by renowned researchers, professors, and well-known practitioners in India and abroad. The specialized contributions from the authors bring a complementary and multidisciplinary management and research on the role of CRP in diseases. Illustrations, figures, and tables have been added at different levels to enrich the text. In order to keep the discussions focused, we have not gone into some of the detailed pathophysiology and management.

As we all know, the field of clinical biology and medicine is not static but constantly changing, so it is not possible to define something very standard but to guide only in clinical practice and research. Editors and authors worked as a team for more than one year to ensure the acceptability of each chapter of this book.

To visualize the age-old CRP, as an innovative therapeutic target, in different new drug discoveries, using molecular modeling for pharmaceutical detailing, designing easy accessible antibodies to target CRP in different disease state, in nanoscience and other research fields can be a promising tool of application in clinical medicine. The expanding research works and functionality of CRP makes it a hot bubbling "Critically Relevant Protein" in a wide range of inflammatory and infectious diseases.

As an educational venture, we just tried to touch and glaze some of the functionality of CRP and its hypervolume in health and diseases. We are transitioning through a pandemic phase caused by Coronavirus, where the vulnerable and affected individuals, their families, financial constrain, swing our dependencies on virology research, upcoming vaccines, their clinical trials, and effectiveness causing an upsurge in demand for healthcare so that we can move from new-normal to normal. Hope to bloom our thoughts with new research updates, findings, thoughts, and experiments in the field of clinical medicine.

Kolkata, West Bengal, India Waliza Ansar
NISER Bhubaneswar, Odisha, India Shyamasree Ghosh

The original version of this Front Matter was revised: Preface, Acknowledgement, Dedication and About the Editors sections have been included.

Acknowledgments

Two years back, to sharpen our educational edges, to provide innovative, thoughtful, and confident strokes in the field of science, we thought of writing this book. It is our second endeavor, a very small step for the scientific community but it is a "giant leap" from our side. At the threshold of completing this book, to achieve this goal, we are fortunate enough to get assistance from all quarters and to remember that we take this special moment to express our praise and sincere appreciation for all those who have contributed both emotionally and professionally in shaping this book.

At the very outset, we would like to express our deepest gratitude to Dr. Chitra Mandal, SERB Distinguished fellow at Indian Institute of Chemical Biology, CSIR (CSIR-IICB), Kolkata; our research supervisor for giving us the opportunity to implement our ideas by channeling our thoughts in the field of immunology.

Indeed, we are grateful to all our teachers, friends, and colleagues starting from our research days to our academic career. Faculty apart, we shall forever remain indebted to those friends who enriched us with their scientific thoughts.

Our heartfelt thanks to all the Directors and Ex-Directors, Vice Chancellor, and Principal of various institutes we were attached.

We are very much grateful and thankful to Prof. Dr. Sharmila Mitra, Principal, Behala College, Kolkata, India, for the whole-hearted support and help. Director, NISER Bhubaneswar, is acknowledged.

We express our heartfelt gratitude to Dr. Sheikh Hasan Habib, Demonstrator, K. P.C. Medical College and Hospital, Kolkata, and Critical Care specialist, for his continuous patience and support.

Special thanks to all the chapter authors for contributing their piece of research, their expertise and thoughts for this book. To write the manuscript, to provide clinical data, and to correct and proofread the work is a tedious, thoughtful, and patient job. Thanks to all authors for extending their support in this journey. A special thanks to them for allowing editors the liberty to edit the chapters freely in order to maintain uniformity throughout the book.

We are grateful to our family for their uncompromised love and affection, sustained support, and encouragement throughout this process.

We are so fortunate to have our children, Saptarshi and Tanaz for their patience, support, and cooperation. Their innocence and consideration is our creative inspiration.

We owe to various authors of different articles from which we gathered so much information that helped in developing confidence and clear vision.

We are grateful to the Council for Scientific and Industrial Research (CSIR), India; Indian Institute of Chemical Biology (IICB), Kolkata; Behala College, Kolkata; Diamond Harbor Women's University; and Asutosh College, Kolkata for providing us the infrastructure to work with all recent facilities.

Our special thanks to the reviewers who reviewed the proposal submitted to Springer, we are thankful for their inputs. We are thankful to the editorial team of Springer from its gestational period for their multiple helpful suggestions in finalizing this book.

Thank you all for your constant help and cooperation, which is what has made us complete this work.

Contents

1 **Multiple Faces of C-Reactive Protein: Structure–Function Relationships** 1
Waliza Ansar

2 **C-Reactive Protein and Neurodegenerative Diseases** 35
Inês Lopes Cardoso and Fernanda Leal

3 **C-Reactive Protein in Inflammatory Bowel Disease** 59
Sayan Malakar

4 **C-Reactive Protein (CRP) and Markers of Oxidative Stress in Acute Myocardial Infarction** 95
Mriganka Baruah

5 **Role of CRP in Monitoring of Acute Pancreatitis** 117
Jawaid Ahmed Khan

6 **Role of SNP in CRP and Biology of Cancer** 175
Rishav Dasgupta and Shyamasree Ghosh

7 **Role of C-Reactive Protein in Tropical Infectious Diseases** 193
Junaid Jibran Jawed

8 **Acute Respiratory Ailments in Pediatric Age Group and Role of CRP in Diagnosis and Management** 213
Chandra Shekhar Das

9 **Role of C-Reactive Protein (CRP) in Sepsis: Severity and Outcome** 249
Sheikh Hasan Habib and Waliza Ansar

10 **C-Reactive Protein: Is Early Prognostic Marker?** 291
S. Yogeshpriya and P. Selvaraj

About the Editors

Waliza Ansar holds a Bachelor's degree in Zoology (Honors) with Chemistry and Botany and a Master's degree in Zoology from Calcutta University. She was awarded the National Scholarship in 1994. She was awarded her PhD degree (2009) from CSIR-Indian Institute of Chemical Biology, Kolkata, Govt. of India, after qualifying for the State Level Eligibility Test (SLET) and National Eligibility Test (NET) conducted by the Ministry of CSIR-HRDG, Govt. of India. In her PhD work, she demonstrated various biological roles of C-reactive protein (CRP) in diseases like malaria, leishmaniasis, and tuberculosis. She also completed 3 years of postdoctoral research at CSIR-IICB, India. She is currently an Assistant Professor of Zoology at Behala College, Kolkata, where she was the Head of the Department of Zoology from 2015 to 2017. She has also worked at Asutosh College, Kolkata, as a lecturer in Environmental Science and Zoology. She has also worked as lecturer in Presidency College, Kolkata; Diamond Harbour Women's University, 24 Parganas (South), India, and Rabindra Bharati Open University. She has conducted research on forensic science, biochemistry, and glycoimmunology. She has published papers on CRP, dipteran flies, and therapeutic applications of monoclonal antibodies in journals such as *Glycoconjugate*, *Cellular Physiology and Biochemistry*, *Immunology Research*, the *Indian Journal of Clinical Medicine*, *Journal of Stem Cells*, etc. She has also published a book chapter on nanomedicine and a book on C-reactive protein (Springer). She has presented her work at various international and national symposiums.

Shyamasree Ghosh is currently a Scientific Officer (F) at the School of Biological Sciences, National Institute of Science Education and Research (NISER), Bhubaneswar. She graduated from Presidency College, Kolkata, in 1998 with Hons in Zoology with Chemistry and Physiology and was awarded a prestigious National Scholarship from the Govt. of India. She received her postgraduate degree in Biotechnology from Calcutta University, ranking second at the University in the year 2000. She completed her PhD in the field of childhood acute lymphoblastic leukemia (ALL) at the Indian Institute of Chemical Biology, under CSIR, Kolkata, after qualifying for the National Eligibility Test conducted by the Ministry of HRD, Govt. of India. Her work was accepted in various national and international journals of repute and is cited by young researchers worldwide. She worked in the field of stem cells and nanotechnology in her postdoctoral research. Later she joined as a faculty in a postgraduate college affiliated to Bangalore University. She has worked and published in Computational Immunology and Machine Learning. She has authored several publications that include books and chapters from international publishers of repute.

Multiple Faces of C-Reactive Protein: Structure–Function Relationships

1

Waliza Ansar

Contents

1.1 Introduction ... 3
1.2 Structure of C-Reactive Protein ... 5
1.3 Biological Role of CRP ... 12
1.4 CRP Concentration in Circulation ... 15
1.5 Structure–Function Relation of CRP ... 18
1.6 CRP and Inflammation ... 19
1.7 CRP in Disease Pathology ... 22
1.8 Discussion ... 25
References ... 25

Abstract

C-reactive protein (CRP) is a fascinating acute phase protein which plays a range of functions in different physiological or pathophysiological states. Since its discovery, it has been regarded as a useful biomarker of tissue injury, infection, and inflammation. The native CRP (nCRP) and the modified isoforms (mCRP) appeared in due course of different researches. The mCRP is formed when CRP is exposed to acidic pH and redox conditions (conditions mimicking inflammatory microenvironments). The nCRP is the native pentameric structure. CRP is functional in both native (nCRP) and its non-native pentameric (mCRP) structural conformations. The functional roles of native CRP are quite different from mCRP. The mCRP binds to PC and also to immobilized (in vitro) factor H, the complement inhibitor for beneficial complement-resistant activity. Thus, nCRP is protective only against early-stage infection, while mCRP is protective against both early- and late-stage infections. As nCRP exhibits PC-independent anti-pneumococcal activity, it is quite feasible that CRP functions as a general

W. Ansar (✉)
Department of Zoology, Behala College, Kolkata, West Bengal, India

W. Ansar, S. Ghosh (eds.), *Clinical Significance of C-reactive Protein*,
https://doi.org/10.1007/978-981-15-6787-2_1

antibacterial molecule. They have different impacts in diverse pathophysiological states. CRP as a biomarker of inflammation, is a clinically significant predictor of potential episodes of cardiovascular disease, independent of other risk factors. Researchers suggested the structure–function relationship between the mCRP isoforms and nCRP. Attempts had been made to recognize the possible significance between the diversity of structures and their opposing roles. This chapter discusses the biochemical facets of CRP biology, emphasizing on the supposed application between the structural biology of CRP isoforms, their physiological relevance, differentiation, and condition-based pathophysiological roles.

Keywords

C-reactive protein · CRP isoforms · Pentraxins · Structural biology · Acute phase response · Bioinformatics · Monomeric CRP · Molecular modeling · Inflammation · FcγR

Abbreviations

APP	Acute phase protein
APR	Acute phase response
CHD	Congenital heart defect
CHD	Congenital heart disease
CR1	Complement receptor 1
CR3	Complement receptor 3
CRP	C-reactive protein
EDTA	Ethylenediaminetetraacetic acid
ESR	Erythrocyte sedimentation rate
FcγR	Fc-gamma receptors
fMLP	*N*-formyl-L-methionyl-L-leucyl-phenylalanine
Glu	Glutamate
HRT	Hormone replacement therapy
HSA	Human serum albumin
IgG	Immunoglobulin
IL-1	Interleukin-1
IL-10	Interleukin 10
IL-12	Interleukin 12
IL-1R	Interleukin-1 receptor
IL-6	Interleukin-6
LDL	Low-density lipoprotein
LPG	Lipophosphoglycan
Lys	Lysine
Mal	Malaria
MALDI-TOF	Matrix-assisted laser desorption/ionization
mCRP	Monomeric CRP
MMDB	Molecular Modeling Database
neo-CRP	neo-antigen CRP

oxLDL	Oxidized low-density lipoproteins
PAF	Paroxysmal atrial fibrillation
PAF	Platelet-activating factor
PBMC	Peripheral blood mononuclear cells
PC	Phosphocholine
PCT	Procalcitonin
PDB	Protein Data Bank
Phe	Phenylalanine
PL	Poly-L-lysine
PMA	Phorbol myristate acetate
RCSB	Research Collaboratory for Structural Bioinformatics
SAA	Serum amyloid A
SAP	Serum amyloid P component
TB	Tuberculosis
Thr	Threonine
VL	Visceral leishmaniasis

Highlight

1. CRP is a member of the pentraxin superfamily, first discovered by Tillett and Francis in 1930. It is the first pattern recognition molecule discovered. The first characterization of this protein was based on its property to precipitate the "C" polysaccharide derived from the pneumococcus cell wall. Subsequently, it was described as an acute phase reactant as CRP levels were increased in patients with a range of inflammatory conditions. CRP is a cyclic protein comprised of five identical non-covalently attached subunits.
2. Apart from other ligands of CRP, phosphocholine (PC) was shown to be the major ligand for CRP binding within the pneumococcal cell wall. PC is found on the cell surface of a number of pathogens.
3. Today, CRP is widely used in the clinic as a marker of inflammation. From different prospective clinical trials, it is shown that CRP levels may serve as a predictor of cardiovascular events, independent of other risk factors.
4. This chapter discusses how momentum gained into the different structural isoforms of CRP have led to an enormous valuation of its prothrombotic and pro-inflammatory role, which is relevant to a wide spectrum of disease states.

1.1 Introduction

C-reactive protein (CRP) is a homo-pentameric classical acute phase inflammatory protein. It is a highly conserved plasma protein known for its ability to bind and precipitate the pneumococcal cell wall C-polysaccharide. In 1930, it was initially discovered by Tillet and Francis during the investigation of the sera of patients suffering from the acute condition of *Pneumococcus* infection. It was then named for its reaction capacity (for precipitation) with the bacterial cell wall somatic capsular (C)-polysaccharide of *Streptococcus pneumoniae* (Tillet and Francis 1930). CRP was the first acute phase protein to be known and is a completely disease-sensitive

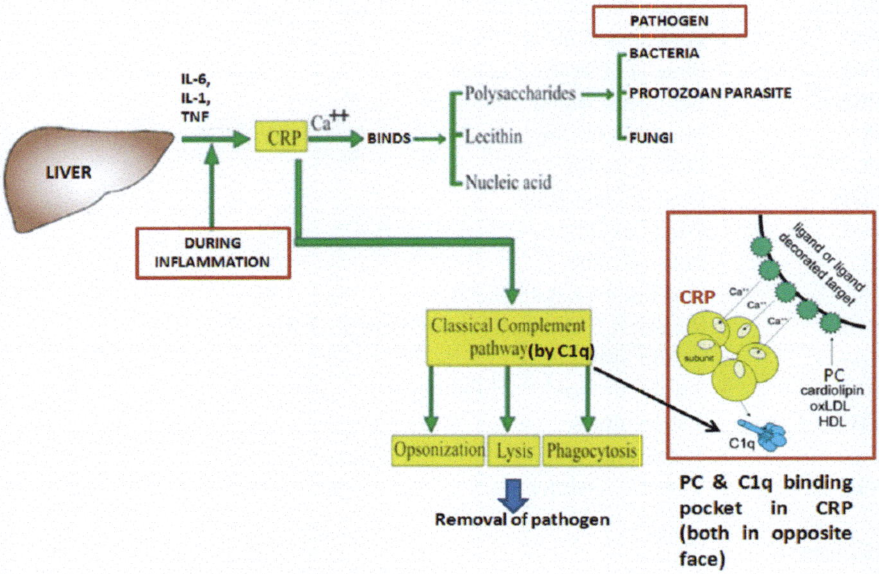

Fig. 1.1 Activation of classical complement pathway by CRP by binding with C1q. CRP also binds (other face) with pathogen through phosphocholine (PC) ligand and calcium (Ca^{++}) ions. Other ligands of CRP are HDL (high-density lipoprotein) and oxLDL (oxidized low density lipoprotein)

biomarker of tissue damage and systemic inflammation (Pepys and Baltz 1983). Since its discovery, CRP is a subject of great clinical interest in acute inflammation, pre- and post-surgery, trauma, state of infection, different heart diseases and in other pathophysiological conditions.

In the presence of calcium, CRP binds to polysaccharides like phosphocholine (PC) on microbes and activates the classical complement pathway of the innate immune system by triggering (by binding with) C1q (Volanakis 2001) (Fig. 1.1). CRP has many homologs in some invertebrates and vertebrates (Black et al. 2004), from arthropod horseshoe crab to humans. CRP is a member of the highly conserved pentraxin family, which includes other structurally related members such as serum amyloid A (SAA), fibronectin, and others (Gewurz et al. 1982). Transcriptional stimulation of the CRP gene chromosome 1(1q23.2) (CRP, NCBI, US 2020) mainly occurs in the liver hepatocyte cell in response to increased concentration of inflammatory cytokines like interleukin-6 (IL-6) (Boras et al. 2014). It is reported that IL-6 is the main inducer of *CRP* gene expression, but it is stimulated by other cytokines also. Other cytokines like IL-1 enhances the secretion of CRP (Szalai et al. 1998). However, IL-6 is necessary for CRP gene induction, it is not sufficient for IL-6 to achieve this alone (Weinhold et al. 1997).

The nonspecific physiological and biochemical response of most endothermic animals toward infection, inflammation, tissue damage, and malignant neoplasia is known as the acute phase response (APR). A number of cytokines are triggered causing activation of liver hepatocyte cells to secrete some proteins (acute phase proteins, APPs) in the bloodstream to be upregulated at the site of pathology. Among array of APPs, the group includes some coagulation, complement, and transport proteins; proteinase inhibitors and other miscellaneous groups. Some behave as positive APPs and some behave as negative APPs. But, CRP, SAA, and procalcitonin (PCT) are some APPs with high sensitivity, response speed, and dynamic functional range (Pepys and Baltz 1983; Pepys and Hirschfield 2003).

The functional role of CRP in host defense lies in its activation of classical complement cascade to enhance phagocytosis and final clearance of pathogen. These roles have been attributed by the binding of CRP to PC (Pepys 1981; Gewurz et al. 1982). CRP also reacts with cells at the sites of tissue injury, binds to nuclear antigens, apoptotic cells, damaged cell membranes, clearance of apoptotic or injured cells, removal of cell debris/material released from these damaged cells. Similarly, to serum amyloid P component (SAP), the functions of CRP is not restricted in identification of microbial antigens but also in maintaining body homeostasis (Du Clos 2003). Under physiological or pathological conditions, the native pentameric CRP (nCRP) has a half-life of about 19 h in plasma. Circulating CRP concentration is determined by its synthesis rate which is solely dependent on the host responses to different pathological process(es). These pathological conditions stimulate the production of CRP. When the stimulus for increased synthesis completely dies down, the circulating CRP level quickly falls to the same level as the CRP clearance rate (Pepys and Hirschfield 2003).

1.2 Structure of C-Reactive Protein

CRP (MW ~120 kDa) belongs to the pentraxins family. The members of the family tree of pentraxins are highly conserved throughout the phylogeny. Although, CRP is conserved in vertebrates, from mouse to man (Ciliberto et al. 1987; Murphy et al. 1995), it has homologues starting from arthropods (Black et al. 2004) and rolling to its high relevance in clinical science as human CRP. Pentraxins have a cyclical multimeric structure. One face of the pentraxin has calcium-dependent ligand-binding site. The prominent ligand for CRP includes PC which is present on the surface of most microbes. CRP binds with PC in a calcium-dependent manner. CRP has also binding site for complement protein C1q and can activate the classical complement pathway (Fig. 1.1). The details of the structure and topology of CRP have been thoroughly investigated (Agrawal and Volanakis 1994). CRP is a multi-functional molecule of the innate immune system in humans.

In CRP trangenic mice injected with endotoxin, acute phase concentrations of human CRP reaches 1 mg/ml in the blood within 24 h of injection and, like in humans, transcription of the CRP gene occurs mainly in the liver (Szalai et al. 2002).

CRP consists of five identical non-covalently bound protomers arranged in cyclic symmetrical pattern(Shrive et al. 1996). CRP as such is a pattern recognition molecule composed of five identical subunits (Shrive et al. 1996). Molecular weight of each subunit is ~23 kDa and each subunit has an intradisulfide bond. In the presence of Ca^{2+}, each subunit binds with very high affinity to PC (Anderson et al.

1978). As all the five ligand-binding sites are all on the same face of the pentamer (homopentamer) (Shrive et al. 1996), CRP binds with very high avidity to ligand-ornamented targets like bacteria or other microbes. On the opposite side of each subunit of CRP is an effector molecule-binding site that arbitrates the protein's interaction with C1q (Agrawal and Volanakis 1994), FcγRI (Marnell et al. 1995), and FcγRII (Bharadwaj et al. 1999).

The diverse range of biological functions of CRP is attributed by a vivid range of ligand with which it interacts in the human body. The classical ligands can be grouped into three major classes. They are (i) compounds that contain PC or some related structures, (ii) polycationic compounds like protamine sulfate and poly-L-lysine, and (iii) carbohydrates that contain D-galactose-related molecular structures (Kilpatrick and Volanakis 1991; Gewurz et al. 1995; Szalai et al. 1997; Lee et al. 2002).

CRP belong to the pentraxin family and, thus, it has a ring-shaped homopentamer (data obtained from crystallographic studies) (Shrive et al. 1996; Lee et al. 2002) whose subunits assemble non-covalently side by side to form a wreath-like structure (Shrive et al. 1996). As all subunits face the same direction, the surface of each side of the wreath has distinctive binding characteristics. X-ray structural analysis of human CRP revealed that each subunit is a β-barrel, and all the five subunits are assembled together in such a way that they face the same direction, thus representing two distinct ring-shaped surfaces (Shrive et al. 1996).

It is reported that the conformation of CRP changes depending on the presence or absence of calcium (Kilpatrick et al. 1982). In total, the soluble form of CRP and CRP bound to a cell surface (known as neo-CRP) or plastic surface express different epitopes (Ying et al. 1989), and therefore, there is a possibility for altered binding specificity (Kempka et al. 1990).

As known before, PC is the most abundant ligand of CRP. It has a single PC-binding site per subunit facing the same direction. The CRP-PC binding required the presence of calcium ions, whereas a polycation (such as poly-L-lysine) needs the dependency of either EDTA (enhances the binding) or calcium for its binding interaction with the CRP. In the presence of multiple PC residues present on a macromolecular structure (pathogen), the affinity of ligand PC toward CRP increased tremendously. As CRP is a doughnut-shaped pentamer with a PC-binding site present on each subunit, a polyvalent ligand that can bind at multiple PC-binding sites should have much increased affinity than the monovalent ligand. The crystallographic structure of ligand-complexed human CRP revealed that PC is bound in a very shallow groove near the calcium ion-binding site. Phosphate group of PC coordinates directly to the bound calcium ions.

Two calcium ions are bound close to each other on one of the surfaces. The phosphate group of PC is directly bound to the calcium ions (Thompson et al. 1999). Two other well-known ligand of CRP are poly-L-lysine (PL) and protamine sulfate belonging to polycationic compounds group and the other group is galactose and related structures (Kilpatrick and Volanakis 1991). All three ligand groups (PC, polycationic compounds and galactose group) appear to bind (shown by X-ray study) on the same side of CRP. It is the same side on which calcium ions are present (Lee et al. 2002).

Calcium ions are critical for the binding of ligands, as well as for continuing the integrity of the CRP. In the presence of calcium ions, human CRP may defy denaturation induced by chemical and physical factors like heat, urea, and reducing agents (Potempa et al. 1983; Wang et al. 2011). The binding of calcium to CRP molecule may protect CRP from proteolytic denaturation and degradation in blood circulation during the acute phase response of infection (Du Clos 2013; Ramadan et al. 2002; Boncler et al. 2019). Although biochemical, physical, and immunological researches on CRP structure showed that the protein might be modified by the changed microenvironment in disease state or in any acute inflammatory condition. Thus, CRP can exist in three isoforms: (i) a multimeric form composed of ten and more subunits, (ii) a native pentameric form, and (iii) a monomeric CRP form, with single subunit. All these isoforms of CRP assigned different biological function and is of special interest to researchers (Boncler et al. 2019).

Apart from the pentameric native CRP, McFadyen et al. 2018, mentioned the presence of highly pro-inflammatory structural isoforms, pCRP* (developed from pentameric CRP, pCRP, or nCRP) which ultimately transformed to mCRP. As pCRP* is structurally more relaxed than pCRP, it expresses more neoepitopes for complement fixation and immune activation. Both pCRP* and mCRP can effect activation of leukocytes, platelets, endothelial cells, and complement components (McFadyen et al. 2018).

Apart from PC-related structural molecules, certain sugar phosphates (such as galactose 6-phosphate) are also bound near the PC-binding site. One of the sugar hydroxyl groups appears to interact with CRP. The small ligands for the polycationic-binding site were Lys-Lys, Lys 4 where polylysine has tremendously enhanced affinity due to the presence of multiple Lys-Lys sequences. Lee et al. suggested experimentally that the PC and polycationic-binding sites do not overlap (Lee et al. 2002).

CRP binds a diverse group of nuclear components like chromatin, histones, nucleosome, and ribonucleoprotein (Robey et al. 1984; Du Clos et al. 1988). CRP clears and subsequently dissolve the potential self-antigens from the site of tissue injury by binding with chromatin and nucleosome (Robey et al. 1985). Histones an integral components of chromatin and nucleosome also bind to CRP (Du Clos et al. 1988). Lee et al. reported that histone binds at the polycationic site and not at the PC-binding site. The ligand PC can bind at the PC-binding site and at the polycationic-binding site and PC effectively compete and inhibit at the polycationic-binding site (Lee et al. 2002).

Das et al. demonstrated that CRPs induced in different pathological conditions differ both in their amino acid and carbohydrate sequences (Das et al. 2003). Differential lectin-like properties, sugar-binding attribute, and immunological reactivity of invertebrate CRP (Mandal et al. 1991) and human CRPs (Köttgen et al. 1992) also revealed differences in their structures and functions. Das et al. 2003 reported first time (70 years after the discovery of CRP) that human CRP is glycosylated. Previously, it was reported to be non-glycosylated. The author also reported that disease-induced molecular variants of human CRP in nearly 16 different pathological conditions.

Preferential glycosylation of CRP is possibly an altered/modulated post-translational modification, which may allow some relevance in the clearance of the inducing agents. Thus, these differences in sequences (amino acid and carbohydrate sequences) may be explicated without invoking the presence of various gene-induced splicing differences at post-transcriptional level (Das et al. 2003).

The induction of some glycosyltransferases needed for post-translational enzyme regulation possibly could induce the molecular variants of CRP. The preferential expression of certain genes in the presence of toxic reagents may also facilitate the differential induction of glycosylation during post-translational modification. In this regard, it may be stated that the CRPs are altered differently, keeping their native pentraxin structures almost the same, but some slight changes in their amino acid sequence and glycosylation make them differ chemically (shown by changes in electrophoretic mobilities, lectin-binding assays, MALDI-TOF, etc). Induced glycosylated molecular variants may be essential for their suitable biological function. Such specific molecular variants of human CRP are of immense clinical importance, helpful for monitoring the acute phase responses, in diagnosis and monitoring of various clinical samples obtained from different pathological conditions (Das et al. 2003).

Apart from the glycosylated molecular variants (Das et al. 2003), another form of CRP, known as mCRP (monomeric CRP) or neo-CRP is synthesized under mildly denaturing conditions (such as acidic environment or 8 M urea or immobilization on plastic surface), like low pH (<3) and surface adsorption (Potempa et al. 1987). It actually dissociates into monomers with conformational changes and newly expressed antigenic determinants. It is known that human CRP circulates in blood in soluble form as stable pentamer. The neo-CRP or mCRP has poor solubility in water (aqueous medium) and quickly assemble to form aggregates.

The mCRP is a physiologically related molecular species, found at the site of inflammation, in vascular tissues (Diehl et al. 2000), attached in inflammatory tissues (Rees et al. 1988), and is also known to exercise stimulatory effects on immunologically relevant cells (monocytes, neutrophils, and platelets) (Potempa et al. 1988) (Fig. 1.2). Neo-CRP/mCRP expresses on its surface a new set of antigenic determinants and has thus undergone considerable conformational changes (Ying et al. 1989).

Köttgen et al. (1992) showed binding of surface immobilized CRP (neo-CRP) to lactosylated HSA (human serum albumin). This neo-CRP does not bind with other lactosylated derivatives containing terminal sugars like sialic acid, mannose, glucose, and N-acetylglucosamine. Lee et al. 2002 also noted that neo-CRP binds better with carbohydrate ligands in the absence of calcium. In contrast, the PC ligand binds with CRP in the presence of calcium ions. Distinct glycosylated molecular variants of CRP binds with different sugar groups and lectins through the PC-binding pocket in the presence of calcium (Das et al. 2003, 2004a). Another ligand of CRP, Protein A has also been reported. Protein A binds differentially with glycosylated CRP (also non-glycosylated CRP) in the absence of calcium. Protein A and PC were reported to be different ligand for CRP as they have separate binding sites (for PC And Protein A) with different binding constants (Das et al. 2004a, b). Glycosylated molecular

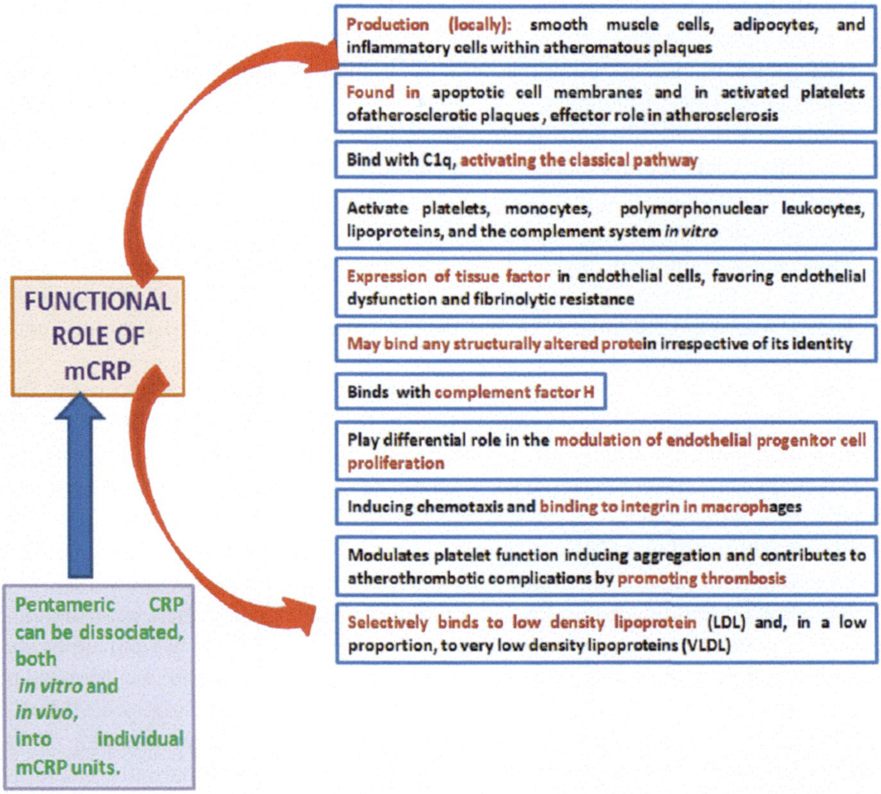

Fig. 1.2 Some pathobiological role of the isoform monomeric CRP (mCRP) (Salazar et al. 2014a, b)

variant of CRP induced in different pathological conditions were first reported by Das et al. in 2003. The binding characteristics of these molecular variants were thoroughly investigated along with structural entity also. The functional relevance of these glycosylated molecular variants were investigated in three very prevalent and common diseases, which are really the health concern in India, viz. malaria (Mal), visceral leishmaniasis (VL), and tuberculosis (TB) (Ansar et al. 2009a, b). Ansar et al. (2009a, b) showed that CRP purified from the serum (obtained from active patients suffering from these three diseases) of these diseases was differentially glycosylated using different protein chemistry tool. It is known that anemia is a common clinical manifestation in malaria, VL, and TB, where erythrocytes of active patients are damaged in disease condition. Glycosylated CRP as compared to deglycosylated CRP after binding with those damaged erythrocytes (erythrocytes were obtained from active malaria, VL, and TB patients) further altered the erythrocyte membrane characteristics, trigger complement cascade (by binding with C1q) and finally remove them from the circulation by lysis. So CRP was glycosylated differentially in disease condition to function as a protective innate immune molecule (Ansar et al. 2009a, b).

CRP bears functional similarity with IgG, likewise binds with Fc receptors (Marnell et al. 1995; Bharadwaj et al. 1999) and also clear damaged erythrocytes (of malaria, VL, and TB patients) by opsonophagocytosis. Thus, disease-induced glycosylated CRP removes damaged erythrocytes from the circulation both by complement activation and by phagocytosis. Therefore, CRP acts as a host immune protective molecule (Ansar et al. 2009a, b).

Thus, the specificity of neo-CRP (as mentioned in the above paragraphs) with reverence to the sugar structure is broad (Lee et al. 2002). Also glycosylated molecular variants of human CRP has different binding characteristics and functional relevance (Das et al. 2003; Ansar et al. 2009a, b).

The serum level of CRP has been a powerful clinical indicator in assessing the presence/absence of infection and inflammation. It was also shown in studies that the increased concentration of serum CRP signifies increased future risk of coronary events (Pepys and Berger 2001). CRP is found attached to atherosclerotic lesion and other sites of inflammation (Reynolds and Vance 1987). At inflammatory sites, the antigenic epitopes of neo-CRP are often expressed. No clear-cut mechanistic process could explain the cause of accumulation of neo-CRP at the inflammation sites.

Numerous three-dimensional X-ray structures of CRP, either the free or ligand-bound form, have been placed in the Protein Data Bank (RCSB PDB) and are freely available for education and research at PDB (http://rcsb.org) and the Molecular Modeling Database (MMDB; https://www.ncbi.nlm.nih.gov/Structure/MMDB/mmdb.shtml). Crystallization of CRP was originally reported in the year 1947 (McCarthy 1947). Human CRP X-ray structures including the partially calcium-loaded CRP, fully calcified.

CRP (in the absence or presence of bound PC), and calcium-depleted CRP was also studied (Ramadan et al. 2002; Thompson et al. 1999; Shrive et al. 1996). The molecular structure of CRP, along with the non-glycosylated polypeptide subunit of monomeric CRP was noted (Salazar et al. 2014a, b).

Sequence analysis of glycosylated CRPs (Das et al. 2003) from three patients using MALDI-TOF revealed the absence of short segments from the N-terminus and C-terminal end. Glycosylated molecular variants of CRP has the same structural motifs (by molecular modeling) like non-glycosylated CRP (Thompson et al. 1999; Shrive et al. 1996). But on the surface opposite the PC-Ca^{2+}-binding region, a modified cleft containing Asn158 or Asn160 was reported. The well-known N-glycosylation motif (Asn-Xaa-Ser/Thr, where Xaa is any amino acid other than proline) is indicated by molecular modeling, which contains the Asn-Val-Asn (amino acids 158–160) sequence (Das et al. 2003).

The increased synthesis of CRP in response to stimulus is quite nonspecific and is triggered by various pathological conditions or disorders (Pepys and Hirschfield 2003; Karasahin et al. 2018; Ryu et al. 2015). Apart from this, in most diseases, the circulating level of CRP reflects the ongoing inflammatory process(es) and/or tissue damage during the acute phase reaction. CRP does these much more accurately than other laboratory parameters like erythrocyte sedimentation rate (ESR) and plasma viscosity (Pepys and Hirschfield 2003; Bray et al. 2016; Ticinesi et al. 2017; Vermeire et al. 2005; Sproston and Ashworth 2018; Chandrashekara 2014). After an acute phase stimulus, the median CRP concentration can spike up nearly 1000-fold from the 0.8–3 mg/L level in normal healthy young adult blood donors. Though, in stress or other intermittent physiological conditions, some normal individuals can have a CRP concentration of nearly 10 mg/L (not exceeding 10 mg/L) (Salazar et al. 2014a, b). Stable CRP concentrations is a characteristic for each individual in the general population (Simental-Mendia et al. 2017), without seasonal variation.

The binding sites for PC calcium and ions was designated as the "recognition face" of CRP protomers, whereas the binding pocket for the complement component C1q (binds the globular domain of the C1q) and Fc receptors are assigned as "effector face." The recognition face binds bacterial cell walls and apoptotic cell membranes (Thompson et al. 1999).

Phe-66 and Glu-81 (also Thr76) also are the two key residues in the hydrophobic-binding pocket necessary for PC-mediating binding. Glu-81 binds with the positively charged nitrogen and Phe-66 maintains hydrophobic interactions with the methyl group of PC. All these binding pockets on the pentameric CRP emerges to be obscure, thus supporting the notion that pCRP does not possess intrinsic pro-inflammatory properties (Thompson et al. 1999; McFadyen et al. 2018; Ngwa and Agrawal 2019).

CRP has pleiotropic effects on human cells and has both pro-inflammatory and anti-inflammatory functions. The results were yielded by studies on CRP based on some contradictory findings, methodological aspects, and the existence of different isoforms (at the site of inflammation, in in vitro studies also) of CRP (Ji et al. 2006; Khreiss et al. 2004).

CRP may also dissociate into subunits (apart from mCRP formed in acidic condition or neo-CRP formed by adherence with plastic surface) during any experimental study by the use of contaminated commercial CRP preparations containing bacteriostatic sodium azide (Molins et al. 2011), LPS (Pepys et al. 2005) or traces of mCRP (Heuertz et al. 2005).

The anti-inflammatory functions of CRP may be associated with its ability to stimulate the expression of interleukin-1 receptor (IL-1R) antagonist in human peripheral blood mononuclear cells (PBMCs) (Tilg et al. 1993).

CRP also can alter the cytokine contour of mouse macrophages by downregulating the secretion of IL-12 (Sproston and Ashworth 2018) and inducing the release of the anti-inflammatory cytokine IL-10. CRP binds with neutrophils to exert stimulatory effects on superoxide production, degranulation, or chemotaxis (Devaraj et al. 2009; Noren Hooten et al. 2012). CRP can exert inhibitory (Sproston and Ashworth 2018; Filep and Foldes-Filep 1989; Heuertz et al. 1999; Kew et al. 1990a, b; Ling et al. 2014) effects on cell after interaction with neutrophils such as inhibiting the respiratory burst of neutrophils in response to a variety of agonists like PAF, fMLP, and PMA (Mortensen and Zhong 2000). Thus, CRP may play a crucial protective role during the early phases of the inflammatory reaction. The nCRP is protective only against early-stage infection, while mCRP is protective against both early- and late-stage infections. CRP bound to various ligands or microbial polysaccharides expressed on damaged cells surface and activates the classical complement cascade by binding with C1q protein. Thus, native CRP plays both pro- and anti-inflammatory role by its interactions with blood cells.

At the site of the injured endothelium, circulating activated platelets accumulates circulating cell-derived microparticles that undergo dissociation from pentameric CRP to mCRP, which exerts inflammatory response by promoting pro-inflammatory effects.

1.3 Biological Role of CRP

The earliest and most fundamental host response to inflammatory injury/infection or stress is the acute phase response, a very transitory (typically) and rapid tuning in the concentration of different plasma APPs. All these proteins, whether the positive APPs or negative APPs has been cited in human and in other animals for their role in disease pathologies. Of these, human CRP, a positive acute phase protein is the prototype (Abernethy and Avery 1941). Thus, adjunct with inflammation and in response to various cytokines, complement activation end products, and hormones, the liver produces a large amount of CRP into the blood circulation. Levels of the CRP in blood concomitantly rises so fast and so dramatically that, in the clinical laboratory, blood CRP level is frequently monitored to determine the intensity of inflammation in patients.

CRP is extensively used as a marker of disease activity. But all the functional attributes of CRP are still to be revealed. The primary function of CRP remains in enhancement of phagocytosis and activation of complement (Fig. 1.3). Human CRP binds with pneumococci and other pathogens or different damaged cells to activate complement cascade and then enhance phagocytosis.

The main role of CRP in innate immune responses appears to be activating complement pathways, trigger opsonization of targeted pathogen and to protect against infection (Mold et al. 1981). The other biological role of CRP include clearance of damaged cells and cell debris, wound healing, and have some immunomodulatory effect (Kilpatrick and Volanakis 1991; Gewurz et al. 1995; Szalai et al. 1997).

In mice, CRP opsonizes and helps in ensuing uptake of *Streptococcus pneumoniae* for increased bacterial clearance and protection against the *Streptococcus* (Szalai et al. 1996). CRP may be believed as one of several plasma proteins contributing in development of host acquired responses (Fearon and Locksley 1996).

CRP helps in clearing damaged tissues as it has been localized at such sites (Kushner and Kaplan 1961). Local production of CRP has been reported earlier (Gould and Weiser 2001). Some extrahepatic tissues like atherosclerotic plaques, vascular smooth muscle, and intracardial tissues also synthesize CRP locally (Salazar et al. 2014a, b).

CRP has been shown to bind small nuclear ribonuclear proteins and histones (Pepys et al. 1994). CRP is further proposed to inhibit chemotaxis of neutrophils (Kew et al. 1990a, b; Ahmed et al. 1996). CRP also binds to PC substituted

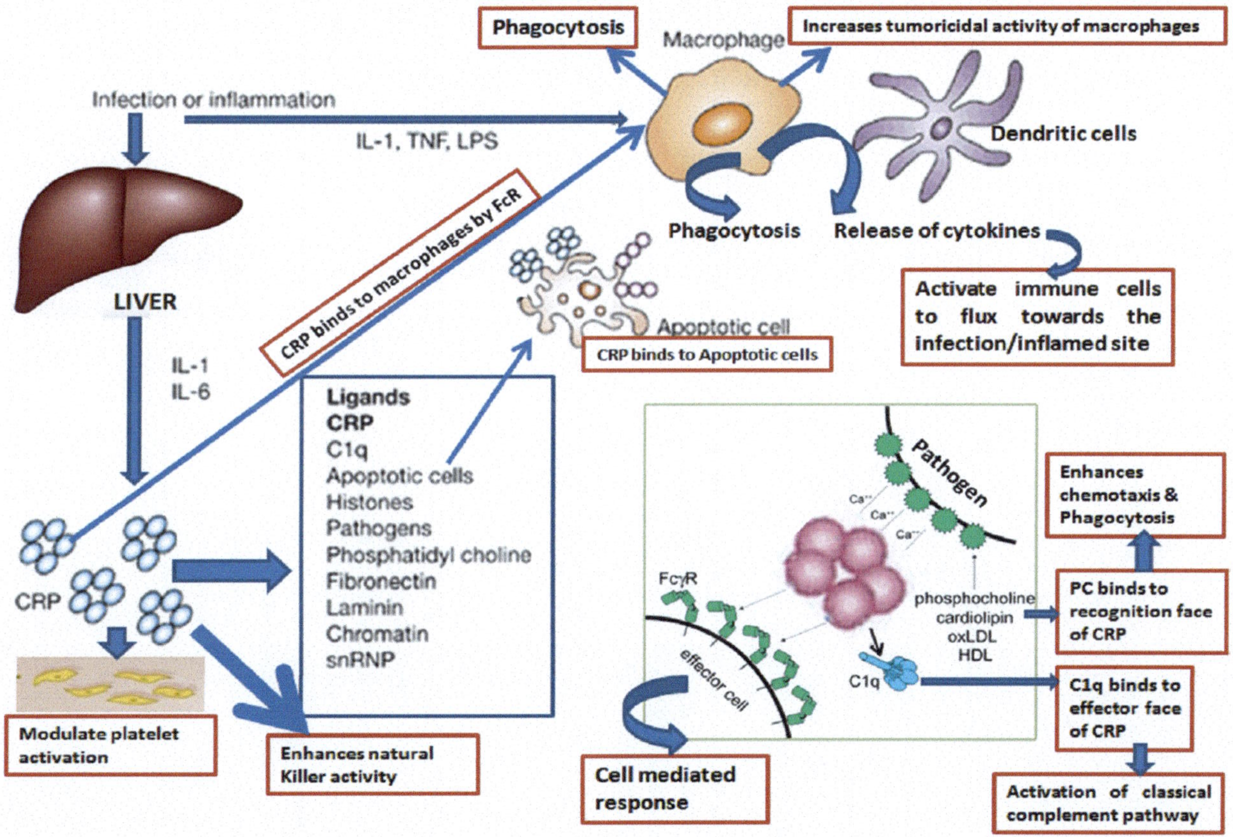

Fig. 1.3 Biological role of CRP

carbohydrate on the surface of microbes like *S. pneumoniae* and with the repeating phosphorylated carbohydratelLipophosphoglycan (LPG) on the upper surface of metacyclic *L. donovani* (Culley et al. 2000).

CRP had been demonstrated to bind with *L. mexicana* metacyclic promastigotes and triggers transformation to a amastigote-like stage, independent of pH and temperature alterations (Bee et al. 2001). CRP effects have also been accounted on neutrophils, macrophages, and endothelial cells.

In neutrophils, CRP balances neutrophil responses by downregulating it at high concentrations and activates responses at low CRP concentrations (Mortensen and Zhong 2000). Upregulation of chemokines and adhesion molecules by CRP has been reported in endothelial cells (Pasceri et al. 2000, 2001).

Due to variation in macrophage function obtained from different sources/species, functional attributes of CRP also changes. Native pentameric CRP stimulates respiratory burst in rat macrophages, whereas other complex response patterns were also reported. The release of a range of cytokines in macrophages to succumb infection after pro-inflammatory activation of CRP has been suggested (Miyagawa et al. 1988 Ballou and Lozanski 1992; Galve et al. 1993).

CRP is known to act as an opsonin. It increases the uptake of an array of particles by both neutrophils and macrophages. CRP was recently shown to bind to LPG present on the surface of Leishmania species (Culley et al. 1996). Opsonization of metacyclic *L. donovani* promastigotes with CRP increases uptake of both differentiated U937 cells and Peripheral Blood Mononuclear cells (PBMCs) with increased parasite survival (Bodman-Smith et al. 2002a, b). CRP binds with various types of Fc receptors (Marnell et al. 1995; Stein et al. 2000; Lu et al. 2018; Black et al. 2004) present on the surface of macrophages or indirectly to complement receptors CR1 and CR3 (Mold et al. 1999).

Various biological roles of CRP in malaria, VL, and TB was demonstrated (Ansar et al. 2006, 2009a, b). Glycosylated CRP (purified from serum of acute patients suffering from visceral leishmaniasis, malaria, and tuberculosis) as compared to deglycosylated CRP by binding with damaged erythrocytes of Leishmania, malaria, and tuberculosis patients further altered the membrane characteristics, triggered complement cascade, and finally removed them from the circulation by lysis. So CRP was glycosylated differentially in disease condition to function as a protective innate immune molecule. Actually, glycosylated CRP altered the mosaic of configuration of complement regulatory proteins on erythrocytes by reducing their numbers and finally hemolyzed the damaged erythrocytes from the circulation. Thus, CRP contributes a major percentage in removing erythrocytes apart from other pathways.

A group of CRP has been purified to apparent homogeneity by calcium-dependent affinity chromatography on PC-Sepharose column from 16 different clinical conditions. Enhanced CRP level was observed in all cases studied. CRP shows a diverse binding interaction as a component of innate immune system and their multiple immune-regulatory functions involved a wide range of biological macromolecules. Variation in glycosylation and amino acid composition of CRP affect their binding interactions. These glycosylated molecular variants were capable of modulating their binding activities with different lectins, sugars, human IgG, and plasma glycoproteins (Das et al. 2003).

1.4 CRP Concentration in Circulation

In healthy individuals, it circulates at low concentrations (nearly 10 mg/L), but its levels increase drastically in response to infections, inflammation, and tissue injury (Pepys and Hirschfield 2003). After a trigger from an acute phase stimulus, level may increase from less than 50 μg/L to even more than 500 mg/L. The rise from normal level may be 10-fold to nearly 10,000-fold. Although CRP is produced mainly by liver hepatocytes, predominantly under the transcriptional control by the cytokine IL-6 (other cytokines like IL-1, TNF-α have enhancing effect), some local CRP synthesis and other sites of possibly CRP secretion have also been noted (Gould and Weiser 2001; Das et al. 2003). The extrahepatic synthesis of CRP although do not substantially influence CRP plasma levels (Black et al. 2004; Agassandian et al. 2014). So in different diseases or pathological conditions, only plasma CRP level is monitored. It may increase to nearly 1000-fold within 24 h of its induction. So CRP is a marker of inflammation to gauge the inflammatory status of a patient. Increased production of CRP is stimulated by certain cytokines such as IL-6, IL-1 growth factor β, and some other extracellular signaling molecules (Weinhold et al. 1997; Volanakis 2001).

Unlike its human equivalent, mouse CRP is produced in only trace amounts and it is not an acute phase reactant (Pepys et al. 1979). As a result, CRP transgenic mice serve as an exclusive and convenient tool for studying the different biologic functions of CRP expressed in vivo (Szalai et al. 1995).

Each healthy individual tend to have stable CRP concentrations which will only spike in case of trauma, inflammation, or any clinical/subclinical infections. In a population of healthy individual, each one slightly differs from the other in their CRP concentration. CRP level may rise also with increasing age (Pepys and Hirschfield 2003; Tang et al. 2017). The presence of subclinical pathologies, stress may be correlated with the variation in individual CRP level of healthy subjects. Baseline CRP has no reported relation with seasonal variation. Significantly high baseline CRP values as a hereditable constituent was also noted. The association between genetic polymorphisms in IL-6 and IL-1 with CRP production has been suggested. Genetic polymorphisms of IL-6 and IL-1, genetic polymorphisms in *CRP* gene itself can be correlated with the near baseline CRP values in systemic lupus erythematosus patients or in susceptible patients with invasive pneumococcal disease. The correlation of baseline CRP concentrations in some disease condition depends on the polymorphism of genetic constituents (Il-6, IL-1, and CRP gene), state of acute phase response, level of CRP production after stimulation, clinical outcome of the subject, and the host defensive role in using CRP in inflammatory disease. Circulating level of CRP more accurately reflects ongoing inflammation than other laboratory parameters used in acute phase response (Pepys and Baltz 1983; Pepys and Hirschfield 2003; Hutchinson et al. 2000; Szalai et al. 2002).

Szalai et al. showed that CRP transgenic mice disclosed a sexually dimorphic pattern of expression, i.e., normal level (baseline) and endotoxin-induced expression of human CRP in CRP transgenic males is appreciably higher than the corresponding level of females (Szalai et al. 1995). This effect is endorsed toward testosterone, as baseline blood CRP level in males is lessened following castration and because consequent reconstitution with testosterone fully re-establishes the ability of castrated males to express normal (baseline) blood CRP. Importantly, castration and testosterone replacement has no outcome on the concentration of CRP articulated during the endotoxin-induced acute phase response. This finding is corroborated to humans, as in a study shows that baseline blood CRP concentration in men are appreciably higher than in women (Yamada et al. 2001). In another studies, Szalai et al. reported that the complement protein C5a also attributes to regulation of CRP gene expression (Szalai et al. 2000a). These data revealed that under basal and acute phase responses, sex hormones and C5a send a synchronized signal to liver hepatocytes to guarantee appropriate IL-6-induced production of CRP (Szalai et al. 2002).

De novo hepatic synthesis of CRP starts very rapidly after a single trigger from a stimulus, the serum level rising above 5 mg/L by about 6 h and spiking a peak at around 48 h. The plasma half-life of CRP is nearly 19 h. The intensity of the pathological process(es) determines the synthesis rate and is constant under all conditions of health and disease. So the stimulation and production of circulating CRP directly reflects the health condition of the individual (Pepys and Hirschfield 2003; Vigushin et al. 1993; Boncler et al. 2019). When the stimulus for amplified production completely ceases, the level of circulating CRP rapidly falls (like erythrocyte sedimentation rate, ESR) and cleared from circulation. Significantly, acute phase CRP level are unaffected by food eating habit and show no rhythmic/nonrhythmic diurnal variation in its concentration. Liver failure or liver cirrhosis impairs production of CRP. CRP values remain mostly unaffected by drugs. Some drugs affecting the intercurrent pathologies or underlying pathology providing the acute phase stimulus may reduce the CRP level, which is rear. Thus, CRP is a very useful nonspecific biochemical marker of inflammation. Measurement of CRP in clinics is importantly contributed toward (a) detection/diagnosis of underlying infection in immune-suppressed individuals in (b) screening of the disease (c) monitoring of the therapeutic response to treatment of infection and inflammation. Interestingly, CRP level spikes to modest level or even absent in some specific diseases showing low or baseline acute phase responses. The concept of very minor or low elevation of CRP production in diseases like systemic lupus erythematosus (SLE) despite evident sign of inflammation and tissue damage is unclear. The intact CRP response to intercurrent infection in individuals suffering from such conditions is also not understandable (Pepys and Baltz 1983; Pepys and Hirschfield 2003).

Elevated expression of CRP was exhibited during infection, different inflammatory conditions like cardiovascular diseases rheumatoid arthritis and many others (Du Clos and Mold 2004). As an acute phase protein, the plasma level of CRP fluctuates by at least 25% during inflammatory conditions (Gabay and Kushner 1999). In some bacterial infections, CRP serum concentrations may spike to a highest level of nearly 1000-fold (Thompson et al. 1999). After the end of the stimuli, CRP level exponentially decreases within 18–20 h, a value close to the half-life of CRP (Ridker et al. 2003). CRP plasma concentration may increase from nearly 1 μg/mL to over 500 μg/mL within 24–72 h of trauma, progressive cancer, or severe tissue damage (Ciubotaru et al. 2005).

We know from different studies and a specific research on CRP transgenic mice showed unequivocal substantiation that human CRP expressed during the natural process of infection surges to host defense. For illustration, in response to infection with *Streptococcus pneumoniae*, CRP transgenic mice reveal an acute phase human CRP response and nearly a 400-fold decrease in bacteremia compared to infected wild type animals (Szalai et al. 1995). This early protection eventually increases the survival rate of CRP transgenic mice. Human CRP is suggestively not protective against pneumococci in CRP transgenic ×Factor B deficient mice (Szalai 2001), revealing that CRP-mediated defense in vivo requires at least the alternative cascade of complement activation. CRP also could not activate alternate complement pathway during clearance of malaria-parasitized RBC. But could trigger the classical pathway in the same disease (Ansar et al. 2009a, b).

CRP binds with *Leishmania* promastigotes through the glycoprotein lipophosphoglycan (LPG) present on the surface of parasite (Culley et al. 1996; Raynes et al. 1994). CRP though facilitates the entry of parasites in macrophages (Culley et al. 1996), it also triggers macrophage responses by activating the release of nitric oxide and other pro-inflammatory cytokines like IL-10, IL-4, and IFN-γ (Bodman-Smith et al. 2002a, b). It is reported that CRP binds with macrophages through Fc receptors (Lu et al. 2018; Black et al. 2004). In contrast, CRP/FcγRI interaction did not dampen significantly CRP's ability to protect CRP transgenic mice (Szalai 2001). CRP transgenic mice also could resist infection with *Salmonella enterica* (Szalai et al. 2000b). CRP also enhances the adaptive immune response in CRP transgenic mice by increasing the serum titer of IgG anti-*S. enterica* antibodies (nearly ten-fold higher) than non-transgenic mice.

Many of the research experiments on CRP uses the commercial preparations of CRP, whose integrity or purity or characterization may give some erroneous result of the protein, and few included adequate controls. All commercial CRP preparation contains the toxic bacteriostatic sodium azide, which inevitably contains the bacterial endotoxin (lipopolysaccharide) as CRP produced by recombinant *Escherichia coli*. Though recent studies claimed the removal of lipopolysaccharide, still using commercial CRP should be the starting point for crucial functional studies. So authentic pure CRP isolated from human samples (serum, pleural fluid, ascetic fluid), free of bacterial contamination, should be used for researches having wide-ranging direct cellular functions of CRP (Danesh and Pepys 2009).

1.5 Structure–Function Relation of CRP

Cyclic pentameric CRP comprises five identical non-covalently attached subunits.

The molecular weight of each subunit is ~23 kDa. Each subunit has an intradisulfide bond (Shrive et al. 1996; Ngwa and Agrawal 2019). In the homopentamer, a PC-binding site is situated on the same face of each subunit. The amino acids Glu81, Thr76, and Phe66 in CRP are crucial for the formation of the PC-binding site (Agrawal et al. 1997, 2002; Thompson et al. 1999). CRP-PC complex activates the classical complement system (Kaplan and Volanakis 1974).

The level of CRP in human serum rises during systemic pneumococcal infection by up to several 100-fold as a component of the acute phase response (Kushner 1982; Tillet and Francis 1930). PC present on the pneumococcal cell wall C-polysaccharide binds with CRP to form CRP-PC complex through Ca^{2+}-dependent interaction (Abernethy and Avery 1941; Volanakis and Kaplan 1971) (Figs. 1.1 and 1.3). Importantly, CRP is only a minor acute phase protein in mice. So, the functions of human CRP in vivo can be studied better in mice (Whitehead et al. 1990). In murine models of pneumococcal infection, different experiments revealed that the host defense function of CRP is to protect against *Streptococcus pneumoniae* infection. Reduction in bacteremia and enhanced survival of mice poses the protective effect of CRP. The structure of CRP has a distinct relationship with its anti-pneumococcal function (Ngwa and Agrawal 2019).

CRP is functional both in its native pentameric and non-native structural conformations. In the native conformation, CRP binds to *Streptococcus* through the PC molecules present on the C-polysaccharide of the *Streptococcus.* The anti-pneumococcal function of CRP probably engrosses the known ability of pathogen-complexed CRP to trigger the complement pathway (Ngwa and Agrawal 2019).

In the native pentameric structure–function linkage, CRP is protective only when given to mice within a few (6 h before to nearly 2 h after the administration of bacteria pneumococci) hours of the administration of pneumococci. CRP was not protective when mice administered with CRP 24 h after infection. Thus, CRP is protective during the early stages of infection but not in the late stage of infection. Both the PC-dependent and PC-independent binding of CRP with pathogens protects the mice against early stages of infection (Ngwa and Agrawal 2019; Gang et al. 2012, 2015). Human CRP are able to activate the lectin and classical systems of complement in human serum and able to trigger the lectin pathway (by interacting with L-Ficolin and M-Ficolin) in murine serum (Suresh et al. 2006). CRP binds with ficolins and galactose-binding proteins present on bacterial surface in the hemolymph of horseshoe crab. Thus, CRP can identify a diverse group of bacteria and channelize them for clearance by recognizing an array of bacterial surface ligands like PC, different sugars, lectins, and some related plasma proteins (Ngwa and Agrawal 2019).

During inflammatory microenvironments (a mimicking condition, in vitro) like redox conditions or acidic pH, the non-native pentameric conformation of CRP is formed. In the non-native conformation (monomeric CRP or mCRP), CRP binds to immobilized factor H (complement inhibitor) in addition to binding with PC

(Fig. 1.2). Recent studies on CRP mutants propose that the factor H-binding function of non-native CRP is protective during late-stage pneumococcal infection. CRP has been reported to bind with various choline-binding proteins present on pneumococci in a Ca^{2+} -independent manner (Ngwa and Agrawal 2019; Frolet et al. 2010).

There is a structural alteration in CRP after binding with PC. This structurally altered CRP then binds with factor H on the pneumococci surface. Now this pneumococci may not be resistant to complement trigger any longer (Ngwa 2016; Ngwa and Agrawal 2019; Agrawal et al. 2008). Therefore, native CRP is protective only during early-stage infection, but non-native CRP is protective against both early- and late-stage infections. CRP binds with various pathogenic serotypes of S. pneumonia (De Beaufort et al., 1997) through PC molecule present both in cell wall and capsular polysaccharides (Edwards et al. 1982). CRP behaves like lectins and binds with polysaccharides that do not contain PC (Kottgen et al. 1992).

CRP binds as an opsonin on Gram-positive bacteria such as *Bacillus subtilis, Staphylococcus aureus, Streptococcus pyogenes,* and *Streptococcus agalactiae* and causes agglutination and lysis of these bacteria (Kindmark 1972; Mold et al. 1982). CRP also binds with Gram-negative bacteria like *Salmonella enterica* and protects mice from the early stages of infection. But CRP does not bind in vitro with *Salmonella enterica* in the absence of serum. PC-positive variants (on the cell exposed outer LPS) of *Haemophilus influenzae* binds with serum CRP with subsequent complement activation for bactericidal activity (Weiser et al. 1998). Murine models of *H. influenzae* infection have not been established yet to determine whether CRP protects against infection with *H. influenzae* (Simons et al. 2014). CRP also binds with commensal *Neisseriae* spp. through the ligand PC (present on the LPS) in the presence of calcium. The protective role of CRP against *Neisseriae* spp. infection in mice has not been evaluated. The protective role of CRP against lethal toxicity of LPS has also been suggested (Hirschfield et al. 2003). CRP binds with different varieties of bacteria in the hemolymph of horseshoe crab (Ngwa and Agrawal 2019; Ng et al. 2007). Thus, CRP plays the role of an antibacterial molecule as non-native CRP shows PC-independent anti-pneumococcal activity. To pursue this, CRP changes its pentameric conformation. CRP need help from some other serum proteins also to figure out pathogen recognition complexes.

1.6 CRP and Inflammation

The body's immediate response to damage its cells and tissues by microbial pathogens, by deadly stimuli-like physical or chemical injury is called inflammation (Fig. 1.4). Leukocytes infiltrate the damaged organ (s) removing the stimulus healing and repairing the tissue are the short-term endeavor of acute inflammation. In contrast, the prolonged, maladaptive and dysregulated chronic inflammation involves active inflammation, response to tissue destruction, and tissue renovations.

Persistent chronic inflammation is associated with different chronic human diseases and conditions such as atherosclerosis, cancer, allergy, arthritis, and some autoimmune diseases. Acute inflammation processes are well defined with its

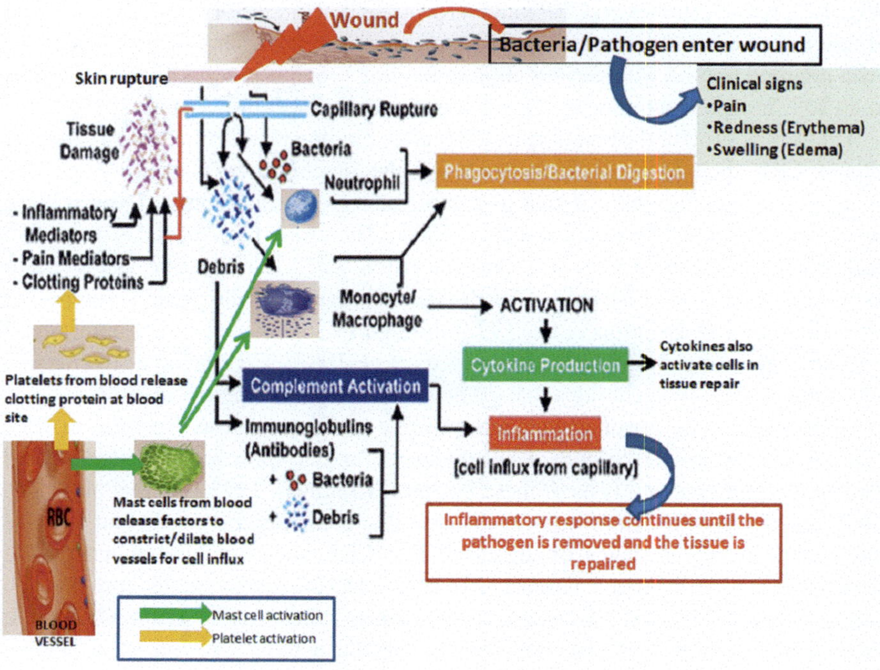

Fig. 1.4 Mechanism of inflammation

initiation and development process, but much less is known about the cellular and molecular pathways or the associated causes of chronic inflammation (Medzhitov 2008). Regarding acute and chronic inflammation, there are two schools of thought. Some biologists most probably firmly believed that acute and chronic inflammation are the two distinct domains of the inflammatory reaction; others feel that inflammation as a single aspect, which cannot be divided on the basis of its just time course.

CRP levels are known to spike dramatically in response to infection, injury, and inflammation. CRP is mainly an acute phase marker of inflammation. It is the first reported pattern recognition receptor (Shrive et al. 1996). Several researches confirmed the important roles that CRP plays in inflammation. CRP is one of the principal downstream mediator of the acute phase response after an inflammatory event and is mainly produced by IL-6-dependent hepatic biosynthesis (Pradhan et al. 2001; Baumeister et al. 2016). CRP binds with the pathogen and activates the C1q molecule leading to flow of complement pathway. It also opsonizes the pathogens (Figs. 1.1 and 1.2). Thus, CRP plays a contributory role in inflammation. CRP can balance the host defense mechanism by activating the complement pathway as well as activating the cell-mediated pathways by binding with Fc receptors of IgG (Pradhan et al. 2001). CRP binding with Fc receptors causes the release of pro-inflammatory cytokines (Du Clos 2000). CRP can also differentiate self and foreign molecules based on pattern recognition ability, which was not achieved by IgG. IgG can only recognize molecules based on their distinct antigenic epitopes

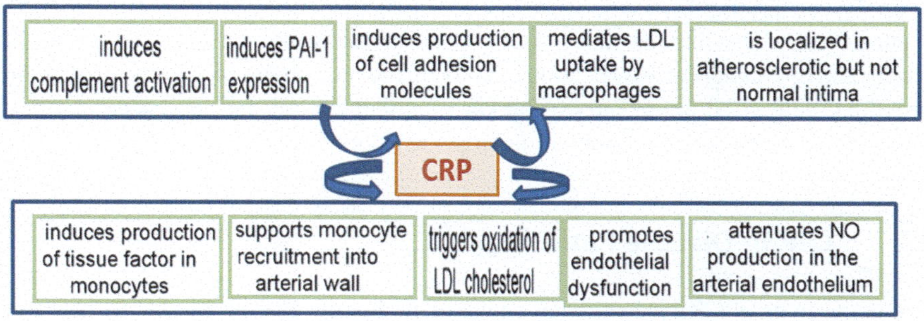

Fig. 1.5 Pathobiological role of CRP

(Du Clos 2000). CRP and IgG bears functional similarity with each other (Marnell et al. 1995; Bharadwaj et al. 1999). Several studies suggest that CRP binds to damaged cell membranes and contributes to the inflammatory loop (Kaplan and Volanakis 1974). CRP was found in atherosclerotic lesions after association with terminal complement complexes (Torzewski et al. 1998), provided the evidence that CRP also promotes local complement activation after localizing in infarcted heart tissue (Lagrand et al. 1997) causing further damage to the heart tissue. Extrahepatic local synthesis of CRP should be cited (Gould and Weiser 2001) in this regard. CRP was also localized to the cell nuclei within the synovium of rheumatoid arthritis patients, but at the same time the cell type was not identified by Gitlin et al. (1977). However, other researches indicate no significant local deposition or localization of CRP in a number of disease state/pathologies, concluding that CRP is found mainly in the fluid phase of blood rather than becoming localized in tissues at the sites of injury or inflammation (Vigushin et al. 1993). The localization of CRP in neutrophil infiltrates, mostly in allergic encephalomyelitis and lesions of vasculitis has been highly correlated (Du Clos et al. 1981; Parish 1971).

CRP levels elevated in many inflammatory disorders/disease state has been clinically used to predict disease outcomes. Thus, CRP level not only reflect the degree of underlying inflammatory response but also provide a useful tool to measure immune injury to tissues (Du Clos 2000). Another fascinating possibility is that CRP could straightforwardly participate in amplifying the immune response, thus leading to further tissue damage. CRP has been known to bind to damaged tissues, lipoproteins, nuclear antigens, and to apoptotic cells for clearance. CRP is also known to participate in complement activation in damaged tissue. Interestingly, CRP is not present in normal blood vessel wall but localizes in atherosclerotic plaques (Fig. 1.5). It has been shown that the migration and appearance of monocytes may be preceded by the deposition of CRP in early atherosclerotic lesions (Lagrand et al., 1997; Reynolds and Vance 1987; Torzewski et al. 2000). Recently, we found that CRP is also known to induce the expression of chemokines and various adhesion molecules in human endothelial cells. As monocytes bind with chemokines on the endothelial cells, so, indirectly CRP helps in stimulating monocytes to migrate to the tissues. The activation of endothelial cells by lipopolysaccharide acts in synergy with CRP molecule. CRP and lipopolysaccharide synergistically also induce tissue factor production by monocytes (Pasceri et al. 2000, 2001; Nakagomi et al. 2000). Thus, from the chain of events, it can be visualized that CRP is not only a marker of inflammation, but also acts as a key molecule in propagating inflammation amplifier loop also (Yeh et al. 2001).

1.7 CRP in Disease Pathology

The major hype of CRP research has centered on the role of CRP and its isoforms on cardiovascular disease and stroke. CRP is used in laboratories as a clinical biomarker of inflammation. Its elevated serum concentration is a strong independent predictor/indicator of cardiovascular disease in mostly asymptomatic individuals. The active role of CRP in the pathophysiology of different cardiovascular disease has been linked in the prognosis of patients suffering from atrial fibrillation, atherosclerotic disease, aortic valve disease, congestive heart failure, heart transplantation, and myocarditis . Different high-throughput assays are used to detect baseline levels of CRP in patients who are at risk of having cardiovascular disease. A person with a CRP level greater than 3 mg/L has an elevated risk of coronary heart disease, and this risk also increases in those individuals with type 2 diabetes . Increased concentration of CRP has been found in individuals with appendicitis, pancreatitis, cholecystitis, and meningitis, suggesting that CRP is a more sensitive marker than erythrocyte sedimentation rate (Pepys and Hirschfield 2003) (Fig. 1.6).

In postmenopausal women, oral hormone replacement therapy (HRT) causes increase in background levels of circulating CRP leading to the risk of thrombotic events like clots (Kurtz et al. 2011; Sproston and Ashworth 2018). A combination of oxidized low-density lipoproteins (oxLDLs) and estrogen causes increased CRP expression in a model of coronary heart disease in both postmenopausal women and older men (Corcoran et al. 2010). Same experiment when repeated with testosterone in place of estrogen supplementation, no effect on CRP expression was seen. Nearly twofold increased levels of circulating CRP was also noticed by Ridker et al. (1999a, b) in healthy postmenopausal women who were taking oral HRT. Several studies have established the predictive role of CRP in cardiovascular disease and also confirmed that HRT used in postmenopausal women increases the risk of blood clots and stroke (Ridker et al. 1999a, b; Decensi et al. 2002; Vongpatanasin et al. 2003).

Persuasive researches confirmed that CRP as a powerful independent clinical predictor of future cardiovascular risk (Fig. 1.6). CRP envisages cardiovascular risk in a wide variety of clinical situations, including men and women without obvious cardiovascular disease, in patients with acute coronary syndromes or presenting stable angina, cardiovascular death, incident myocardial infarction, in patients with post-myocardial infarction, and even those with the metabolic syndrome (Ridker et al. 2000a, b, 2003; Biasucci et al. 1999;Liuzzo et al. 1994). In addition, CRP also predicts the risk of sudden cardiac death, ischemic stroke, incident peripheral and restenosis after percutaneous coronary intervention artery disease (Ridker 2002;

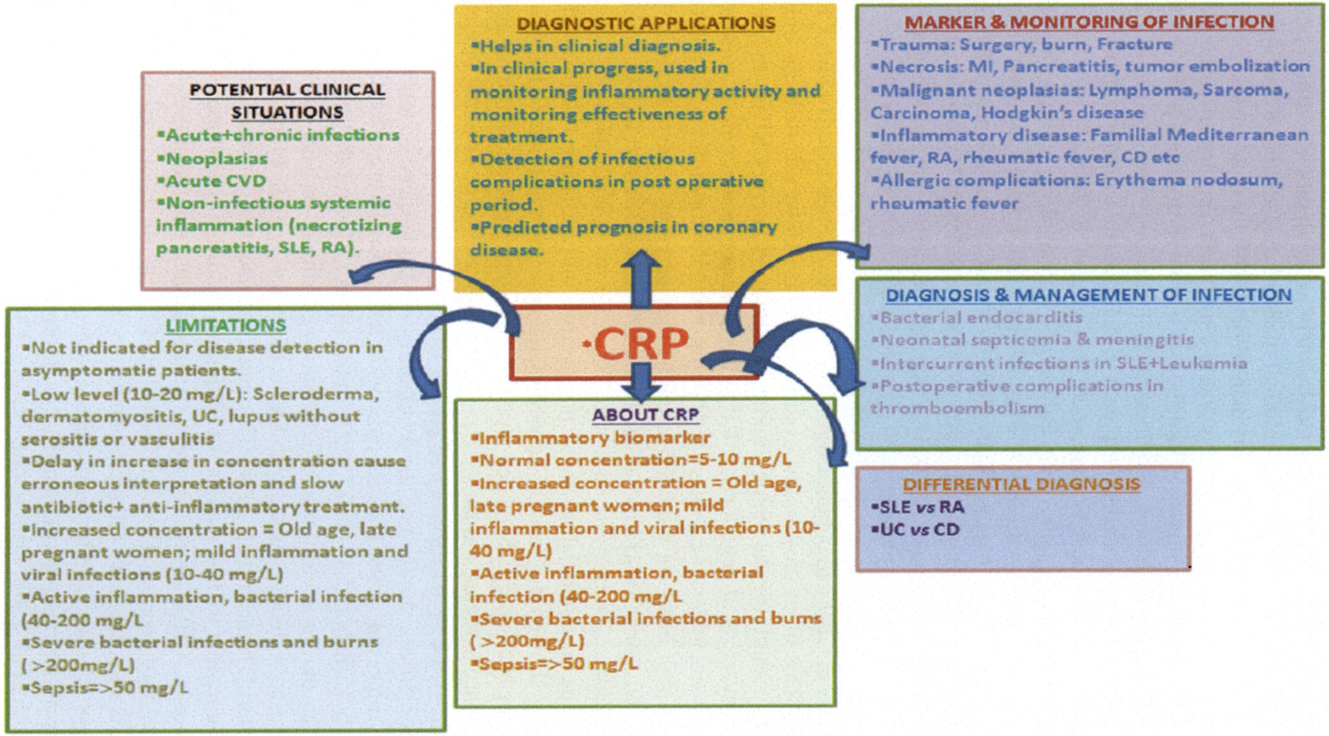

Fig. 1.6 Clinical relevance of CRP. The diagnostic, clinical indicator, prognosis of CRP in different disease condition was mentioned along with its limitations. *SLE* systemic Lupus erythromatosus, *CD* Crohn's disease, *UC* ulcerative colitis, *RA* rheumatoid arthritis, *MI* myocardial infarction, *CVD* cardiovascular diseases (Aguiar et al. 2013; Pepys and Hirschfield 2003)

Albert et al. 2002; Ridker et al. 2001; Blake and Ridker 2002). CRP was found to be the stronger predictor than LDL cholesterol in the incident of cardiovascular events (Ridker et al. 2002). Thus, CRP was used widely as an analyte in clinical practice (Fig. 1.6). CRP can be an appropriate clinical marker because it has a long half-life. CRP levels do not fluctuate with circadian rhythm and its concentration remains stable over time without any variability and fasting blood samples are not required. Currently, CRP levels <1 mg/L, 1–3 mg/L, and > 3 mg/L are used to denote low-, moderate-, and high risk groups. Modified CRP showed a greater pro-inflammatory phenotype and hence exhibited even greater risk for adverse cardiovascular events (Khreiss et al. 2004; Yeh and Willerson 2003; Verma et al. 2004).

Normal CRP level may spike up in individuals with smoking habit, with obesity, diabetes, or in individuals with lack of exercise habit. All these factors may increase the risk of cardiovascular events. So by lifestyle modification, elevated CRP levels can be normalized (Tchernof et al. 2002; Rohde et al. 1999; Haffner et al. 2002; Esposito et al. 2003; Verma et al. 2004). Other risk-reduction strategies include uses of statins, and aspirin in patients with elevated CRP level to lower cardiac risk. Studies revealed that apparently healthy individuals with low levels of LDL but with elevated levels of CRP will benefit from statin therapy. As mentioned before, CRP is a better predictor of cardiovascular events than LDL. However, this can only be elucidated by identifying the major binding sites for monomeric CRP on endothelial cells. The conversion of pentameric CRP to the more active monomeric form at the inflammatory site may serve to limit CRP's pro-inflammatory effects. Reports also demonstrated that blocking the CRP receptors may serve as a novel tactic to limit endothelial cell activation (Ridker et al. 1997, 1999a, b; Plenge et al. 2002; Albert et al. 2001; Ridker 2003a, b; Khreiss et al. 2004; Verma et al. 2004).

In 1996, the first population-based prospective study relating to observational epidemiology showed the relation between CRP concentration and CHD risk. By 2009, 50 more such studies had been reported. Each year more and more researches are coming out validating the same. The speedy appearance of different epidemiological studies on CRP has been aided greatly by the relative stability of CRP during long-duration frozen storage and the availability of standardized quantitative assay methods (Kuller et al. 1996; Ridker et al. 1997, 2002; Danesh et al. 2004; Emerging Risk Factors Collaboration et al. 2010). Interestingly, despite the spikes in CRP concentration during acute phase inflammatory condition, its year-to-year consistency within an individual is comparable to those for systolic blood pressure and total cholesterol (Danesh et al. 2004). Through researches continuous associations of CRP concentration with a variety of different conditions or disease state like ischemic stroke, CHD, several cancers causing deaths, external injury or trauma, and chronic lung disease. The significance of CRP to such a varied spectrum of conditions is quite unclear. Other inflammatory biomarkers like leukocyte count, fibrinogen, or albumin have been previously shown to have correlations with a range of both nonvascular and vascular conditions (Fibrinogen Studies Collaboration et al. 2005; Danesh et al. 1998; Danesh and Pepys 2009).

1.8 Discussion

CRP, after its identification in 1930, was later considered to be an "acute phase protein." It is an early indicator of inflammatory or infectious conditions. An elevated level of CRP serves as an early marker of the intensity of inflammation in processes as dissimilar as myocardial infarction and appendicitis. It has been studied, since then, as a screening tool for inflammation. It acts as a diagnostic adjunct and a marker for disease activity. Different advanced methods of quantifying CRP have led to amplified application to clinical medicine. In the emergency departments of hospitals, CRP must be implemented in the clinical context of various diseases. No single value of CRP (or other biomarkers) can be used to rule in or rule out a specific disease diagnosis. Therefore, CRP has limited utility in the emergency departments. It may be a useful accessory to serial examinations in evasive presentations of appendicitis in those primary health cares without equipped access to computed tomography (CT) scan. It may be elevated with impediments or treatment failures in patients with pancreatitis, pneumonia, urinary tract infections, and pelvic inflammatory disease. The concentration of circulating CRP associates with endovascular diseases. It may serve to recognize otherwise asymptomatic patients at critical cardiovascular risk to necessitate aggressive therapy. CRP is usually elevated in patients with occult bacteremia, meningitis, and neonatal sepsis.

References

Abernethy TJ, Avery OT (1941) The occurrence during acute infections of a protein not normally present in the blood: I. Distribution of the reactive protein in patients' sera and the effect of calcium on the flocculation reaction with C polysaccharide of pneumococcus. J Exp Med 73:173–182

Agassandian M, Shurin GV, Ma Y, Shurin MR (2014) C-reactive protein and lung diseases. Int J Biochem Cell Biol 53:77–88. M

Agrawal A, Volanakis JE (1994) Probing the C1q-binding site on human C-reactive protein by site-directed mutagenesis. J Immunol 152:5404–5410

Agrawal A, Lee S, Carson M, Narayana SVL, Greenhough TJ, Volanakis JE (1997) Site-directed mutagenesis of the phosphocholine-binding site of human C-reactive protein: role of Thr76 and Trp67. J Immunol 158:345–350

Agrawal A, Simpson MJ, Black S, Carey MP, Samols D (2002) A C reactive protein mutant that does not bind to phosphocholine and pneumococcal C-polysaccharide. J Immunol 169:3217–3222. https://doi.org/10.4049/jimmunol.169.6.3217

Agrawal A, Suresh MV, Singh SK, Ferguson DA Jr (2008) The protective function of human C-reactive protein in mouse models of Streptococcus pneumoniae infection. Endocr Metab Immune Disord Drug Targets 8:231–237. https://doi.org/10.2174/187153008786848321

Aguiar FJ, Ferreira-Júnior M, Sales MM, Cruz-Neto LM, Fonseca LA, Sumita NM, Duarte NJ, Lichtenstein A, Duarte AJ (2013) C-reactive protein: clinical applications and proposals for a rational use. Rev Assoc Med Bras (1992) 59(1):85–92

Ahmed N, Thorley R, Xia D, Samols D, Webster RO (1996) Transgenic mice expressing CRP exhibit diminished chemotactic factor induced alveolitis. Am J Respir Crit Care Med 153:1141–1147

Albert MA, Danielson E, Rifai N, Ridker PM, PRINCE Investigators (2001) Effect of statin therapy on C-reactive protein levels: the pravastatin inflammation/CRP evaluation (PRINCE): a randomized trial and cohort study. JAMA 286(1):64–70

Albert CM, Ma J, Rifai N, Stampfer MJ, Ridker PM (2002) Prospective study of C-reactive protein, homocysteine, and plasma lipid levels as predictors of sudden cardiac death. Circulation 105 (22):2595–2599

Anderson JK, Stroud RM, Volanakis JE (1978) Studies on the binding specificity of human C-reactive protein for phosphorylcholine. Fed Proc 37:1495

Ansar W, Bandyopadhyay SM, Chowdhury S, Habib SH, Mandal C (2006) Role of C-reactive protein in complement-mediated hemolysis in Malaria. Glycoconj J 23(3–4):233–240

Ansar W, Mukhopadhyay S, Habib SK, Basu S, Saha B, Sen AK, Mandal CN, Mandal C (2009a) Disease-associated glycosylated molecular variants of human C-reactive protein activate complement-mediated hemolysis of erythrocytes in tuberculosis and Indian visceral leishmaniasis. Glycoconj J 26(9):1151–1169. https://doi.org/10.1007/s10719-009-9236-y

Ansar W, Habib SK, Roy S, Mandal C, Mandal C (2009b) Unraveling the C-reactive protein complement-cascade in destruction of red blood cells: potential pathological implications in Plasmodium falciparum malaria. Cell Physiol Biochem 23(1–3):175–190. https://doi.org/10.1159/000204106

Ballou SP, Lozanski G (1992) Induction of inflammatory cytokine release from cultured human monocytes by C-reactive protein. Cytokine 4:361–368

Baumeister D, Akhtar R, Ciufolini S, Pariante CM, Mondelli V (2016) Childhood trauma and adulthood inflammation: a meta-analysis of peripheral C-reactive protein, inetleukin-6 and tumour necrosis factor-α. Mol Psychiatry 21:642–649. https://doi.org/10.1038/mp.2015.67

Bee A, Culley FJ, Alkhalife IS, Bodman-Smith KB, Raynes JG, Bates PA (2001) Transformation of Leishmania mexicana metacyclic promastigotes to amastigote-like forms mediated by binding of C -reactive protein. Parasitology 122:521–529

Bharadwaj D, Stein M-P, Volzer M, Mold C, DuClos TW (1999) The major receptor for C-reactive protein on leukocytes is Fcg receptor II. J Exp Med 190:585–590

Biasucci LM, Liuzzo G, Grillo RL, Caligiuri G, Rebuzzi AG, Buffon A, Summaria F, Ginnetti F, Fadda G, Maseri A (1999) Elevated levels of C-reactive protein at discharge in patients with unstable angina predict recurrent instability. Circulation 99(7):855–860

Black S, Kushner I, Samols D (2004) C-reactive protein. J Biol Chem 279(47):48487–48490. https://doi.org/10.1074/jbc.R400025200

Blake GJ, Ridker PM (2002) C-reactive protein and prognosis after percutaneous coronary intervention. Eur Heart J 23:923–925

Bodman-Smith KB, Melendez AJ, Campbell I, Harrison PT, Allen JM, Raynes JG (2002a) C-reactive protein-mediated phagocytosis and phospholipase D signalling through the high-affinity receptor for immunoglobulin G (FcγRI). Immunology 107:252–260

Bodman-Smith KB, Mbuchi M, Culley FJ, Bates PA, Raynes JG (2002b) C-reactive protein-mediated phagocytosis of Leishmania donovani promastigotes does not alter parasite survival or macrophage responses. Parasite Immunol 24(9–10):447

Boncler M, Wu Y, Watala C (2019) The multiple faces of C-reactive protein—physiological and pathophysiological implications in cardiovascular disease. Molecules 2062:24. https://doi.org/10.3390/molecules24112062

Boras E, Slevin M, Alexander MY, Aljohi A, Gilmore W, Ashworth J, Krupinski J, Potempa LA, Al Abdulkareem I, Elobeid A, Matou-Nasri S (2014) Monomeric C-reactive protein and Notch-3 co-operatively increase angiogenesis through PI3K signalling pathway. Cytokine 69 (2):165–179. https://doi.org/10.1016/j.cyto.2014.05.027

Bray C, Bell LN, Liang H, Haykal R, Kaiksow F, Mazza JJ, Yale SH (2016) Erythrocyte sedimentation rate and C-reactive protein measurements and their relevance in clinical medicine. Wis Med J 115:317–321

Chandrashekara S (2014) C-reactive protein: An inflammatory marker with specific role in physiology, pathology, and diagnosis. J Rheumatol Clin Immunol 2:S1

Ciliberto G, Arcone R, Wagner EF, Ru¨ther U (1987) Inducible and tissue-specific expression of human C-reactive protein in transgenic mice. EMBO J 6:4017–4022

Ciubotaru I, Potempa LA, Wander RC (2005) Production of modified C-reactive protein in U937-derived macrophages. Exp Biol Med 230(10):762–770. https://doi.org/10.1177/153537020

Corcoran MP, Meydani M, Lichtenstein AH, Schaefer EJ, Dillard A, Lamon-Fava S (2010) Sex hormone modulation of proinflammatory cytokine and CRP expression in macrophages from older men and postmenopausal women. J Endocrinol 206(2):217–224. https://doi.org/10.1677/JOE-10-0057

CRP C-Reactive Protein (Homo sapiens) (2020) Entrez gene. National Center for Biotechnology Information (NCBI), U.S. National Library of Medicine.

Culley FJ, Harris RA, Kaye PM, McAdam PWJ, Raynes JG (1996) C-reactive protein binds to a novel ligand on Leishmania donovani and increases uptake into human macrophages. J Immunol 156:4691–4696

Culley FJ, Bodman-Smith KB, Ferguson MAJ, Nikolaev AV, Shantilal N, Raynes JG (2000) C-reactive protein binds to phosphorylated carbohydrates. Glycobiology 10:59–65

Danesh J, Pepys MB (2009) C-reactive protein and coronary disease: is there a causal link? Circulation 120:2036–2039

Danesh J, Collins R, Appleby P, Peto R (1998) Association of fibrinogen, C-reactive protein, albumin, or leukocyte count with coronary heart disease: meta-analyses of prospective studies. JAMA 279:1477–1482

Danesh J, Wheeler JG, Hirschfield GM, Eda S, Eiriksdottir G, Rumley A, Lowe GD, Pepys MB, Gudnason V (2004) C-reactive protein and other circulating markers of inflammation in the prediction of coronary heart disease. N Engl J Med 350:1387–1397

Das T, Sen AK, Kempf T, Pramanik SR, Mandal C, Mandal C (2003) Induction of glycosylation in human C-reactive protein under different pathological conditions. Biochem J 373 (Pt 2):345–355

Das T, Mandal C, Mandal C (2004a) Variations in binding characteristics of glycosylated human C-reactive proteins in different pathological conditions. Glycoconj J 20:537–543

Das T, Mandal C, Mandal C (2004b) Protein A-a new ligand for human C-reactive protein. FEBS Lett 576:107–113

De Beaufort AJ, Langermans JAM, Matze-van der Lans AM, Hiemstra PS, Vossen JM, van Furth R (1997) Difference in binding of killed and live Streptococcus pneumoniae serotypes by C-reactive protein. Scand J Immunol 46:597–600. https://doi.org/10.1046/j.1365-3083.1997.d01-171.x

Decensi A, Omodei U, Robertson C, Bonanni B, Guerrieri-Gonzaga A, Ramazzotto F et al (2002) Effect of transdermal estradiol and oral conju-gated estrogen on C-reactive protein in retinoid-placebo trial in healthy women. Circulation 106(10):1224–1228. https://doi.org/10.1161/01.CIR.0000028463.74880.EA

Devaraj S, Dasu MR, Singh U, Rao LV, Jialal I (2009) C-reactive protein stimulates superoxide anion release and tissue factor activity in vivo. Atherosclerosis 203:67–74

Diehl EE, Haines GK 3rd, Radosevich JA, Potempa LA (2000) Immunohistochemical localization of modified C-reactive protein antigen in normal vascular tissue. Am J Med Sci 319(2):79–83

Du Clos TW (2000) Function of C-reactive protein. Ann Med 32(4):274–278. https://doi.org/10.3109/07853890009011772

Du Clos TW (2003) C-reactive protein as a regulator of autoimmunity and inflammation. Arthritis Rheum 48:1475–1477

Du Clos TW (2013) Pentraxins: Structure, function, and role in inflammation. ISRN Inflamm 2013:379,040

Du Clos TW, Mold C (2004) C-reactive protein: an activator of innate immunity and a modulator of adaptive immunity. Immunol Res 30(3):261–277. https://doi.org/10.1385/IR:30:3:261

Du Clos TW, Mold C, Paterson PY, Alroy J, Gewurz H (1981) Localization of C-reactive protein in inflammatory lesions of experimental allergic enceph-alomyelitis. Clin Exp Immunol 43:565–573

Du Clos TW, Zlock LT, Rubin RL (1988) Analysis of the binding of C-reactive protein to histones and chromatin. J Immunol 141(12):4266–4270

Edwards KM, Gewurz H, Lint TF, Mold C (1982) A role for C-reactive protein in the complement-mediated stimulation of human neutrophils by type 27 Streptococcus pneumoniae. J Immunol 128:2493–2496

Emerging Risk Factors Collaboration, Kaptoge S, Di Angelantonio E, Lowe G, Pepys MB, Thompson SG, Collins R, Danesh J (2010) C-reactive protein concentration and risk of coronary heart disease, stroke, and mortality: an individual participant meta-analysis. Lancet 375 (9709):132–140. https://doi.org/10.1016/S0140-6736(09)61717-7

Esposito K, Pontillo A, Di Palo C et al (2003) Effect of weight loss and lifestyle changes on vascular inflammatory markers in obese women: a randomized trial. JAMA 289:1799–1804

Fearon DT, Locksley RM (1996) The instructive role of innate immunity in the acquired immune response. Science 272(5258):50–53

Fibrinogen Studies Collaboration, Danesh J, Lewington S, Thompson SG, Lowe GD, Collins R, Kostis JB, Wilson AC, Folsom AR, Wu K, Benderly M, Goldbourt U, Willeit J, Kiechl S, Yarnell JW, Sweetnam PM, Elwood PC, Cushman M, Psaty BM, Tracy RP, Tybjaerg-Hansen-A, Haverkate F, de Maat MP, Fowkes FG, Lee AJ, Smith FB, Salomaa V, Harald K, Rasi R, Vahtera E, Jousilahti P, Pekkanen J, D'Agostino R, Kannel WB, Wilson PW, Tofler G, Arocha-Piñango CL, Rodriguez-Larralde A, Nagy E, Mijares M, Espinosa R, Rodriquez-Roa E, Ryder E, Diez-Ewald MP, Campos G, Fernandez V, Torres E, Marchioli R, Valagussa F, Rosengren A, Wilhelmsen L, Lappas G, Eriksson H, Cremer P, Nagel D, Curb JD, Rodriguez B, Yano K, Salonen JT, Nyyssönen K, Tuomainen TP, Hedblad B, Lind P, Loewel H, Koenig W, Meade TW, Cooper JA, De Stavola B, Knottenbelt C, Miller GJ, Cooper JA, Bauer KA, Rosenberg RD, Sato S, Kitamura A, Naito Y, Palosuo T, Ducimetiere P, Amouyel P, Arveiler D, Evans AE, Ferrieres J, Juhan-Vague I, Bingham A, Schulte H, Assmann G, Cantin B, Lamarche B, Després JP, Dagenais GR, Tunstall-Pedoe H, Woodward M, Ben-Shlomo Y, Davey Smith G, Palmieri V, Yeh JL, Rudnicka A, Ridker P, Rodeghiero F, Tosetto A, Shepherd J, Ford I, Robertson M, Brunner E, Shipley M, Feskens EJ, Kromhout D, Dickinson A, Ireland B, Juzwishin K, Kaptoge S, Lewington S, Memon A, Sarwar N, Walker M, Wheeler J, White I, Wood A (2005) Plasma fibrinogen level and the risk of major cardiovascular diseases and nonvascular mortality: an individual participant meta-analysis. JAMA 294(14):1799–1809

Filep J, Foldes-Filep E (1989) E_ects of C-reactive protein on human neutrophil granulocytes challenged with N-formyl-methionyl-leucyl-phenylalanine and platelet-activating factor. Life Sci 44:517–524

Frolet C, Beniazza M, Roux L, Gallet B, Noirclerc-Savoye M, Vernet T, Di Guilmi AM (2010) New adhesin functions of surface-exposed pneumococcal proteins. BMC Microbiol 10:190. https://doi.org/10.1186/1471-2180-10-190

Gabay C, Kushner I (1999) Acute-phase proteins and other systemic responses to inflammation. N Engl J Med 340(6):448–454. https://doi.org/10.1056/NEJM199902113400607

Galve-de Rochemonteix B, Wiktorowicz K, Kushner I, Dayer JM (1993) C-reactive protein increases production of IL-1 alpha, IL-1 beta, and TNF-alpha, and expression of mRNA by human alveolar macrophages. J Leukoc Biol 53:439–445

Gang TB, Hammond DJ Jr, Singh SK, Ferguson DA Jr, Mishra VK, Agrawal A (2012) The phosphocholine-binding pocket on C-reactive protein is necessary for initial protection of mice against pneumococcal infection. J Biol Chem 287:43116–43125. https://doi.org/10.1074/jbc.M112.427310

Gang TB, Hanley GA, Agrawal A (2015) C-reactive protein protects mice against pneumococcal infection via both phosphocholine-dependent and phosphocholine-independent mechanisms. Infect Immun 83:1845–1852. https://doi.org/10.1128/IAI.03058-14

Gewurz H, Mold C, Siegel J, Fiedel B (1982) C-reactive protein and the acute phase response. Adv Intern Med 27:345–372

Gewurz H, Zhang XH, Lint TF (1995) Structure and function of the pentraxins. Curr Opin Immunol 7(1):54–64
Gitlin JD, Gitlin JI, Gitlin D (1977) Localizing of C-reactive protein in synovium of patients with rheumatoid arthritis. Arthritis Rheum 20(8):1491–1499. https://doi.org/10.1002/art.1780200808
Gould JM, Weiser JN (2001) Expression of Creactive protein in the human respiratory tract. Infect Immun 69:1747–1175
Haffner SM, Greenberg AS, Weston WM et al (2002) Effect of rosiglitazone treatment on nontraditional markers of cardiovascular disease in patients with type 2 diabetes mellitus. Circulation 106:679–684
Heuertz RM, Tricomi SM, Ezekiel UR, Webster RO (1999) C-reactive protein inhibits chemotactic peptide-induced p38 mitogen-activated protein kinase activity and human neutrophil movement. J Biol Chem 274:17,968–17,974
Heuertz RM, Schneider GP, Potempa LA, Webster RO (2005) Native and modified C-reactive protein bind different receptors on human neutrophils. Int J Biochem Cell Biol 37:320–335
Hirschfield GM, Herbert J, Kahan MC, Pepys MB (2003) Human C-reactive protein does not protect against acute lipopolysaccharide challenge in mice. J Immunol 171:6046–6051. https://doi.org/10.4049/jimmunol.171.11.6046
Hutchinson WL, Koenig W, Fröhlich M, Sund M, Lowe GD, Pepys MB (2000) Immunoradiometric assay of circulating C-reactive protein: age-related values in the adult general population. Clin Chem 46(7):934–938
Ji SR, Wu Y, Potempa LA, Qiu Q, Zhao J (2006) Interactions of C-reactive protein with low-density lipoproteins: Implications for an active role of modified C-reactive protein in atherosclerosis. Int J Biochem Cell Biol 38:648–661
Kaplan MH, Volanakis JE (1974) Interaction of C-reactive protein complexes with the complement system. I. Consumption of human complement associated with the reaction of C-reactive protein with pneumococcal C-polysaccharide and with the choline phosphatides, lecithin, and sphingomyelin. J Immunol 112:2135–2147
Karasahin O, Tasar PT, Timur O, Yildirim F, Binici DN, Sahin S (2018) The value of C-reactive protein in infection diagnosis and prognosis in elderly patients. Aging Clin Exp Res 30:555–562
Kempka G, Roos PH, Kolb-Bachofen V (1990) A membrane-associated form of C-reactive protein is the galactose-specific particle receptor on rat liver macrophages. J Immunol 144 (3):1004–1009
Kew RR, Hyers TM, Webster RO (1990a) Human C-reactive protein inhibits neutrophil chemotaxis in vitro: Possible implications for the adult respiratory distress syndrome. J Lab Clin Med 115:339–345
Kew RR, Hyers TM, Webster RO (1990b) Human CRP inhibits neutrophil chemotaxis in vitro; possible implications for ARDS. J Lab Clin Med 115:339–345
Khreiss T, József L, Potempa LA, Filep JG (2004) Conformational rearrangement in C-reactive protein is required for proinflammatory actions on human endothelial cells. Circulation 109 (16):2016–2022
Kilpatrick JM, Volanakis JE (1991) Molecular genetics, structure, and function of C-reactive protein. Immunol Res 10(1):43–53
Kilpatrick JM, Kearney JF, Volanakis JE (1982) Demonstration of calcium-induced conformational change(s) in C-reactive protein by using monoclonal antibodies. Mol Immunol 19:1159–1165
Kindmark CO (1972) In vitro binding of human C-reactive protein by some pathogenic bacteria and zymosan. Clin Exp Immunol 11:283–289
Köttgen E, Hell B, Kage A, Tauber R (1992) Lectin specificity and binding characteristics of human C-reactive protein. J Immunol 149:445–453
Kuller LH, Tracy RP, Shaten J, Meilahn EN (1996) Relation of C-reactive protein and coronary heart disease in the MRFIT nested case-control study: Multiple Risk Factor Intervention Trial. Am J Epidemiol 144:537–547

Kurtz EG, Ridker PM, Rose LM, Cook NR, Everett BM, Buring JE et al (2011) Oral postmenopausal hormone therapy, C-reactive protein and cardiovascular out-comes. Menopause 18 (1):23–29. https://doi.org/10.1097/gme.0b013e3181e750dd

Kushner I (1982) The phenomenon of the acute phase response. Ann NY Acad Sci USA 389:39–48. https://doi.org/10.1111/j.1749-6632.1982.tb22124.x

Kushner I, Kaplan MH (1961) Studies of acute phase proteins. I. An immunohistochemical method for the localisation of CRP in rabbits. Association with necrosis in local inflammatory lesions. J Exp Med 114:961–979

Lagrand WK, Niessen HW, Wolbink GJ, Jaspars LH, Visser CA, Verheugt FW et al (1997) C-reactive protein colocalizes with complement in human hearts during acute myocardial infarction. Circulation 95(1):97–103. https://doi.org/10.1161/01.CIR.95.1.97

Lee RT, Takagahara I, Lee YC (2002) Mapping the binding areas of human C-reactive protein for phosphorylcholine and polycationic compounds. Relationship between the two types of binding sites. J Biol Chem 277(1):225–232

Ling MR, Chapple IL, Creese AJ, Matthews JB (2014) E_ects of C-reactive protein on the neutrophil respiratory burst in vitro. Innate Immun 20:339–349

Liuzzo G, Biasucci LM, Gallimore JR, Grillo RL, Rebuzzi AG, Pepys MB, Maseri A (1994) The prognostic value of C-reactive protein and serum amyloid a protein in severe unstable angina. N Engl J Med 331(7):417–424

Lu J, Mold C, Du Clos TW, Sun PD (2018) Pentraxins and Fc Receptor-Mediated Immune Responses. Front Immunol 9:2607

Mandal C, Biswas M, Nagpurkar A, Mookerjea S (1991) Isolation of a phosphoryl choline-binding protein from the hemolymph of the snail, Achatina fulica. Dev Comp Immunol 15:227–239

Marnell LL, Mold C, Volzer MA, Burlingame RW, DuClos TW (1995) C-reactive protein binds to FcgR1 in transfected COS cells. J Immunol 155:2185–2193

McCarthy MI (1947) The occurrence during acute infections of a protein not normally present in the blood. J Exp Med 85:491–498

McFadyen JD, Kiefer J, Braig D, Loseff-Silver J, Potempa LA, Eisenhardt SU, Peter K (2018) Dissociation of C-reactive protein localizes and amplifies inflammation: evidence for a direct biological role of C-reactive protein and its conformational changes. Front Immunol 9:1351. https://doi.org/10.3389/fimmu.2018.01351. eCollection 2018

Medzhitov R (2008) Origin and physiological roles of inflammation. Nature 454(7203):428–435

Miyagawa N, Okamoto Y, Nakano H (1988) Effect of C-reactive protein on peritoneal macrophages. II. Human C-reactive protein activates peritoneal macrophages of guinea pigs to release superoxide anion in vitro. Microbiol Immunol 32:721–723

Mold C, Nakayama S, Holzer TJ, Gewurz H, Du Clos TW (1981) C-reactive protein is protective against Streptococcus pneumoniae infection in mice. J Exp Med 154(5):1703–1708

Mold C, Rodgers CP, Kaplan RL, Gewurz H (1982) Binding of human C-reactive protein to bacteria. Infect Immun 38:392–395

Mold C, Gewurz H, Du Clos TW (1999) Regulation of complement activation by c-reactive protein. Immunopharmacology 42:23–30

Molins B, Pena E, de la Torre R, Badimon L (2011) Monomeric C-reactive protein is prothrombotic and dissociates from circulating pentameric C-reactive protein on adhered activated platelets under flow. Cardiovasc Res 92:328–337

Mortensen RF, Zhong W (2000) Regulation of phagocytic leukocyte activities by C-reactive protein. J Leukoc Biol 67:495–500

Murphy C, Beckers J, Rüther U (1995) Regulation of the human C-reactive protein gene in transgenic mice. J Biol Chem 270:704–708

Nakagomi A, Freedman SB, Geczy CL (2000) Interferon-gamma and lipopolysaccharide potentiate monocyte tissue factor induction by C-reactive protein: relationship with age, sex, and hormone replacement treatment. Circulation 101:1785–1791

Ng PML, Le Saux A, Lee CM, Tan NS, Lu J, Thiel S et al (2007) C-reactive protein collaborates with plasma lectins to boost immune response against bacteria. EMBO J 26:3431–3440. https://doi.org/10.1038/sj.emboj.7601762

Ngwa DN (2016) Comparison of anti-pneumococcal functions of native and modified forms of C-reactive protein. Dissertation/Master's thesis. East Tennessee States University, Johnson City, TN

Ngwa DN, Agrawal A (2019) Structure-function relationships of C-reactive protein in bacterial infection. Front Immunol 10:166

Noren Hooten N, Ejiogu N, Zonderman AB, Evans MK (2012) Association of oxidative DNA damage and C-reactive protein in women at risk for cardiovascular disease. Arterioscler Thromb Vasc Biol 32:2776–2784

Parish WE (1971) Studies on vasculitis. I. Immunoglobulins, 1C, C-reactive protein, and bacterial antigens in cutaneous vasculitis lesions. Clin Allergy 1:97–109. https://doi.org/10.1111/j.1365-2222.1971.tb02451.x

Pasceri V, Willerson JT, Yeh ET (2000) Direct proinflammatory effect of C-reactive protein on human endothelial cells. Circulation 102:2165–2168

Pasceri V, Cheng JS, Willerson JT, Yeh ET, Chang J (2001) Modulation of C-reactive protien-mediated monocyte chemoattractant protein-1 induction in human endothelial cells by anti-atherosclerosis drugs. Circulation 103:2531–2534

Pepys MB (1981) C-reactive protein fifty years on. Lancet 1:653–657

Pepys MB, Baltz ML (1983) Acute phase proteins with special reference to C-reactive protein and related proteins (pentaxins) and serum amyloid A protein. Adv Immunol 34:141–212

Pepys MB, Berger A (2001) The renaissance of C reactive protein. BMJ 322(7277):4–5

Pepys MB, Hirschfield GM (2003) C-reactive protein: a critical update. J Clin Invest 111:1805–1812. https://doi.org/10.1172/JCI200318921

Pepys MB, Baltz ML, Gomer K, Davies AJS, Doenhoff M (1979) Serum amyloid P component is an acute-phase reactant in the mouse. Nature 278:259–261

Pepys MB, Booth SE, Tennet GA, Butler PJG, Williams DG (1994) Binding of pentraxins to different nuclear structures. CRP binds to small nuclear ribonuclearprotein particles and serum amyloid P binds to chromatin and nucleoli. Clin Exp Immunol 97:152–157

Pepys MB, Hawkins PN, Kahan MC, Tennent GA, Gallimore JR, Graham D, Sabin CA, Zychlinsky A, De Diego J (2005) Pro-inflammatory Effects of Bacterial Recombinant Human C-Reactive Protein are Caused by Contamination with Bacterial Products not by C-Reactive Protein Itself. Circ Res 97:97–103

Plenge JK, Hernandez TL, Weil KM et al (2002) Simvastatin lowers C-reactive protein within 14 days: an effect independent of low-density lipoprotein cholesterol reduction. Circulation 106:1447–1452

Potempa LA, Maldonado BA, Laurent P, Zemel ES, Gewurz H (1983) Antigenic, electrophoretic and binding alterations of human C-reactive protein modified selectively in the absence of calcium. Mol Immunol 20:1165–1175

Potempa LA, Siegel JN, Fiedel BA, Potempa RT, Gewurz H (1987) Expression, detection and assay of a neoantigen (Neo-CRP) associated with a free, human C-reactive protein subunit. Mol Immunol 24(5):531–541

Potempa LA, Zeller JM, Fiedel BA, Kinoshita CM, Gewurz H (1988) Stimulation of human neutrophils, monocytes, and platelets by modified C-reactive protein (CRP) expressing a neoantigenic specificity. Inflammation 12(4):391–405

Pradhan AD, Manson JE, Rifai N, Buring JE, Ridker PM (2001) C-reactive protein, interleukin 6, and risk of developing type 2 diabetes mellitus. JAMA 286(3):327–334. https://doi.org/10.1001/jama.286.3.327

Ramadan MA, Shrive AK, Holden D, Myles DA, Volanakis JE, DeLucas LJ, Greenhough TJ (2002) The three-dimensional structure of calcium-depleted human C-reactive protein from perfectly twinned crystals. Acta Crystallogr D Biol Crystallogr 58:992–1001

Raynes JG, Curry A, Harris RA (1994) Binding of C-reactive protein to Leishmania. Biochem Soc Trans 22(1):3S

Rees RF, Gewurz H, Siegel JN, Coon J, Potempa LA (1988) Expression of a C-reactive protein neoantigen (neo-CRP) in inflamed rabbit liver and muscle. Clin Immunol Immunopathol 48 (1):95–107

Reynolds GD, Vance RP (1987) C-reactive protein immunohistochemical localization in normal and atherosclerotic human aortas. Arch Pathol Lab Med 111(3):265–269

Ridker PM (2002) Inflammatory biomarkers, statins, and the risk of stroke: cracking a clinical conundrum. Circulation 105:2583–2585

Ridker PM (2003a) Rosuvastatin in the primary prevention of cardiovascular disease among patients with low levels of low-density lipoprotein cholesterol and elevated high-sensitivity C-reactive protein: rationale and design of the JUPITER trial. Circulation 108:2292–2297

Ridker PM (2003b) Clinical application of C-reactive protein for cardiovascular disease detection and prevention. Circulation 107:363–369. https://doi.org/10.1161/01.cir.0000053730.47739.3c

Ridker PM, Cushman M, Stampfer MJ et al (1997) Inflammation, aspirin, and the risk of cardiovascular disease in apparently healthy men. N Engl J Med 336:973–999

Ridker PM, Hennekens CH, Rifai N, Buring JE, Manson JE (1999a) Hormone replacement therapy and increased plasma concentration of C-reactive protein. Circulation 100(7):713–716. https://doi.org/10.1161/01.CIR.100.3.230

Ridker PM, Rifai N, Pfeffer MA et al (1999b) Long-term effects of pravastatin on plasma concentration of C-reactive protein: the Cholesterol And Recurrent Events (CARE) Investigators. Circulation 100:230–235

Ridker PM, Hennekens CH, Buring JE, Rifai N (2000a) C-reactive protein and other markers of inflammation in the prediction of cardiovascular disease in women. N Engl J Med 342 (12):836–843. https://doi.org/10.1056/NEJM200003233421202

Ridker PM, Rifai N, Stampfer MJ, Hennekens CH (2000b) Plasma concentration of interleukin-6 and the risk of future myocardial infarction among apparently healthy men. Circulation 101 (15):1767–1772

Ridker PM, Stampfer MJ, Rifai N (2001) Novel risk factors for systemic atherosclerosis: a comparison of C-reactive protein, fibrinogen, homocysteine, lipoprotein(a), and standard cholesterol screening as predictors of peripheral arterial disease. JAMA 285:2481–2485

Ridker PM, Rifai N, Rose L, Buring JE, Cook NR (2002) Comparison of C-reactive protein and low-density lipoprotein cholesterol levels in the prediction of first cardiovascular events. N Engl J Med 347:1557–1565

Ridker PM, Buring JE, Cook NR, Rifai N (2003) C-reactive protein, the metabolic syndrome, and risk of incident cardiovascular events: an 8-year follow-up of 14,719 initially healthy American women. Circulation 107(3):391–397

Robey FA, Jones KD, Tanaka T, Liu TY (1984) Binding of C-reactive protein to chromatin and nucleosome core particles. A possible physiological role of C-reactive protein. J Biol Chem 259 (11):7311–7316

Robey FA, Jones KD, Steinberg AD (1985) C-reactive protein mediates the solubilization of nuclear DNA by complement in vitro. J Exp Med 161(6):1344–1356

Rohde LE, Hennekens CH, Ridker PM (1999) Survey of C-reactive protein and cardiovascular risk factors in apparently healthy men. Am J Cardiol 84:1018–1022

Ryu JA, Yang JH, Lee D, Park CM, Suh GY, Jeon K, Cho J, Baek SY, Carriere KC, Chung CR (2015) Clinical usefulness of procalcitonin and c-reactive protein as outcome predictors in critically ill patients with severe sepsis and septic shock. PLoS One 10(9):e0138150

Salazar J, Martinez MS, Mervin Chavez M, Toledo A, Anez R, Yaquelín Torres Y et al (2014a) C-reactive protein: clinical and epidemiological perspectives. Cardiol Res Pract 2014:605810. https://doi.org/10.1155/2014/605810

Salazar J, Martínez MS, Chávez-Castillo M, Núñez V, Añez R, Torres Y, Toledo A, Chacín M, Silva C, Pacheco E, Rojas J, Bermúdez V (2014b) C-reactive protein: an in-depth look into structure, function, and regulation. Int Sch Res Notices 2014:653045. https://doi.org/10.1155/2014/653045. eCollection 2014

Shrive AK, Cheetham GMT, Holden D, Myles DAA, Turnell WG, Volanakis JE, Pepys MB, Bloomer AC, Greenhough TJ (1996) Nat Struct Biol 3:346–354. https://doi.org/10.1038/nsb0496-346

Simental-Mendia LE, Sahebkar A, Rodriguez-Moran M, Zambrano-Galvan G, Guerrero-Romero F (2017) Effect of magnesium supplementation on plasma C-reactive protein concentrations: a systematic review and meta-analysis of randomized controlled trials. Curr Pharm Des 23:4678–4686

Simons JP, Loeffler JM, Al-Shawi R, Ellmerich S, Hutchinson WL, Tennent GA, Petrie A, Raynes JG, de Souza JB, Lawrence RA, Read KD, Pepys MB (2014) C-reactive protein is essential for innate resistance to pneumococcal infection. Immunology 142(3):414–420. https://doi.org/10.1111/imm.12266

Sproston NR, Ashworth JJ (2018) Role of C-Reactive Protein at Sites of Inflammation and Infection. Front Immunol 9:754

Stein MP, Edberg JC, Kimberley PP et al (2000) CRP binding to Fc RIIa on human monocytes and neutrophils is allele specific. J Clin Invest 105:369–376

Suresh MV, Singh SK, Ferguson DA Jr, Agrawal A (2006) Role of the property of C-reactive protein to activate the classical pathway of complement in protecting mice from pneumococcal infection. J Immunol 176:4369–4374. https://doi.org/10.4049/jimmunol.176.7.4369

Szalai AJ (2001) The antimicrobial activity of C-reactive protein. Microbes Infect 4:201–205

Szalai AJ, Briles DE, Volanakis JE (1995) Human C-reactive protein is protective against fatal Streptococcus pneumoniae infection in transgenic mice. J Immunol 155:2557–2563

Szalai AJ, Briles DE, Volanakis JE (1996) Role of complement in C-reactive-protein-mediated protection of mice from Streptococcus pneumoniae. Infect Immun 64(11):4850–4853

Szalai AJ, Agrawal A, Greenhough TJ, Volanakis JE (1997) C-reactive protein: structural biology, gene expression, and host defense function. Immunol Res 16(2):127–136

Szalai AJ, van Ginkel FW, Dalrymple SA, Murray R, McGhee JR, Volankis JE (1998) Testosterone and IL-6 requirements for human C-reactive protein gene expression in transgenic mice. J Immunol 160(11):5294–5299

Szalai AJ, van Ginkel FW, Wang Y, McGhee JR, Volanakis JE (2000a) Complement-dependent acute-phase expression of C-reactive protein and serum amyloid P-component. J Immunol 165:1030–1035

Szalai AJ, VanCott JL, McGhee JR, Volanakis JE, Benjamin WH Jr (2000b) Human C-reactive protein is protective against fatal Salmonella enterica serovar typhimurium infection in transgenic mice. Infect Immun 68:5652–5656

Szalai AJ, McCrory HA, Cooper GS, Wu J, Kimberly RP (2002) Association between baseline levels of C-reactive protein (CRP) and a dinucleotide repeat polymorphism in the intron of the CRP gene. Genes Immun 3:14–19

Tang Y, Fung E, Xu A, Lan HY (2017) C-reactive protein and ageing. Clin Exp Pharmacol Physiol 44:9–14

Tchernof A, Nolan A, Sites CK et al (2002) Weight loss reduces C-reactive protein levels in obese postmenopausal women. Circulation 105:564–569

Thompson D, Pepys MB, Wood SP (1999) The physiological structure of human C-reactive protein and its complex with phosphocholine. Structure 7:169–177. https://doi.org/10.1016/S0969-2126(99)80023-9

Ticinesi A, Lauretani F, Nouvenne A, Porro E, Fanelli G, Maggio M, Meschi T (2017) C-reactive protein (CRP) measurement in geriatric patients hospitalized for acute infection. Eur J Intern Med 37:7–12

Tilg H, Vannier E, Vachino G, Dinarello CA, Mier JW (1993) Antiinflammatory Properties of Hepatic Acute Phase Proteins: Preferential Induction of Interleukin 1 (IL-1) Receptor Antagonist over ID113 Synthesis by Human Peripheral Blood Mononuclear Cells. J Exp Med 178:1629–1636

Tillet WS, Francis T (1930) Serological reactions in pneumonia with a non- protein somatic fraction of Pneumococcus. J Exp Med 52(4):561–571. https://doi.org/10.1084/jem.52.4.561

Torzewski J, Torzewski M, Bowyer DE, Fröhlich M, Koenig W, Waltenberger J et al (1998) C-reactive protein frequently colocalizes with the terminal complement complex in the intima of early atherosclerotic lesions of human coronary arteries. Arterioscler Thromb Vasc Biol 18 (9):1386–1392. https://doi.org/10.1161/01.ATV.18.9.1386

Torzewski M, Rist C, Mortensen RF et al (2000) C-reactive protein in the arterial intima: role of C-reactive protein receptor-dependent monocyte recruitment in atherogenesis. Arterioscler Thromb Vasc Biol 20:2094–2099

Verma S, Szmitko PE, Yeh ETH (2004) C-reactive protein structure affects function. Circulation 109:1914–1917

Vermeire S, Van Assche G, Rutgeerts P (2005) The role of C-reactive protein as an inflammatory marker in gastrointestinal diseases. Nat Clin Pract Gastr 2:580–586

Vigushin DM, Pepys MB, Hawkins PN (1993) Metabolic and scintigraphic studies of radioiodinated human C-reactive protein in health and disease. J Clin Invest 91:1351–1357. https://doi.org/10.1172/JCI116336

Volanakis JE (2001) Human C-reactive protein: expression, structure, and function. Mol Immunol 38:189–197

Volanakis JE, Kaplan MH (1971) Specificity of C-reactive protein for choline phosphate residues of pneumococcal C-polysaccharide. Proc Soc Exp Biol Med 136:612–614. https://doi.org/10.3181/00379727-136-35,323

Vongpatanasin W, Tuncel M, Wang Z, Arbique D, Mehrad B, Jialal I (2003) Differential effects of oral versus transdermal estrogen replacement therapy on C-reactive protein in postmenopausal women. J Am Coll Cardiol 41(8):1358–1363. https://doi.org/10.1016/S0735-1097(03)00156-6

Wang MY, Ji SR, Bai CJ, El Kebir D, Li HY, Shi JM, Zhu W, Costantino S, Zhou HH, Potempa LA et al (2011) A redox switch in C-reactive protein modulates activation of endothelial cells. FASEB J 25:3186–3196

Weinhold B, Bader A, Valeria POLI, Rütehr U (1997) Interleukin-6 is necessary, but not sufficient, for induction of the human C-reactive protein gene in vivo. Biochem J 325(3):617–621. https://doi.org/10.1042/bj3250617

Weiser JN, Pan N, McGowan KL, Musher D, Martin A, Richards J (1998) Phosphorylcholine on the lipopolysaccharide of Haemophilus influenza contributes to persistence in the respiratory tract and sensitivity to serum killing mediated by C-reactive protein. J Exp Med 187:631–640. https://doi.org/10.1084/jem.187.4.631

Whitehead AS, Zahedi K, Rits M, Mortensen RF, Lelias JM (1990) Mouse C-reactive protein: generation of cDNA clones, structural analysis, and induction of mRNA during inflammation. Biochem J 266:283–290. https://doi.org/10.1042/bj2660283

Yamada S, Gotoh T, Nakashima Y, Kayaba K, Ishikawa S, Nago N, Nakamura Y, Itoh Y, Kajii E (2001) Distribution of serum C-reactive protein and its association with atherosclerosis risk factors in a Japanese population. Am J Epidemiol 153:1183–1190

Yeh ET, Willerson JT (2003) Coming of age of C-reactive protein: using inflammation markers in cardiology. Circulation 107:370–372

Yeh ETH, Anderson HV, Pasceri V, Willerson JT (2001) C-reactive protein linking inflammation to cardiovascular complications. Circulation 104:974–975

Ying S-C, Gewurz H, Kinoshita CM, Potempa LA, Siegel JN (1989) Identification and partial characterization of multiple native and neoantigenic epitopes of human C-reactive protein by using monoclonal antibodies. J Immunol 143:221–228

C-Reactive Protein and Neurodegenerative Diseases

2

Inês Lopes Cardoso and Fernanda Leal

Contents

2.1 Introduction 37
2.1.1 CRP Characterization and Structure 37
2.1.2 Biosynthesis and Metabolism of CRP 38
2.2 CRP and Alzheimer's Disease 40
2.2.1 Alzheimer's Disease Etiology 40
2.2.2 Role of CRP in Alzheimer's Disease 42
2.3 CRP and Parkinson's Disease 43
2.3.1 Parkinson's Disease Etiology 43
2.3.2 Role of CRP in Parkinson's Disease 44
2.4 CRP and Huntington's Disease 45
2.4.1 Huntington's Disease Etiology 45
2.4.2 Role of CRP in Huntington's Disease 46
2.5 CRP and ALS 47
2.5.1 ALS Etiology 47
2.5.2 Role of CRP in ALS 47
2.6 CRP and Prion's Disease 48
2.6.1 Prion's Disease Etiology 48
2.6.2 Role of CRP in Prion's Diseases 49
2.7 Overview 49
References 50

Abstract

Inflammatory biomarkers are a very useful tool used to detect and monitor inflammatory processes and in the selection of the therapy strategy to use.

I. L. Cardoso (✉) · F. Leal
Health Sciences Faculty, University Fernando Pessoa, Porto, Portugal

FP-ENAS/CEBIMED, Energy, Environment and Health Research Unit/Biomedical Research Centre, University Fernando Pessoa, Porto, Portugal
e-mail: mic@ufp.edu.pt

W. Ansar, S. Ghosh (eds.), *Clinical Significance of C-reactive Protein*,
https://doi.org/10.1007/978-981-15-6787-2_2

C-reactive protein (CRP) is an nonspecific (yet highly useful) clinical biomarker of inflammation. The importance of CRP results from the lack of large fluctuations during the day or season in its plasma levels. CRP is produced by hepatocytes during the inflammatory process, as a result of the action of pro-inflammatory cytokines. Several studies refer CRP as an acute inflammation biomarker, however, this protein has also been associated with chronic inflammatory processes.

Neurodegenerative diseases are increasing worldwide. This type of conditions is characterized by the gradual loss of neuronal structure and function. Although the leading causes of neurodegeneration have not been identified yet, several studies have related the inflammatory process with several neurodegenerative pathways, associated with increased risk of developing dementia and depression, consequences of neurodegenerative diseases.

Several studies have observed changes in the levels of CRP in patients suffering from one type of neurodegenerative diseases. However, concerning Alzheimer's disease, the results stated in those studies are still not consistent. Some of them have reported increased CRP levels, others a decrease, and some have not even observed significant changes in this protein. On the other hand, data from several studies show higher CRP levels in patients suffering from Parkinson's disease and other neurodegenerative diseases, such as Huntington's disease, Amyotrophic lateral sclerosis (ALS), and Prion's disease, when compared with controls.

Thus, the goal of this study is to clarify the role of CRP in the development of neurodegenerative diseases and its possible role on the pathogenesis of these disorders, as well as its use in their diagnosis.

Keywords

C-reactive protein · Neurodegenerative diseases · Alzheimer's disease · Parkinson's disease · Inflammation · Biomarkers

Abbreviations

ALS	Amyotrophic lateral sclerosis
ALSFRS-R	ALS Functional Rating Scale-Revised
ApoE4	Apolipoprotein E isoform 4
APP	βA precursor protein
CAG	Cytosine-adenine-guanine
CRP	C-reactive protein
Da	Daltons
FcyRs	Receptors for the Fc portion of immunoglobulin G
hs-CRP	High-sensitivity CRP
Htt	Huntingtin protein
IL	Interleukin
LDL	Low density lipoprotein

mCRP	Monomeric C-reactive protein
mHtt	Mutant Huntingtin
NSE	Neuron-specific enolase
pCRP	Pentameric C-reactive protein
PolyQ	Polyglutamine
PrP	Prion protein
PrPC	Normal cellular prion protein
PrPRes	Prion protein resistant to proteinase K digestion
PrPSc	Scrapie prion protein
ROS	Reactive oxygen species
SOD1	Superoxide dismutase 1
TFC	Total functioning capacity
TNF	Tumor necrosis factor
βA	β-amyloid

Highlights

- Changes in CRP levels are indicative of acute or chronic inflammation.
- Potential use of CRP as biomarker of neurodegeneration.
- CRP involvement in the process of neurodegeneration.
- Are changes in CRP level cause or the outcome of neurodegenerative diseases?
- CRP role in deposition of protein tangles characteristic of neurodegeneration.

2.1 Introduction

The main characteristic of neurodegenerative diseases is the gradual loss of neuronal structure and function and are increasing worldwide. Although all biochemical and/or physiological processes are not fully understood, it is known that the main neurodegenerative diseases, such as the Alzheimer's, Parkinson's, Huntington's, ALS, and Prion's, have similarities, particularly regarding intracellular or extracellular deposition of protein tangles (Leal et al. 2012).

Several studies point out CRP as being, in some way, involved in the pathogenesis of neurodegenerative disorders. The main goal of this chapter is to discuss and clarify the possible involvement of CRP in neurodegenerative diseases.

2.1.1 CRP Characterization and Structure

The first description of CRP is attributed to Tillet and Francis, 1930 through their work on patients with *Streptococcus pneumoniae* infections. These researchers observed that, during the acute symptoms phase, the patient's serum contained a substance capable of precipitating the C polysaccharide of the cell wall of pneumococcus. For this reason, the name given to this substance was C-reactive. As stated by several studies, this protein cannot be detected in healthy patients, however, can reach very high amounts in patients carrying infections. Plasma levels drop to undetectable levels after patient's recovery (Marnell et al. 2005).

CRP, with a molecular weight of 115,135 Da, belongs to the pentraxins family of proteins, being composed of five copies of a non-glycosylated polypeptide chain (23,027 Da) with 206 amino acids. These monomers are bonded to each other non-covalently, leading to a very stable discoid structure that is resistant to proteolysis (Xia and Samols 1997). Each monomeric chain has two faces, the recognition/binding face and the effector face. The binding face, associated with two calcium ions, is responsible for recognition of the phosphocholine residues of the C polysaccharide of *Streptococcus pneumoniae* (Thompson et al. 1999; Stone and Kazil 2014), and the effector face has affinity for the complement factor 1q and FcγRs (Thompson et al. 1999).

CRP can exist in two isoforms with different conformations. The native CRP is a pentameric form (pCRP) and the other is the monomeric (mCRP) form. The pCRP form has pro-inflammatory and anti-inflammatory properties (Wu et al. 2015). These activities are context dependent. On the other hand, mCRP has strong pro-inflammatory activity on endothelial cells, leukocytes, and platelets and is probably involved in the amplification of the inflammatory response (Wu et al. 2015). There might be a direct link between the dissociation of pCRP in mCRP and the inflammatory process (Wu et al. 2015).

2.1.2 Biosynthesis and Metabolism of CRP

The production of CRP, an acute phase protein, in liver results from the stimulation by pro-inflammatory cytokines derived from monocytes/macrophages or adipose tissue (Silva and Lacerda 2012; Jialal and Devaraj 2001). Liver cells are mainly stimulated by interleukin (IL)-6, and this synthesis is synergistically increased by IL-1 (Marnell et al. 2005; Haverkate et al. 1997). It is well known that pro-inflammatory risk factors such as oxidized low density lipoprotein (LDL) and infectious agents like *Chlamydia pneumoniae* activate a pro-inflammatory response (Inoue et al. 2005). This response increases the secretion of IL-1β and α-tumor necrosis factor (TNF-α) that will further lead to the release of IL-6. Moreover, the binding of IL-6 to liver receptors activates the synthesis and the release of CRP and amyloid A serum protein (Jialal and Devaraj 2001).

Besides being produced in liver, several studies state that CRP is also synthesized in the atherosclerotic lesion (particularly in smooth muscle and macrophages), in kidneys, neurons, and alveolar macrophages (Calabro et al. 2003; Dong and Wright 1996). Moreover, some data report activation of CRP synthesis by lipid peroxidation and infection by cytomegalovirus that triggers a cascade of pro-inflammatory cytokines (Hajsadeghi et al. 2015). Chang et al. (2002) showed that CRP does not attach to native LDL. Instead, increased levels of reactive oxygen species oxidize LDL, and subsequently CRP stimulates its entrance in macrophages. CRP can attach to oxidized phosphatidylcholine present at the surface of LDL (Chang et al. 2002).

Healthy individuals exhibit low CRP plasma levels (<0.5 mg/dL). However, during inflammation, its serum concentration can increase one hundred to one thousand times. CRP production initiates 4–6 h after stimulation, duplicates every

8 h and reaches its maximum level after 50 h (approximately 2 days). In only 48 h, CRP levels can arrive to 300 mg/dL. These facts make CRP an important clinical marker due to its stability, high sensitivity, good reproductivity, and precision.

This protein has a 19 h half-life, even after stimulus, trauma, or surgery, and basal levels can be reached after several days (Bezerra et al. 2008; Denardi et al. 2008). If no chronic stimulus is present, plasma CRP content can normalize in 3–4 days. In chronic inflammatory periods, CRP might persist high forever (Bezerra et al. 2008).

CRP plasma levels seem also to be related to CRP genetic variants. Work of Michopoulos et al. (2015) stated the presence of higher CRP plasma levels in individuals with the rs1130864 CRP polymorphism.

Moreover, another study detected mean CRP levels considerably higher in men carrying one copy of the rs3093075 and rs1417938 alleles when compared to those with no copies (Markt et al. 2014). On the other hand, these researchers observed that men with one copy of the variant allele rs1800947 had significantly lower mean CRP levels. No relationship was observed between the rs2808630 polymorphism and circulating CRP (Markt et al. 2014).

Plasma CRP level differs among individuals. Each individual has its particular basal level that remains stable over the years without big changes during the day or season. However, CRP levels tend to suffer a slight rise with age in healthy individuals, probably due to the presence of diseases in subclinical stages (Denardi et al. 2008).

In the same way, diet does not interfere with plasma CRP levels. However, levels of CRP seem to vary with ethnicity ancestry. Shah et al. (2010) observed higher CRP content in individuals of Hispanic origin.

Among existing markers, CRP stands out due to its high sensitivity in identifying inflammatory states, by showing a considerable rise as a response to diverse inflammatory and infectious stimuli. Quantification of plasma CRP aids in the clinical identification of inflammation, as well as in the determination of its extension and response to therapy (Bezerra et al. 2008).

Role of CRP in Cellular Metabolism

The anti-inflammatory activity of CRP seems to be associated with its action on the induction of the expression of the antagonist of IL-1 receptor (Galea et al. 2018). Moreover, CRP can induce the secretion of IL-10, an anti-inflammatory cytokine, and, at the same time, reduce IL-12 production in macrophages (Mold et al. 2002).

On the other hand, CRP effect in neutrophils is to exert inhibitory or excitatory effects on cell chemotaxis (Ling et al. 2014) or to promote superoxide production (Devaraj et al. 2009). Another effect of CRP in neutrophils is the inhibition of the respiratory burst (Mortensen and Zhong 2000).

All these findings suggest that CRP probably has a protective role during the first stages of inflammation (Boncler et al. 2019).

In most diseases, changes in the plasma level of CRP indicates the presence of inflammation or tissue damage. This measurement is much more accurate than other acute phase response parameters, like erythrocyte sedimentation rate, since, as previously mentioned, it does not suffer diurnal variation and is not altered by food intake.

CRP synthesis is only affected by liver failure and very few drugs lead to the reduction in CRP production. This can only be achieved with drugs that also interfere with the acute phase stimulus that triggered the pathology.

Therefore, the CRP level is a very useful nonspecific biochemical marker of inflammation that has been routinely used in the diagnosis of organic diseases and in the follow-up of treatments of inflammatory conditions (Pepys and Baltz 1983). It has also been used in the detection of infection in a small number of specific diseases with modest or absent acute phase responses (Pepys and Baltz 1983). Moreover, it also plays an important role in monitorization of bacterial infection, inflammation, and consequent neurodegeneration (Luan and Yao 2018).

2.2 CRP and Alzheimer's Disease

2.2.1 Alzheimer's Disease Etiology

The Alzheimer's disease is a progressive and irreversible disease that affects memory and, with evolution, eventually affects other mental functions. Most cases occur in the elderly. After diagnosis of the disease, an Alzheimer's patient resists on average 8 years. It may arise from a genetic cause or sporadically, having in this case several factors responsible for its appearance, such as altered cholinergic transmission, amyloid cascade, apolipoprotein E isoform 4 (ApoE4), and tau protein (Rauk 2009; Zatta et al. 2009).

In Alzheimer's disease, histopathologically, there is deposition of extracellular senile plaques, β-amyloid (βA) peptide aggregates, and of intracellular fibrils of the polymerized (hyperphosphorylated) tau protein. The βA peptide results from the proteolysis of the βA precursor protein (APP) by α-, β-, and γ-secretases (Sola Vigo et al. 2009).

The proteolytic process that occurs in APP, with the consequent formation of βA, is not homogeneous, since this production gives rise to different lengths of βA. The βA can exist in different forms such as a monomer, dimer, oligomer, fibrils, and fibril aggregates. The most frequent forms of βA consist of 40 and 42 amino acids. The βA isoforms that have a greater length are associate with faster aggregation, oligomerization, and fibril formation. The oligomeric form is the most neurotoxic. The active mechanism for removing βA oligomers occurs by promoting its enzymatic breakdown into monomers (Rauk 2009).

Several studies have suggested that free radicals are involved in βA fibrillation and neurofibrillary tangle formation in Alzheimer's patients. The temporal coincidence of the increase of free radicals and antioxidant enzymes with the onset of amyloid plaques deposition is still evidencing that aging induces oxidative changes in the brain, which are an important factor in improving βA production and deposition in the Alzheimer's patient brain. This type of deposition can be seen as a protection mechanism that inactivates the extreme toxicity of oligomeric species.

Since Alzheimer's disease is a chronic disorder, it is likely that neurons are undergoing near-lethal levels of oxidative stress for prolonged periods (Silvestrelli et al. 2006).

Axons are mainly composed of tau protein and low molecular weight microtubules. This tau protein plays an important role in stabilizing microtubules and inducing their formation.

βA deposition triggers hyperphosphorylation of the tau protein, leading to pathological effects that promote neurotoxicity. Oligomers formed from the βA peptide cause instability in the tau protein and its accumulation as neurofibrillary tangles (Selkoe and Hardy 2016). This leads to the activation of inflammatory cells, initiation of oxidative stress, rapid dissociation of microtubules, and breakdown of the axon structure that together end in the decline of synaptic and neuronal health and consequent neurodegeneration (Selkoe and Hardy 2016). The process culminates in the stablishment of a synapse processing disorder and subsequently death of neurons, which affects the hippocampus and neocortex zones (Maccioni et al. 2010).

The neurodegeneration observed in Alzheimer's disease is associated with the described accumulation of soluble oligomeric βA in patient's brain. The βA peptides are key mediators of synaptotoxicity and neurotoxicity (Jin et al. 2011).

The pathogenesis of this disorder seems also to be induced by the accumulation of inflammatory proteins outside of the brain. In fact, Engelhart et al. (2004) observed high levels of these molecules in brains of individuals having Alzheimer's disease. The slight increase in inflammatory markers is responsible for the development of a low-grade inflammation. In this way, the body experiences a mild chronic elevation of these molecules, but not enough to lead to an acute inflammation (Eklund 2009). Among these biomarkers is CRP, a marker of low-grade inflammation, that has been found moderately increased in neurodegenerative disorders (Woloshin and Schwartz 2005; Kushner et al. 2006).

This low-grade inflammation has been considered an important contributor to the pathophysiology of hypertension and cardiovascular diseases (Nosalski et al. 2017). In the same way, several studies have related the existence of a low-grade peripheral inflammation with the development of Alzheimer's disease (Michaud et al. 2013; Heneka et al. 2015).

However, Heppner et al. (2015) stated that, besides the pathophysiological events that have been associated with the development of Alzheimer's disease, also the immune system seems to be an important contributor to the pathogenesis of Alzheimer's disease. As mentioned before, after βA deposition there is the formation of neurofibrillary tangles by hyperphosphorylation of the tau protein. Studies showed that the production of the neurofibrillary tangles is, in fact, activated by microglial cells, that drive the neuroinflammation process (Kitazawa et al. 2005). This local microglial cell-driven immune response might be enough for activation of tau protein pathology (Heneka et al. 2014).

2.2.2 Role of CRP in Alzheimer's Disease

CRP is considered an excellent biomarker of chronic inflammation and, although the biochemical mechanisms are still not fully understood, it is known that CRP is involved directly in the development of associated pathologies such as neurodegenerative diseases (Vachatova et al. 2016).

In most neurodegenerative diseases, an increased level of CRP is observed. This is related to the presence of an inflammatory process that culminates in the development of neurodegeneration. However, concerning Alzheimer's disease no consensus exists about this relationship.

First studies saw the existence of CRP in senile plaques of Alzheimer's patients, suggesting that the formation of these plaques probably involves an acute phase inflammation and/or the production of CRP (Iwamoto et al. 1994).

It seems that βA plaques may induce the dissociation of pCRP into the monomeric form. In fact, Strang et al. (2012) observed association of mCRP and the βA plaques. Slevin et al. (2015) stated that the increased level of mCRP was associated with reduced expression of βA/tau protein, suggesting a possible role of mCRP in dementia promotion.

Several studies observed a significant association between CRP and Alzheimer's disease (Engelhart et al. 2004), together with enhanced content of IL-1, IL-6, and α-1-antichymotrypsin (Licastro et al. 2000). Work of Gupta et al. (2005) also observed increased CRP content in these patients. Moreover, Desikan et al. (2015) correlated high CRP, high serum cholesterol, and phospholipids with increased risk of Alzheimer's development.

Supporting mentioned data, Schmidt et al. (2002) observed that men with increased levels of CRP had a three times higher chance of developing any sort of dementia. This study observed moderate elevated CRP levels in individuals developing Alzheimer's disease later in life (Schmidt et al. 2002). In the same way, Tomita (2011) stated that a mid-life increase in plasma CRP levels is a risk factor for Alzheimer's disease. The high CRP level was even associated with increased odds of all causes of dementia in women (Kravitz et al. 2009).

Another study detected significantly increased serum levels of adhesion molecules and CRP in dementia patients (Dimopoulos et al. 2006). These results suggest the presence of accelerated atherosclerosis in these patients. Moreover, the observed low-grade chronic inflammation might probably lead to cognitive impairment (Dimopoulos et al. 2006). Other researchers have also related the high CRP level with cognitive decline (Schmidt et al. 2002; Yaffe et al. 2003; Teunissen et al. 2003; Engelhart et al. 2004).

However, other studies did not observe that adverse correlation. Instead, no association between CRP and Alzheimer's disease or any sort of mild cognitive impairment was observed (Dik et al. 2005; Kim et al. 2015). Nevertheless, Kim et al. (2015) were able to detect some tendency for higher CRP in Alzheimer's patients. Moreover, in these studies CRP levels did not significantly correlate with age or level of education. In addition, Devcic et al. (2014) did not find a role for CRP or homocysteine in Alzheimer's development.

Several researchers report the existence of an inverse correlation (Nilsson et al. 2008; O'Bryant et al. 2010). These studies stated a lowered CRP content in patients with stabilized Alzheimer's disease. On the other way around, increased CRP was associated with decreased risk of Alzheimer's disease and all types of dementia (Tan et al. 2007; Sundelof et al. 2009; van Himbergen et al. 2012; Silverman et al. 2012; Yarchoan et al. 2013; Wichmann et al. 2014).

A recent study analyzed the relationship between CRP and Alzheimer's disease in different age groups (Gabin et al. 2018). Data from this work showed the adverse association between these two factors in individuals in the age range 60–70.5, while an inverse association was observed in senior individuals (>70.6 years of age) (Gabin et al. 2018).

Moreover, Locascio et al. (2008) related the low plasma CRP with faster progression of the disease, manifested as cognitive and functional decline. Nilsson et al. (2011) associated high CRP with shorter survival time, while low content of CRP was present in Alzheimer's patients. In opposition to these studies, Hazen et al. (2020) observed no association between increased levels of inflammatory markers and faster progression of cognitive impairment or Alzheimer's dementia.

Gong et al. (2016) observed no significant difference in CRP plasma content between Alzheimer's patients and control individuals. However, the same study saw lower CRP levels in patients with moderate or mild Alzheimer's disease. This study suggests that CRP content can be used in the diagnosis of mild or moderate Alzheimer's disease (Gong et al. 2016).

A recent study analyzed the relationship of CRP and the genotype for ApoE (Martiskainen et al. 2018). This work observed that the genotypes ε3ε4 and ε4ε4 were significantly linked with decreased levels of high-sensitivity CRP in plasma (hs-CRP). Moreover, carriers of the ε4 allele showed higher total cholesterol levels (Jofre-Monseny et al. 2008; Rebeck 2017). These data are in accordance with the association of the ε4 allele with the development of Alzheimer's disease. Supporting these findings is the work of Hooper et al. (2018) that stated an inverse association of chronically raised CRP with βA peptide. This relationship was found to be specific for the ε4 allele carriers (Hooper et al. 2018).

2.3 CRP and Parkinson's Disease

2.3.1 Parkinson's Disease Etiology

Parkinson's disease is considered the second most usual neurodegenerative disease in the world. It is a neurological disorder of movement with progressive and long-lasting evolution, which leads to disability. The main symptoms are rest tremor, bradykinesia, muscular stiffness, and postural instability. This disease has no cure and affects more often men than women (Teive 2005).

In Parkinson's disease, there are multiple monoaminergic dysfunctions, including dopaminergic, cholinergic, serotonergic, and noradrenergic systems deficits. 90% of the cases are sporadic and without known etiology. However, there are some processes that cause atrophy and death of dopaminergic neurons in the nervous system. The most referred are oxidative stress, excitotoxicity, mitochondrial dysfunction, and protein aggregation (Weintraub et al. 2008).

Parkinson's disease is characterized by specific neurodegeneration of the substantia nigra accompanied by a significant decrease in dopamine in all basal ganglia components and the presence of Lewy bodies, cytoplasmic inclusions composed of a variety of proteins such as α-synuclein, parkin, and ubiquitin. The ubiquitin proteasomal system represents the major mechanism responsible for the elimination of damaged mutant intracellular proteins. This system consists of 26S and 20S proteasomes, which are proteases with multicatalytic action. A loss of α-subunits of the 20S proteasomal system was evidenced in the nigral dopaminergic neurons, leading to instability of the proteasomal system, with loss of function and subsequent triggering of the neurodegenerative process (Togo et al. 2001).

The propensity of Parkinson's disease to occur in elderly patients suggests that the neuronal proteasomal system becomes insufficient in its function of eliminating abnormal proteins that may accumulate over time. Thus, an imbalance between a high abundance of abnormal proteins associated with low activity of the proteasomal system and the presence of low levels of proteasomal activators in the compact part of the midbrain substantia nigra leads to neurodegenerative processes (Teive 2005).

Neuroinflammatory processes can play a role in the sequence of events that cause dopaminergic loss. The accumulation of α-synuclein motivates the secretion of pro-inflammatory cytokines, giving rise to chronic neuroinflammation in the brain of patients with Parkinson's disease. It has been found that acute systemic inflammation worsens the motor symptoms of Parkinson's disease (Codolo et al. 2013; Haghshomar et al. 2019).

In recent decades, hypotheses have been constructed that neuroinflammation by microglia may collaborate to the Parkinson's disease pathogenesis. In addition, studies reveal that non-steroidal anti-inflammatory drugs are neuroprotective and greatly reduce the risk of Parkinson's disease, which strongly supports the referred hypothesis (Bassani et al. 2015; Tansey and Goldberg 2010).

2.3.2 Role of CRP in Parkinson's Disease

CRP is an acute phase protein type and is controlled by pro-inflammatory cytokines. IL-6 is the main cytokine, among others, that increases the expression of the CRP gene in hepatocytes. High levels of CRP are strongly associated with inflammatory process. Inflammation may be linked to cognitive decrease and involved in the Parkinson's disease pathogenesis. CRP may be involved in the atherogenesis process and be able to activate microglia. Therefore, CRP can affect the progression of neurodegenerative processes, specifically in Parkinson's disease (Moghaddam et al. 2018).

Studies show that CRP is linked with the neurological disorders pathogenesis and is considered a factor of risk for Parkinson's disease. A recent meta-analysis (Qiu et al. 2019) revealed that Parkinson's disease is associated with increased CRP. This study reveals that CRP levels in the blood and cerebrospinal fluid were higher in patients with Parkinson's disease compared to controls without disease. Then, CRP may be a factor of risk for Parkinson's disease or Parkinson's disease leading to an inflammatory response.

Another recent study (Moghaddam et al. 2018) confirmed that CRP concentrations in the cerebrospinal fluid correlate with motor and non-motor severity in individuals with Parkinson's disease. These results show that neuroinflammation plays a role in the onset and maybe in the progression of motor and non-motor symptoms of Parkinson's disease, which can provide a better analysis of the pathological mechanisms inherent to Parkinson's disease.

A different study (Choi et al. 2016) suggests a correlation between serum levels of hs-CRP and cognitive function in patients with Parkinson's disease. Serum hs-CRP levels were inversely associated with the Mini-Mental State Examination scores and performance on neuropsychological tests of language, visuospatial function, visual memory, and executive function. Linking hs-CRP levels with cognitive function is important, as it is possible that the handling of inflammation has beneficial effects on the cognition of Parkinson's disease patients.

Previously, a study by Sawada et al. (2015) showed that baseline concentrations of CRP were associated with death risk and life prognosis in Parkinson's disease patients. These associations were shown to be independent of duration, severity, cognitive function, age, and nutritional conditions. In addition, in another study, Umemura et al. (2015) showed that baseline plasma CRP levels were associated with motor deterioration and predicted motor prognosis in Parkinson's disease patients. These associations were independent of gender, age, severity, dementia, and anti-Parkinsonian agents. Both studies suggest the possibility that subclinical systemic inflammation may accelerate neurodegeneration in Parkinson's disease.

An earlier study (Song et al. 2011) also showed that hs-CRP levels in drug-free early Parkinson's disease patient were more elevated than in normal controls, which confirmed the hypothesis that neuroinflammatory reactions play an important role in the Parkinson's disease pathogenesis. In addition, the results cautiously suggested that levels of hs-CRP increased before the clinical start of Parkinson's disease.

2.4 CRP and Huntington's Disease

2.4.1 Huntington's Disease Etiology

Huntington's disease is an autosomal dominant progressive neurodegenerative disorder that affects movement and leads to cognitive impairment and psychiatric disturbances. The most known symptomatology of this condition is an excess of uncoordinated motor movements, called chorea, and gait disturbances. The disease progresses over time and becomes fatal 15 to 20 years after the onset of the first symptoms. The leading cause of patient death is usually infectious or cardiovascular respiratory complications (Walker 2007).

Neuropathologically, Huntington's disease is characterized by gradual neuronal loss of the striatum (caudate nucleus and putamen) and cortex. However, many other nuclei are also affected, including pale globe, thalamus, hypothalamus, subthalamic nucleus, substantia nigra, and cerebellum (Raymond et al. 2011).

Huntington's disease is induced by the unstable expansion of a repeat of a cytosine-adenine-guanine (CAG) trinucleotide in the Huntington's disease gene coding region present in the short arm of chromosome 4. This gene encodes the Huntingtin protein (Htt). Mutations in that gene lead to a mutant Huntingtin (mHtt). Translation of the mutated gene causes the formation of a long-lasting polyglutamine (polyQ) sequence that will particularly lead to the mHtt aggregation. These intracellular mHtt aggregates are called inclusion bodies that result in the striatum and deeper layers of cortex degeneration, which are selectively affected during disease progression (Jones et al. 2016).

Elimination of inclusion bodies can occur through different cellular processes, which, depending on the environment, can aggravate or mitigate disease progression. Mechanisms involved include molecular chaperones, ubiquitin proteasomal system, and autophagy pathways. These networks can detect mHtt as a threat and initiate beneficial defense responses, leading to the delay of disease progression (Maiti et al. 2014).

There is growing evidence that the immune system may be included in the Huntington's disease pathogenesis. An increase in the activation of the components of the innate and adaptive immune system was detected in the central nervous system and in the peripheral tissues of patients with Huntington's disease. Especially, there were observed increased concentrations of TNF-α and IL-6 in plasma of carriers of Huntington's disease mutations in comparison with the controls (Bjorkqvist et al. 2008; Silvestroni et al. 2009).

2.4.2 Role of CRP in Huntington's Disease

TNF-α and IL-6 are pro-inflammatory cytokines that are launched into the bloodstream in response to the acute phase. The levels of IL-6 increase in the plasma of pre-manifested individuals, 16 years before the expected onset of the disease. Curiously, the cytokines that increase earlier in the elapse of the disease are IL-6 and IL-8, both involved in the innate immune response (Bjorkqvist et al. 2008).

CRP synthesis is upregulated in the liver under IL-6 regulation. Since during the acute phase response, CRP levels increase, this protein can be considered a positive acute phase protein. Associations were observed between levels of acute phase protein and Total Functioning Capacity (TFC) Score, severity of apathy, cognitive impairment, and use of antipsychotics. The levels of hs-CRP were higher in patients with mutations who used antipsychotics continuously, when compared to those who never used antipsychotics. This study shows that carriers of Huntington's disease mutations who use antipsychotics are likely to develop an acute phase response (Bouwens et al. 2014).

In another study (Sánchez-López et al. 2012), CRP levels were higher in Huntington's disease mutation carriers with more advanced disease. In addition, it was observed that CRP levels increase significantly in carriers of Huntington's disease mutations in the advanced stage of the disease when compared to controls (Stoy et al. 2005). However, in a different study (Silajdzic et al. 2013), it was found that CRP was reduced at the onset of Huntington's disease.

2.5 CRP and ALS

2.5.1 ALS Etiology

ALS is a rare neurodegenerative disease characterized by progressive loss of motor function and respiratory capacity, which mortality is fundamentally due to respiratory repercussion in the final stages of the disease, with average survival time ranging from 2 to 5 years. Neurodegeneration occurs mainly in the motor cortex, brainstem, and spinal cord and is more common in men than in women. The first symptom is symmetrical weakness in one limb associated with progressive muscular atrophy (Hayashi et al. 2016).

ALS is the major adult-onset motor neuron disease. Approximately 90% of all ALS cases have no genetic component or known cause and are called sporadic. The remaining 10% have a genetic bond and are known as familial ALS. Of these cases, 25% are due to mutations in the gene encoding superoxide dismutase 1 (SOD1), representing the most studied example. SOD1 is a major component of the cell defense mechanism against oxidative damage. One strange feature is why carriers of mutations in SOD1 are apparently healthy by the end of middle age and then undergo rapid neurological decline (Carrì et al. 2017; Keskin et al. 2016).

Mechanisms by which misfolded SOD1 species originate ALS are still poorly understood. However, involvement in mitochondrial function disturbance, stress induction in the endoplasmic reticulum, and axonal transport disorder and aggregation have been suggested. Mutated SOD1 associates or interacts with the endoplasmic reticulum, mitochondria, glial NADPH oxidase (NOX)/Rac1 complex, resulting in endoplasmic reticulum stress, mitochondrial dysfunction, and reactive oxygen species (ROS) overproduction, respectively (Carrì et al. 2017).

2.5.2 Role of CRP in ALS

CRP is an inflammatory response biomarker with meaning prognostic value for various types of tumors, and cardiovascular and rheumatic diseases. Studies have shown that neurons are also able to produce CRP in response to brain damage (Yasojima et al. 2000). CRP-centered studies make sense according to the surprising ALS neuroinflammation literature (Corcia et al. 2018).

In a recent study (Lunetta et al. 2017), serum CRP levels in ALS patients correlated with the severity of functional impairment, as measured by the total

ALS Functional Rating Scale-Revised (ALSFRS-R) Score, and with the patient's survival. These results suggest that patients with ALS and elevated levels of serum CRP progress more rapidly than those with lower levels of CRP. This increase may also mirror a neuroinflammatory state potentially responsive to immune regulators, such as NP001. These results confirm results of previous studies that corroborate the potential usefulness of CRP as a biomarker in ALS. However, another recent study (Nagel et al. 2017) investigated the association of adipokines and hs-CRP with risk and prognosis of ALS, and no association was found.

Before, Ryberg et al. (2010) observed a significant increase in the levels of CRP in the cerebrospinal fluid in patients with ALS, compared with the levels obtained in a control group and with an Alzheimer's cohort. It is important to note that there was no difference between the ALS and multiple sclerosis subgroups in terms of the levels of CRP in the cerebrospinal fluid. This has been clarified as associated to the fact that both conditions exhibit inflammation of the central nervous system. In a group of patients with ALS, Keizman et al. (2009) found a correlation between clinical disability and some sensitive biomarkers of inflammation, including CRP.

2.6 CRP and Prion's Disease

2.6.1 Prion's Disease Etiology

Prion's diseases, or transmissible spongiform encephalopathies, are a set of neurodegenerative diseases that can affect both human and animal species. Prion's diseases identified in humans are Creutzfeldt-Jakob's disease, Gerstmann-Straussler-Scheinker's syndrome, fatal familial insomnia, and Kuru's disease, with Creutzfeldt-Jakob's disease being the most common (Manix et al. 2015).

Prion's diseases result in progressive and/or complete degeneration of nerve cells, resulting in a debilitating and incurable state, leading to death. After acquiring the disease, the individual experiences extrapyramidal motor signs including tremors, ataxia, myoclonus, and postural instability. These disorders are essentially characterized by long incubation periods, neural loss, and failure to induce an inflammatory response by the affected living being (Chen and Dong 2016).

Prion proteins (PrP) are transmissible infectious particles of a protein nature. The function of these normal proteins is not yet fully understood. Normal cellular prion protein (PrPC) is present principally in the central nervous system, lymphatic tissue, and neuromuscular junctions. PrPC plays an important role in neuronal processes such as circadian rhythm, neuroprotection, and neuroplasticity. However, PrPC can become abnormal or infectious, being called scrapie prion protein (PrPSc). Prion's diseases are associated with the conversion of PrPC into the insoluble conformer PrPSc, the latter being associated with disease development (Chen and Dong 2016).

The main event in all Prion's diseases is the change in PrPC conformation that mainly consists of α-helices, in PrPSc, composed essentially of β-sheets. Thus, PrPSc becomes resistant to proteinase K digestion (PrPRes) and more prone to β-sheets accumulation in the brain. The slow and progressive accumulation of

PrPSc in the brain raises little doubt about its role in triggering neurodegeneration and disease (Venneti 2010).

Creutzfeldt-Jakob's disease is a rare, progressive, and fatal neurodegenerative disorder of the central nervous system, with an incidence of about two new cases per million inhabitants per year worldwide. An ultimate diagnosis is possible only based on clinical and neuropathological results. The most specific indication is the identification of PrPSc in brain tissue, but this is only possible by postmortem examination. There are different studies of the detection of plasma S-100, neuron-specific enolase (NSE) and PrPC, and for this reason they have been designated as marker proteins associated with Creutzfeldt-Jakob's disease. The S-100 and the NSE are considered markers of neuronal damage (Völkel et al. 2001).

2.6.2 Role of CRP in Prion's Diseases

Tissue damage is typical of the late stages of Creutzfeldt-Jakob's disease, in addition to neuronal loss. Moreover, during tissue injury, acute phase proteins are generally seen in plasma. CRP is a significant acute phase reactant that is widely used as an indicator of inflammatory status and tissue damage. The association with cell injury may explain the increased concentrations of CRP and IL-6 (as an inducer of CRP) in the plasma of patients with Creutzfeldt-Jakob's disease.

Fratini et al. (2012) analyzed plasma samples from patients with sporadic Creutzfeldt-Jakob's disease to find diagnostic biomarkers. Levels of different acute phase inflammatory proteins in plasma, including CRP, were significantly higher in patients with sporadic Creutzfeldt-Jakob's disease than in neurologic and healthy controls.

In a previous study, Völkel et al. (2001) analyzed plasma samples from patients with Creutzfeldt-Jakob's disease and from a group of control of healthy plasma donors, for CRP and IL-6 as markers of inflammation or tissue injury. To elucidate the value of these proteins in the diagnosis, they were compared with the proteins S-100, NSE, and PrPC. Creutzfeldt-Jakob's disease was identified by the CRP and IL-6 markers in more samples than by the S-100 and PrPC markers because easy-to-use commercial kits are available for the former. No correlation was observed between increased NSE concentrations and the disease. However, none of the marker proteins analyzed was able to recognize all cases of Creutzfeldt-Jakob's disease.

2.7 Overview

Aging of the population is now a worldwide phenomenon and therefore the incidence and prevalence of neurodegenerative diseases is likely to increase in the future. Accordingly, in recent years, neurodegenerative diseases, like Alzheimer's, Parkinson's, Huntington's, ALS, and Prion's, have assumed an extremely important

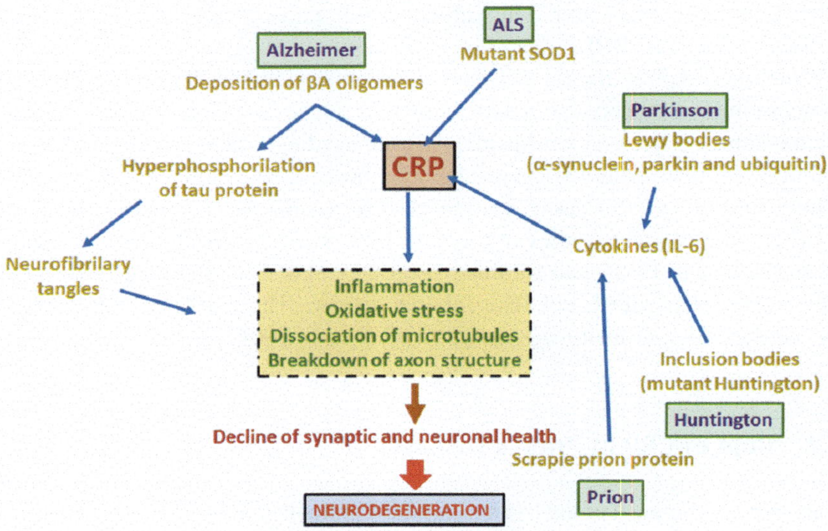

Fig. 2.1 Overview of the role of CRP in the etiology of neurodegenerative diseases

place. Knowledge about the etiology of these type of disorders has increased but still there is a lot to clarify.

As can be concluded from all the studies reported in this work, inflammatory proteins, such as CRP, are most likely involved in the pathogenesis of neurodegenerative disorders since they exhibit an altered profile because of acute or chronic inflammation (Fig. 2.1).

In most neurodegenerative diseases, such as Parkinson's, Huntington's, ALS, and Prion's, an increased level of CRP is observed. CRP is linked to the pathogenesis of these neurological disorders and is considered a factor of risk. Almost all the studies favor the potential utility of CRP as a biomarker. This is related to the presence of an inflammatory process that culminates in the development of neurodegeneration. However, concerning Alzheimer's disease no consensus exists about this relationship.

References

Bassani TB, Vital MA, Rauh LK (2015) Neuroinflammation in the pathophysiology of Parkinson's disease and therapeutic evidence of anti-inflammatory drugs. Arq Neuropsiquiatr 73:616–623. https://doi.org/10.1590/0004-282X20150057

Bezerra CFR, Amaral BA, Camelo FP, Aquino ARL, Seabra EG, Carreiro AFP (2008) O papel da proteína C-reativa na relação entre as doenças periodontais e cardiovasculares. R Periodontia 18 (1):7–13

Bjorkqvist M, Wild EJ, Thiele J, Silvestroni A, Andre R, Lahiri N, Raibon E et al (2008) A novel pathogenic pathway of immune activation detectable before clinical onset in Huntington's disease. J Exp Med 205:1869–1877. https://doi.org/10.1084/jem.20080178

Boncler M, Wu Y, Watala C (2019) The multiple faces of C-reactive protein – physiological and pathophysiological implications in cardiovascular disease. Molecules 24(11):E2062. https://doi.org/10.3390/molecules24112062

Bouwens JA, Hubers AAM, van Duijn E, Cobbaert CM, Roos RAC, van der Mast RC, Giltay EJ (2014) Acute-phase proteins in relation to neuropsychiatric symptoms and use of psychotropic medication in Huntington′s disease. Eur Neuropsychopharmacol 24(8:1248–1256. https://doi.org/10.1016/j.euroneuro.2014.05.004

Calabro P, Willerson JT, Yeh ET (2003) Inflammatory cytokines stimulated C-reactive protein production by human coronary artery smooth muscle cells. Circulation 108:1930–1932. https://doi.org/10.1161/01.CIR.0000096055.62724.C5

Carrì MT, D'Ambrosi N, Cozzolino M (2017) Pathways to mitochondrial dysfunction in ALS pathogenesis. BiochemBiophys Res Commun 483(4):1187–1193. https://doi.org/10.1016/j.bbrc.2016.07.055

Chang MK, Binder CJ, Torzewski M, Witztum JL (2002) C-reactive protein binds to both oxidized LDL and apoptotic cells through recognition of a common ligand: phosphorylcholine of oxidized phospholipids. Proc Natl Acad Sci U S A 99:13043–13048. https://doi.org/10.1073/pnas.192399699

Chen C, Dong XP (2016) Epidemiological characteristics of human prion diseases. Infect Dis Poverty 5(1):47. https://doi.org/10.1186/s40249-016-0143-8

Choi S, Kim BC, Kang KW, Choi K, Nam T, Kim J et al (2016) Relationship between serum high-sensitivity C-reactive protein levels and cognitive function in patients with Parkinson's disease. Neurol Asia 21(4):349–356

Codolo G, Plotegher N, Pozzobon T, Brucale M, Tessari I, Bubacco L, de Bernard M (2013) Triggering of inflammasome by aggregated alpha-synuclein, an inflammatory response in synucleinopathies. PLoS One 8(1):e55375. https://doi.org/10.1371/journal.pone.0055375

Corcia P, Blasco H, Beltran S, Andres C, Vourc'h P (2018) Couratier P.C-reactive protein: a promising biomarker in ALS? Rev Neurol 174(3:104–105. https://doi.org/10.1016/j.neurol.2017.07.001

Denardi CAS, Filho AC, Chagas ACP (2008) A proteína C-reativa na atualidade. Rev Socer J 2:329–334

Desikan RS, Schork AJ, Wang Y, Thompson WK, Dehghan A, Ridker PM, Chasman DI et al (2015) Polygenic overlap between C-reactive protein, plasma lipids and Alzheimer's disease. Circulation 131(23):2061–2069. https://doi.org/10.1161/CIRCULATIONAHA.115.015489

Devaraj S, Dasu MR, Singh U, Rao LV, Jialal I (2009) C-reactive protein stimulates superoxide anion release and tissue factor activity in vivo. Atherosclerosis 203:67–74. https://doi.org/10.1016/j.atherosclerosis.2008.05.060

Devcic S, Glamuzina L, Ruljancic N, Mihanovic M (2014) There are no differences in IL-6, CRP and homocysteine concentrations between women whose mothers had AD and women whose mothers did not have AD. Psychiatry Res 220:970–974. https://doi.org/10.1016/j.psychres.2014.08.059

Dik MG, Jonker C, Hack CE, Smit JH, Comijs HC, Eikelenboom P (2005) Serum inflammatory proteins and cognitive decline in older persons. Neurology 64:1371–1377. https://doi.org/10.1212/01.WNL.0000158281.08946.68

Dimopoulos N, Piperi C, Salonicioti A, Mitropoulos P, Kallai E, Liappas I, Lea RW, Kalofoutis A (2006) Indices of low-grade chronic inflammation correlate with early cognitive deterioration in an elderly Greek population. Neurosci Lett 398:118–123. https://doi.org/10.1016/j.neulet.2005.12.064

Dong Q, Wright JR (1996) Expression of C-reactive protein by alveolar macrophages. J Immunol 156:4815–4820

Eklund CM (2009) Proinflammatory cytokines in CRP baseline regulation. Adv Clin Chem 48:111–136

Engelhart MJ, Geerlings MI, Meijer J, Kiliaan A, Ruitenberg A, van Swieten JC, Stijnen T et al (2004) Inflammatory proteins in plasma and the risk of dementia: the Rotterdam study. Arch Neurol 61(5):668–672. https://doi.org/10.1001/archneur.61.5.668

Fratini F, Principe S, Puopolo M, Ladogana A, Poleggi A, Piscopo P, Bruno G et al (2012) Increased levels of acute-phase inflammatory proteins in plasma of patients with sporadic CJD. Neurology 79:1–7. https://doi.org/10.1212/WNL.0b013e318265a55d

Gabin JM, Saltvedt I, Tambs K, Holmen J (2018) The association of high sensitivity C-reactive protein and incident Alzheimer disease in patients 60 years and older: the HUNT study. Norway Immun Ageing 15:4. https://doi.org/10.1186/s12979-017-0106-3. eCollection 2018

Galea J, Ogungbenro K, Hulme S, Patel H, Scarth S, Hoadley M, Illingworth K et al (2018) Reduction of inflammation after administration of interleukin-1 receptor antagonist following aneurysmal subarachnoid hemorrhage: results of the subcutaneous interleukin-1Ra in SAH (SCIL-SAH) study. J Neurosurg 128:515–523. https://doi.org/10.3171/2016.9.JNS16615

Gong C, Wei D, Wang Y, Ma J, Yuan C, Zhang W, Yu G, Zhao Y (2016) A meta-analysis of C-reactive protein in patients with Alzheimer's disease. Am J Alzheimers Dis Other Demen 31 (3):194–200. https://doi.org/10.1177/1533317515602087

Gupta A, Watkins A, Thomas P, Majer R, Habubi N, Morris G, Pansari K (2005) Coagulation and inflammatory markers in Alzheimer's and vascular dementia. Int J Clin Pract 59:52–57. https://doi.org/10.1111/j.1742-1241.2004.00143.x

Haghshomar M, Rahmani F, HadiAarabi M, Shahjouei S, Sobhani S, Rahmani M (2019) White matter changes correlates of peripheral neuroinflammation in patients with Parkinson's disease. Neuroscience 403:70–78. https://doi.org/10.1016/j.neuroscience.2017.10.050

Hajsadeghi S, Chitsazan M, Chitsazan M, Salehi N, Amin A, Maleki M, Babaali N, Abdi S, Mohsenian M (2015) Changes of high sensitivity C-reactive protein during clopidogrel therapy in patients undergoing percutaneous coronary intervention. Res Cardiovasc Med 5(1):e28997. https://doi.org/10.5812/cardiovascmed28997. eCollection 2016 Feb

Haverkate F, Thompson SG, Pyke SD, Gallimore JR, Pepys MB (1997) Production of C-reactive protein and risk of coronary events in stable and unstable angina. European concerted action on thrombosis and disabilities angina pectoris study group. Lancet 349:462–466. https://doi.org/10.1016/s0140-6736(96)07591-5

Hayashi Y, Homma K, Ichijo H (2016) SOD1 in neurotoxicity and its controversial roles in SOD1 mutation-negative ALS. Adv Biol Regul 60:95–104. https://doi.org/10.1016/j.jbior.2015.10.006

Hazen J, Vistnes M, Barca ML, Eldholm RS, Persson K, Braekhus A, Saltvedt I, Selbaek G, Engedal K, Knapskog AB (2020) The association between circulating inflammatory markers and the progression of Alzheimer disease in Norwegian memory clinic patients with mild cognitive impairment or dementia. Alzheimer Dis Assoc Disord 34(1):47–53. https://doi.org/10.1097/WAD.0000000000000034*2

Heneka MT, Kummer MP, Latz E (2014) Innate immune activation in neurodegenerative disease. Nat Rev Immunol 14:463–477. https://doi.org/10.1038/nri3705

Heneka MT, Carson MJ, El Khoury J, Landreth GE, Brosseron F, Feinstein DL, Jacobs AH et al (2015) Neuroinflammation in Alzheimer's disease. Lancet Neurol 14(4):388–405. https://doi.org/10.1016/S1474-4422(15)70016-5

Heppner FL, Ransohoff RM, Becher B (2015) Immune attack: the role of inflammation in Alzheimer disease. Nat Rev Neurosci 16(6):358–372. https://doi.org/10.1038/nrn3880

van Himbergen TM, Beiser AS, Ai M, Seshadri S, Otokozawa S, Au R, Thongtang N, Wolf PA, Schaefer EJ (2012) Biomarkers for insulin resistance and inflammation and the risk for all-cause dementia and Alzheimer disease: results from the Framingham heart study. Arch Neurol 69 (5):594–600. https://doi.org/10.1001/archneurol.2011.670

Hooper C, De Souto BP, Cantet C, Cesari M, Payoux P, Salabert AS, Vellas B (2018) Chronically raised C-reactive protein is inversely associated with cortical β-amyloid in older adults with subjective memory complaints. Exp Gerontol 108:226–230. https://doi.org/10.1016/j.exger.2018.04.014

Inoue T, Kato T, Uchida T, Sakuma M, Nakajima A, Shibazaki M, Imoto Y, Saito M, Hashimoto S, Hikichi Y, Node K (2005) Local release of C-reactive protein from vulnerable plaque or coronary arterial wall injured by stenting. J Am Coll Cardiol 46:239–245. https://doi.org/10.1016/j.jacc.2005.04.029

Iwamoto N, Nishiyama E, Ohwada J, Arai H (1994) Demonstration of CRP immuno-reactivity in brains of Alzheimer's disease: immunohistochemical study using formic acid pretreatment of tissue sections. Neurosci Lett 177:23–26. https://doi.org/10.1016/0304-3940(94)90035-3

Jialal I, Devaraj S (2001) Inflammation and atherosclerosis: The value of the high-sensitivity C-reactive protein assay as a risk marker. Am J Clin Pathol 116:S108–S115. https://doi.org/10.1309/J63V-5LTH-WYFC-VDR5

Jin M, Shepardson N, Yang T, Chen G, Walsh D, Selkoe DJ (2011) Soluble amyloid betaprotein dimers isolated from Alzheimer cortex directly induce Tau hyperphosphorylation and neuritic degeneration. Proc Natl Acad Sci U S A 108(14):5819–5824. https://doi.org/10.1073/pnas.1017033108

Jofre-Monseny L, Minihane AM, Rimbach G (2008) Impact of apoE genotype on oxidative stress, inflammation and disease risk. Mol Nutr Food Res 52:131–145. https://doi.org/10.1002/mnfr.200700322

Jones U, Busse M, Enright S, Rosser AE (2016) Respiratory decline is integral to disease progression in Huntington's disease. Eur Respir J 48(2):1–3. https://doi.org/10.1183/13993003.02215-2015

Keizman D, Rogowski O, Berliner S, Ish-Shalom M, Maimon N, Nefussy B, Artamonov I, Drory VE (2009) Low-grade systemic inflammation in patients with amyotrophic lateral sclerosis. Acta Neurol Scand 119(6):383–389. https://doi.org/10.1111/j.1600-0404.2008.01112.x

Keskin I, Forsgren E, Lange DJ, Weber M, Birve A, Synofzik M, Gilthorpe JD, Andersen PM, Marklund SL (2016) Effects of cellularpathway disturbances on misfolded superoxide dismutase-1 in fibroblasts derived from ALS patients. PLoS One 11(2):e0150133. https://doi.org/10.1371/journal.pone.0150133. eCollection 2016

Kim H, Shin A, Lee KJ (2015) Differences in C-reactive protein level in patients with Alzheimer's disease and mild cognitive impairment. J Psychiatry 18:1. https://doi.org/10.4172/1994-8220.1000194

Kitazawa M, Oddo S, Yamasaki TR, Green KN, LaFerla FM (2005) Lipopolysaccharide-induced inflammation exacerbates tau pathology by a cyclin dependent kinase 5-mediated pathway in a transgenic model of Alzheimer's disease. J Neurosci 25:8843–8853. https://doi.org/10.1523/JNEUROSCI.2868-05.2005

Kravitz BA, Corrada MM, Kawas CH (2009) Elevated C-reactive protein levels are associated with prevalent dementia in the oldest-old. Alzheimers Dement 5(4):318–323. https://doi.org/10.1016/j.jalz.2009.04.1230

Kushner I, Rzewnicki D, Samols D (2006) What does minor elevation of C-reactive protein signify? Am J Med 119(2):166.e17–166.e28. https://doi.org/10.1016/j.amjmed.2005.06.057

Leal MFC, Catarino RIL, Pimenta AM, Souto MRS, Pinheiro TSN (2012) Speciation of copper and zinc in urine – importance of metals in neurodegenerative diseases. Quim Nova 35(10):1985–1990. https://doi.org/10.1590/S0100-40422012001000018

Licastro F, Pedrini S, Caputo L, Annoni G, Davis LJ, Ferri C, Casadei V, Grimaldi LM (2000) Increased plasma levels of interleukin-1, interleukin-6 and alpha-1- antichymotrypsin in patients with Alzheimer's disease: peripheral inflammation or signals from the brain? J Neuroimmunol 103:97–102. https://doi.org/10.1016/s0165-5728(99)00226-x

Ling MR, Chapple IL, Creese AJ, Matthews JB (2014) Effects of C-reactive protein on the neutrophil respiratory burst in vitro. Innate Immun 20:339–349. https://doi.org/10.1177/1753425913493199

Locascio JJ, Fukumoto H, Yap L, Bottiglieri T, Growdon JH, Hyman BT, Irizarry MC (2008) Plasma amyloid beta-protein and C-reactive protein in relation to the rate of progression of Alzheimer disease. Arch Neurol 65(6):776–785. https://doi.org/10.1001/archneur.65.6.776

Luan Y, Yao Y (2018) The clinical significance and potential role of C-reactive protein in chronic inflammatory and neurodegenerative diseases. Front Immunol 9:1302. https://doi.org/10.3389/fimmu.2018.01302. eCollection 2018

Lunetta C, Lizio A, Maestri E, Sansone VA, Mora G, Miller RG, Appel SH, Chiò A (2017) Serum C-reactive protein as a prognostic biomarker in amyotrophic lateral sclerosis. JAMA Neurol 74 (6):660–667. https://doi.org/10.1001/jamaneurol.2016.6179

Maccioni RB, Farías G, Morales I, Navarrete L (2010) The revitalized tau hypothesis on Alzheimer's disease. Arch Med Res 41(3):226–231. https://doi.org/10.1016/j.arcmed.2010.03.007

Maiti P, Manna J, Veleri S, Frautschy S (2014) Molecular chaperone dysfunction in neurodegenerative diseases and effects of curcumin. Biomed Res Int 2014:495091. https://doi.org/10.1155/2014/495091

Manix M, Kalakoti P, Henry M, Takur J, Menger R, Guthikonda B, Nanda A (2015) Creutzfeldt-Jakob disease: updated diagnostic criteria, treatment algorithm, and the utility of brain biopsy. Neurosurg Focus 39(5):E2. https://doi.org/10.3171/2015.8.FOCUS15328

Markt SC, Rider JR, Penney KL, Schumacher FR, Epstein MM, Fall K, Sesso HD, Stampfer MJ, Mucci LA (2014) Genetic variation across C-reactive protein and risk of prostate cancer. Prostate 74(10):1034–1042. https://doi.org/10.1002/pros.22820

Marnell L, Mold C, Du Clos TW (2005) C-reactive protein: ligands, receptors and role in inflammation. Clin Immunol 117:104–111. https://doi.org/10.1016/j.clim.2005.08.004

Martiskainen H, Takalo M, Solomon A, Stancáková A, Marttinen M, Natunen T (2018) Decreased plasma C-reactive protein levels in APOE ε4 allele carriers. Ann Clin Transl Neurol 5 (10):1229–1240. https://doi.org/10.1002/acn3.639. eCollection 2018 Oct

Michaud M, Balardy L, Moulis G, Gaudin C, Peyrot C, Vellas B, Cesari M, Nourhashemi F (2013) Proinflammatory cytokines, aging, and age-related diseases. J Am Med Dir Assoc 14 (12):877–882. https://doi.org/10.1016/j.jamda.2013.05.009

Michopoulos V, Rothbaum AO, Jovanovic T, Almli LM, Bradley B, Rothbaum BO, Gillespie CF, Ressler KJ (2015) CRP genetic variation and CRP levels are associated with increased PTSD symptoms and physiological responses in a highly traumatized civilian population. Am J Psychiatry 172:353–362. https://doi.org/10.1176/appi.ajp.2014.14020263

Moghaddam HS, Valitabar Z, Ashraf-Ganjouei A, Zadeh MM, Sherbaf FG, Aarabi MH (2018) Cerebrospinal fluid C-reactive protein in Parkinson's disease: associations with motor and non-motor symptoms. Neuromolecular Med 20:376–385. https://doi.org/10.1007/s12017-018-8499-5

Mold C, Rodriguez W, Rodic-Polic B, Du Clos TW (2002) C-Reactive protein mediates protection from lipopolysaccharide through interactions with FcγR. J Immunol 169:7019–7025. https://doi.org/10.4049/jimmunol.169.12.7019

Mortensen RF, Zhong W (2000) Regulation of phagocytic leukocyte activities by C-reactive protein. J Leukoc Biol 67:495–500. https://doi.org/10.1002/jlb.67.4.495

Nagel G, Peter RS, Rosenbohm A, Koenig W, Dupuis L, Rothenbacher D, Ludolph AC (2017) Adipokines, C-reactive protein and Amyotrophic Lateral Sclerosis – results from a population-based ALS registry in Germany. Sci Rep 7:4374. https://doi.org/10.1038/s41598-017-04706-5

Nilsson K, Gustafson L, Hultberg B (2008) C-reactive protein: vascular risk marker in elderly patients with mental illness. Dement Geriatr Cogn Disord 26(3):251–256. https://doi.org/10.1159/000160957

Nilsson K, Gustafson L, Hultberg B (2011) C-reactive protein level is decreased in patients with Alzheimer's disease and related to cognitive function and survival time. Clin Biochem 44 (14–15):1205–1208. https://doi.org/10.1016/j.clinbiochem.2011.07.011

Nosalski R, McGinnigle E, Siedlinski M, Guzik TJ (2017) Novel immune mechanisms in hypertension and cardiovascular risk. Curr Cardiovasc Risk Rep 11(4):12. https://doi.org/10.1007/s12170-017-0537-6

O'Bryant SE, Waring SC, Hobson V, Hall JR, Moore CB, Bottiglieri T, Massman P, Diaz-Arrastia R (2010) Decreased C-reactive protein levels in Alzheimer disease. J Geriatr Psychiatry Neurol 23(1):9–53. https://doi.org/10.1177/0891988709351832

Pepys MB, Baltz ML (1983) Acute phase proteins with special reference to C-reactive protein and related proteins (pentaxins) and serum amyloid a protein. Adv Immunol 34:141–212

Qiu X, Xiao Y, Wu J, Gan L, Huang Y, Wang J (2019) C-reactive protein and risk of Parkinson's disease: asystematic review and meta-analysis. Front Neurol 10:364. https://doi.org/10.3389/fneur.2019.00384. eCollection 2019

Rauk A (2009) The chemistry of Alzheimer's disease. Chem Soc Rev 38:2698–2715. https://doi.org/10.1039/b807980n

Raymond LA, André VM, Cepeda C, Gladding CM, Milnerwood AJ, Levine MS (2011) Pathophysiology of Huntington's disease: time-dependent alterations in synaptic and receptor function. Neuroscience 198:252–273. https://doi.org/10.1016/j.neuroscience.2011.08.052

Rebeck GW (2017) The role of APOE on lipid homeostasis and inflammation in normal brains. J Lipid Res 58:1493–1499. https://doi.org/10.1194/jlr.R075408

Ryberg H, An J, Darko S, Lustgarten JL, Jaffa M, Gopalakrishnan V, Lacomis D, Cudkowicz M, Bowser R (2010) Discovery and verification of amyotrophic lateral sclerosis biomarkers by proteomics. Muscle Nerve 42:104–111. https://doi.org/10.1002/mus.21683

Sánchez-López F, Tasset I, Aguera E, Feijóo M, Fernández-Bolaños R, Sánchez FM, Ruiz MC, Cruz AH, Gascón F, Túnez I (2012) Oxidative stress and inflammation biomarkers in the blood of patients with Huntington's disease. Neurol Res 34(7):721–724. https://doi.org/10.1179/1743132812Y.0000000073

Sawada H, Oeda T, Umemura A, Tomita S, Kohsaka M, Park K, Yamamoto K, Sugiyama H (2015) Baseline C-reactive protein levels and life prognosis in Parkinson disease. PLoS One 10(7): e0134118. https://doi.org/10.1371/journal.pone.0134118. eCollection 2015

Schmidt R, Schimdt H, Curb JD, Masaki K, White LR, Launer LJ (2002) Early inflammation and dementia: a 25-year follow-up of the Honolulu-Asia aging study. Ann Neurol 52(2):168–174. https://doi.org/10.1002/ana.10265

Selkoe DJ, Hardy J (2016) The amyloid hypothesis of Alzheimer's disease at 25 years. EMBO Mol Med 8:595–608. https://doi.org/10.15252/emmm201606210

Shah T, Newcombe P, Smeeth L, Addo J, Casas JP, Whittaker J, Miller MA et al (2010) Ancestry as a determinant of mean population C-reactive protein values: implications for cardiovascular risk prediction. Circulation 3:436–444. https://doi.org/10.1161/CIRCGENETICS.110.957431

Silajdzic E, Rezeli M, Vegvari A, Lahiri N, Andre R, Magnusson-Lind A, Nambron R et al (2013) A critical evaluation of inflammatory markers in Huntington's disease plasma. J Huntingtons Dis 2:125–134. https://doi.org/10.3233/JHD-130049

Silva D, Lacerda AP (2012) Proteína C reativa de alta sensibilidade como biomarcador de risco na doença coronária. Rev Port Cardiol 31:733–745. https://doi.org/10.1016/j.repc.2012.02.018

Silverman JM, Schmeidler J, Beeri MS, Rosendorff C, Sano M, Grossman HT, Carrión-Baralt JR et al (2012) C-reactive protein and familial risk for dementia: a phenotype for successful cognitive aging. Neurology 79(11):1116–1123. https://doi.org/10.1212/WNL.0b013e3182698c89

Silvestrelli G, Lanari A, Parnetti L, Tomassoni D, Amenta F (2006) Treatment of Alzheimer's disease: from pharmacology to a better understanding of disease pathophysiology. Mech Ageing Dev 127:148–157. https://doi.org/10.1016/j.mad.2005.09.018

Silvestroni A, Faull RL, Strand AD, Moller T (2009) Distinct neuroinflammatory profile in post-mortem human Huntington's disease. Neuroreport 20:1098–1103. https://doi.org/10.1097/WNR.0b013e32832e34ee

Slevin M, Matou S, Zeinolabediny Y, Corpas R, Weston R, Liu D, Boras E et al (2015) Monomeric C-reactive protein – a key molecule driving development of Alzheimer's disease associated with brain ischemia? Sci Rep 5:13281. https://doi.org/10.1038/srep13281

Sola Vigo F, Kedikian G, Heredia L, Heredia F, Añel AD, Rosa AL, Lorenzo A (2009) Amyloid-precursor protein mediates neuronal toxicity of amyloid β through go protein activation. Neurobiol Aging 30(9):1379–1392. https://doi.org/10.1016/j.neurobiolaging.2007.11.017

Song I, Chung S, Kim J, Lee K (2011) Association between high-sensitivity C-reactive protein and risk of early idiopathic Parkinson's disease. Neurol Sci 32:31–34. https://doi.org/10.1007/s10072-010-0335-0

Stone PA, Kazil J (2014) The relationships between serum C-reactive protein level and risk and progression of coronary and carotid atherosclerosis. Semin Vasc Surg 27:138–142. https://doi.org/10.1053/j.semvascsurg.2015.04.002

Stoy N, Mackay GM, Forrest CM, Christofides J, Egerton M, Stone TW, Darlington LG (2005) Tryptophan metabolism and oxidative stress in patients with Huntington's disease. J Neurochem 93:611–623. https://doi.org/10.1111/j.1471-4159.2005.03070.x

Strang F, Scheichl A, Chen YC, Wang X, Htun NM, Bassler N, Eisenhardt SU, Habersberger J, Peter K (2012) Amyloid plaques dissociate pentameric to monomeric C-reactive protein: a novel pathomechanism driving cortical inflammation in Alzheimer's disease? Brain Pathol 22 (3):337–346. https://doi.org/10.1111/j.1750-3639.2011.00539.x

Sundelof J, Kilander L, Helmersson J, Larsson A, Ronnemaa E, Degerman-Gunnarsson M, Basun H, Lannfelt L, Basu S (2009) Systemic inflammation and the risk of Alzheimer's disease and dementia: a prospective population-based study. J Alzheimers Dis 18(1):79–87. https://doi.org/10.3233/JAD-2009-1126

Tan ZS, Beiser AS, Vasan RS, Roubenoff R, Dinarello CA, Harris TB, Benjamin EJ et al (2007) Inflammatory markers and the risk of Alzheimer disease: the Framingham study. Neurology 68 (22):1902–1908. https://doi.org/10.1212/01.wnl.0000263217.36439.da

Tansey MG, Goldberg MS (2010) Neuroinflammation in Parkinson's disease: its role in neuronal death and implications for therapeutic intervention. Neurobiol Dis 37:510–518. https://doi.org/10.1016/j.nbd.2009.11.004

Teive HAG (2005) Etiopathogenesis of Parkinson disease. Rev Neurociências 13(4):201–214

Teunissen CE, van Boxtel MP, Bosma H, Bormans E, Delanghe J, de Bruijn C, Wauters A et al (2003) Inflammation markers in relation to cognition in a healthy aging population. J Neuroimmunol 134(1–2):142–150. https://doi.org/10.1016/s0165-5728(02)00398-3

Thompson D, Pepys MB, Wood SP (1999) The physiological structure of human C-reactive protein and its complex with phosphocholine. Structure 7:169–177. https://doi.org/10.1016/S0969-2126(99)80023-9

Togo T, Iseki E, Marui W, Akiyama H, Uéda K, Kosaka K (2001) Glial involvement in the degeneration process of Lewy body bearing neurons and the degradation process of Lewy bodies in brains of dementia with Lewy bodies. J Neurol Sci 184(1):71–75. https://doi.org/10.1016/s0022-510x(00)00498-6

Tomita N (2011) Honolulu-Asia aging study (HAAS). Nihon Rinsho 69(Suppl8):603–606

Umemura A, Oeda T, Yamamoto K, Tomita S, Kohsaka M, Park K, Sugiyama H, Sawada H (2015) Baseline plasma C-reactive protein concentrations and motor prognosis in Parkinson disease. PLoS One 10(8):e0136722. https://doi.org/10.1371/journal.pone.0136722. eCollection 2015

Vachatova S, Andrys C, Krejsek J, Salavec M, Ettler K, Rehacek V, Cermakova E, Fiala Z, Borska L (2016) Metabolic syndrome and selective inflammatory markers in psoriatic patients. J Immunol Res 2016:5380792. https://doi.org/10.1155/2016/5380792

Venneti S (2010) Prion diseases. Clin Lab Med 30(1):293–309. https://doi.org/10.1016/j.cll.2009.11.002

Völkel D, Zimmermann K, Zerr I, Lindner T, Bodemer M, Poser S, Schwarz HP (2001) C-reactive protein and IL-6: new marker proteins for the diagnosis of CJD in plasma? Transfusion 41:1509–1514. https://doi.org/10.1046/j.1537-2995.2001.41121509.x

Walker F (2007) Huntington's disease. Lancet 369:218–228. https://doi.org/10.1016/S0140-6736(07)60111-1

Weintraub D, Comella CL, Horn S (2008) Parkinson's disease – part I: pathophysiology, symptoms, burden, diagnosis, and assessment. Am J Manag Care 14(2):S40–S48

Wichmann MA, Cruickshanks KJ, Carlsson CM, Chappell R, Fischer ME, Klein BE, Tsai MY, Schubert CR (2014) Long-term systemic inflammation and cognitive impairment in a population-based cohort. J Am Geriatr Soc 62(9):1683–1691. https://doi.org/10.1111/jgs.12994

Woloshin S, Schwartz LM (2005) Distribution of C-reactive protein values in the United States. N Engl J Med 352(15):1611–1613. https://doi.org/10.1056/NEJM200504143521525
Wu Y, Potempa LA, El Kebir D (2015) Filep JG.C-reactive protein and inflammation: conformational changes affect function. Biol Chem 396:1181–1197. https://doi.org/10.1515/hsz-2015-0149
Xia D, Samols D (1997) Transgenic mice expressing rabbit C-reactive protein are resistant to endotoxemia. Proc Natl Acad Sci U S A 94:2575–2580. https://doi.org/10.1073/pnas.94.6.2575
Yaffe K, Lindquist K, Penninx BW, Simonsick EM, Pahor M, Kritchevsky S, Launer L, Kuller L, Rubin S, Harris T (2003) Inflammatory markers and cognition in well-functioning African-American and white elders. Neurology 61(1):76–80. https://doi.org/10.1212/01.wnl.0000073620.42047.d7
Yarchoan M, Loureva N, Xie SX, Swenson FJ, Hu W, Soares H, Trojanowski JQ et al (2013) Association of plasma C-reactive protein levels with the diagnosis of Alzheimer's disease. J Neurol Sci 333(1–2):9–12. https://doi.org/10.1016/j.jns.2013.05.028
Yasojima K, Schwab C, McGeer EG, McGeer PL (2000) Human neurons generate C-reactive protein and amyloid P: upregulation in Alzheimer's disease. Brain Res 887(1):80–89. https://doi.org/10.1016/s0006-8993(00)02970-x
Zatta P, Drago D, Bolognin S, Sensi SL (2009) Alzheimer's disease, metal ions and metal homeostatic therapy. Trends Pharmacol Sci 30(7):346–355. https://doi.org/10.1016/j.tips.2009.05.002

C-Reactive Protein in Inflammatory Bowel Disease

3

Sayan Malakar

Contents

3.1 Introduction ... 62
3.2 Incidence and Prevalence of IBD ... 63
3.3 Pathophysiological Consideration ... 65
3.3.1 Gut Mucosal Immunity and Inflammation ... 65
3.3.2 Intestinal Microbiota ... 66
3.3.3 Genetics and Epigenetics in IBD ... 67
3.3.4 Inflammatory Response in IBD ... 68
3.3.5 Role of Innate and Adaptive Immune System in Inflammation in IBD ... 69
3.3.6 Intestinal Adhesion Molecule and Leukocyte Recruitment ... 70
3.3.7 Altered T-Cell Tolerance and IBD ... 71
3.3.8 The Final Inflammatory Cascade in IBD ... 72
3.4 The Link: Inflammation and CRP ... 72
3.5 Histopathology ... 73
3.5.1 Crohn's Disease ... 73
3.5.2 Ulcerative Colitis ... 75
3.6 Clinical Features ... 75
3.6.1 Severity ... 77
3.6.2 Extra-Intestinal Manifestation ... 77
3.7 Extra-Intestinal Manifestation of UC ... 78
3.7.1 Musculo-Skeletal ... 78
3.7.2 Hepatobiliary ... 78
3.7.3 Cutaneous/Oral ... 78
3.7.4 Ophthalmologic ... 78
3.7.5 Hematologic ... 79
3.8 Diagnosis ... 79
3.9 CRP in IBD ... 80
3.9.1 CRP and Endoscopic Correlation ... 81
3.9.2 IBD Relapse and CRP: Any Link? ... 81
3.9.3 CRP to Differentiate Between IBD and IBS ... 83

S. Malakar (✉)
Department of Gastroenterology and Hepatology, GB Pant Institute of Post-graduate Medical, Education and Research, New Delhi, India

W. Ansar, S. Ghosh (eds.), *Clinical Significance of C-reactive Protein*,
https://doi.org/10.1007/978-981-15-6787-2_3

3.10 Malignancy in IBD 83
3.11 IBD and Pregnancy 84
3.12 Management 84
3.12.1 Drugs Used in Management of IBD 85
3.12.2 Treatment Algorithm: Choosing Drugs in Sequential Order Depending on Severity of IBD 86
3.12.3 Nutritional Therapy 87
3.12.4 Surgery 87
3.13 Indications of Surgery in UC and CD is Outlined 87
3.14 Morbidity and Mortality 87
References 87

Abstract

Crohn's disease (CD) and ulcerative colitis (UC) are the two main forms of inflammatory bowel disease (IBD) which are characterized by chronic idiopathic inflammatory damage to the intestine. Though they share some common inflammatory pathways, still they can be distinguished clinically, endoscopically, and histologically. Even after intense research, the etiology for IBD remains unknown. Defective immune regulation, altered gut microbiota, and other environmental factors trigger immune cascade in the background of genetic predisposition in patients with IBD. Activation of helper T cells, i.e., T_H1 and T_H17, and sequential cascade of inflammatory mediators involving IL-1, IL-4, IL-5, IL-6, and TNF and other cytokines play the major role in IBD. Each step of this inflammatory cascade is not only potential target for therapy, the markers of inflammation may guide us for diagnosing, confirming the diagnosis even in follow-up, and prognostication in IBD. Serum biomarkers are mainly markers of inflammation, i.e., C-reactive protein (CRP), ESR, ferritin, or antibodies, i.e., ANCA (anti-neutrophilic cytoplasmic antibody), ASCA anti-OmpC whereas stool markers are mainly leukocyte-derived molecules like fecal calprotectin and lactoferrin. CRP is an acute phase reactant protein secreted by the hepatocyte cells in response to stimulation by circulating IL-6 and TNF-α. CRP level ranges from 5 to 200 mg/L. CRP level sharply rises following any inflammatory condition due to increased recruitment of hepatocytes and decrease of its peak also follows a sharp trend. CRP has been incorporated in the assessment of severity of UC (Truelove and Witts criteria) though various studies have shown that clear increase of CRP is much more prominent in CD than UC, the reason of which is not still very clear and satisfactory. Two explanations for that are higher IL-6 concentration and transmural involvement of intestine in CD. Though not specific, still CRP remains one of the very important biomarkers in IBD.

Keywords

C-reactive protein · Inflammatory bowel disease · Crohn's disease · Ulcerative colitis · Biomarkers

Abbreviation

AGR2	Anterior gradient 2
ANCA	Anti-neutrophilic cytoplasmic antibody
ASA	5-Aminosalicylic acid
ASCA	Anti-saccharomyces cerevisiae antibody
ATG16L1	Autophagy-related 16 like 1
CCR6	C-C chemokine receptor 6
CCR9	C-C chemokine receptor type 9
CD	Crohn's disease
CRC	Colorectal cancer
CRP	C-reactive protein
CT	Computed tomography
CTE	Computed tomography enterography
DC	Dendritic cell
EIMs	Extra-intestinal manifestations
ER	Endoplasmic reticulum
ESR	Erythrocyte sedimentation rate
FOXP3	Forkhead box P3
GC	Glucocorticosteroid
GI	Gastrointestinal
IBD	Inflammatory bowel disease
IBS	Irritable bowel syndrome
IFN	Interferon
IgA	Immunoglobulin A
IgG	Immunoglobulin G
IL	Interleukins
IL23R	Interleukin 23 receptor
IPAA	Ileal pouch/anal anastomosis
IRGM	Immunity-related GTPase family M protein
IV	Intravenous
JAK2	Janus kinase 2
JAK-STAT	Janus-associated kinase signal transducers and activators of transcription proteins
MAP	Mitogen-activated protein
mCRP	Monomeric CRP
MMP	Matrix metalloproteinase
MRE	Magnetic resonance enterography
MRI	Magnetic resonance imaging
MST1	Macrophage-stimulating 1 gene
MTX	Methotrexate
nCRP	Native CRP
NF-κB	Nuclear factor kappa-light-chain-enhancer of activated B cells
NO	Nitric oxide

NOD2	Nucleotide-binding oligomerization domain containing 2
OCP	Oral contraceptive pills
OR	ODDs ratio
pANCA	Perinuclear antineutrophil cytoplasmic antibodies
PTGER4	Prostaglandin receptor 4
PTPN2	Protein tyrosine phosphatase non-receptor type 2
SBFT	Small bowel follow through
TGF	Transforming growth factor
TLR	Toll-like receptor
TNF	Tumor necrosis factor
TNFAIP3	Tumor necrosis factor, alpha-induced protein 3
T-reg	T-regulatory
UC	Ulcerative colitis
WAS	Wiskott–Aldrich syndrome
XLID	X-linked immune dysregulation

Highlights

1. IBD is chronic, relapsing, immune-mediated idiopathic inflammatory intestinal disorder that mainly comprises CD and UC. CD is a granulomatous transmural disease mainly due to T_H1 response but UC is superficial mucosal inflammation mediated by T_H2 cells.
2. Various cytokines come into play to maintain immune cascade, these pathways are not only important for therapeutic challenges faced for the management of IBD, but they may also help us in diagnosing the disease and to prognosticate them.
3. Inflammatory cytokines mainly IL-6 stimulates production of CRP from hepatocytes.
4. Other biological molecules in IBD detection are ANCA, anti-OmpC, anti-Cbir1, ASMA, IL-6, IL-1 haptoglobin, hemopexin, ceruloplasmin, ferritin, etc. But CRP is cheap, reproducible, and easily available though it may increase in other inflammatory or infective conditions.
5. Rise of CRP in UC and CD differs significantly. In UC, the inflammation is superficial, thus rise of CRP is not that sharp but in CD depth of inflammation is transmural with higher systemic level of IL-6 and CRP.

3.1 Introduction

Idiopathic inflammatory bowel disease (IBD) encompasses two major forms, namely the Crohn's disease (CD) and ulcerative colitis (UC). They are mainly characterized by chronic, idiopathic, and relapsing course of diseases due to immune activation leading to intestinal inflammation. Though these two are the major forms of IBD, other chronic intestinal inflammatory states may share some features with IBD, but they have differential identified etiologies. Examples of these disorders are bypass enteropathy, diversion colitis, drug induced colitis and radiation colitis. UC and CD, these two major forms of IBD share some clinical, radiological, pathological, and epidemiological features. And these are the evidences that underlying causes may be similar for those two entities. Approximately 10% patients with Crohn's disease cannot be differentiated from UC on clinical backdrops and hence called as indeterminate colitis although these two diseases have distinct symptoms with divergent treatment and prognosis.

With a background of genetic susceptibility, environmental factors along with altered microbiota play a role for immune dysregulation in gut. Normally, intestinal epithelial cells, microbiota, and immune cells maintain homeostasis in gut maintaining controlled physiological inflammation. Any disruption to this may activate immune system in gut leading to inflammatory cascade activation.

The immune system in gut is usually non-responsive and non-reactive to luminal content due to oral tolerance. Multiple mechanisms to induce oral tolerance have been postulated like deletion or anergy of antigen reactive T cells, induction of CD4 + T cells that suppresses gut inflammation like Foxp3 expressing regulatory T cells. T regulatory cells (T-reg) produce anti-inflammatory cytokines, namely IL-35, IL-10, and TGF-β. In UC and CD inflammatory cascades initiate due to genetic susceptibility that expose gut mucosa to inappropriate innate immune activation, epithelial sensing, and reactivity to commensal bacteria. Which in turn lead to activation of CD4+ and CD8+ T cells leading to cytokines secretion. Excessive release of cytokines overload the pro-inflammatory side compared to T-reg mediated anti-inflammatory counterpart. Immune inflammatory response is initiated and cascade is propagated once abnormally activated innate cells sense bacteria through parenchymal cells or dendritic cells. There is recruitment of activated macrophages, B cells, and neutrophils through integrins. Thus, T_H1, T_H2, and T_H17 come into play and secret various cytokines which spill over into blood. Those cytokines in blood can be assessed along with neutrophil-derived inflammatory markers in stool and can be detected guiding us for diagnosing and prognosticating the diseases.

In 2001, the expression "biomarker" was defined as "*a characteristic molecule or agent that is measured and evaluated as an indicator of normal biological processes, pathogenic processes or pharmacologic responses to a therapeutic agent.*" Leonard and Connor mentioned the perfect biomarker as "one that exhibits accuracy, reproducibility, sensitivity, specificity, and patient acceptability." The greatest impediments in tracing biological biomarkers for IBD are differential expression of symptoms among patients and partial instability of the symptoms in the same suffering patient.

CRP is an acute phase reactant protein secreted by hepatocyte cells in acute condition and chronic inflammation (in some disease condition), stimulated by circulating IL-6 and TNF-α. In comparison to UC, patients with CD shows more clear increase in blood CRP.

3.2 Incidence and Prevalence of IBD

In spite of limitation in gold standard diagnostic modality, geographical and temporal trends have been observed in CD (Table 3.1). CD has been inversely associated with sunlight exposure (Nerich et al. 2011), and the reason might be linked to

Table 3.1 Causes of ulcerative colitis and Crohn's disease

Features	UC	CD
Age of onset	2nd to 5th decade and 7th to 9th decade	2nd to 4th decade and 7th to 9th decade
Female: male	0.51–1.58	0.34–1.65
Smoking	May prevent disease	May cause disease
Oral contraceptive pills	Contraceptive Pills increase doesn't increase the risk	May increase the risk
Appendectomy	Protective	Not protective
Monozygotic twins	6–18% concordance	38–58 % concordance
Dizygotic twins	0–2 % concordance	4% concordance
Antiobiotic used in 1st year of life	Three times increased risk of childhood IBD	Three times increased risk of childhood IBD

vitamin D. And a north-south gradient has been seen in France (Nerich et al. 2006) and the United States (Kappelman et al. 2007; Jacobsen et al. 2006) supporting this evidence. Incidence rate has been reported to be 20.2 and 12.7 per 100,000 persons-year in North America and Europe, respectively. There is slight female preponderance (Kappelman et al. 2007; Bernstein et al. 2006) in CD (female: male 1.3:1); in pediatric population, it is reversed (Kappelman et al. 2007). Mean age of diagnosis in CD is 30, age for most of the diagnosed cases of CD ranges from 15 to 30 years of age (Bernstein et al. 2006; Loftus et al. 2007). In a study from Minnesota, there is a stable incidence of CD up to age group of 20, second peak of disease was seen among the population over 60 years of age (Loftus et al. 2007).

Pathological finding in young and older age group do not vary, but there are some studies showing increased proportion of colonic and distal diseases among older population (Charpentier et al. 2014) and ileocolonic involvement in younger patients (Castro et al. 2008). The prevalence and incidence of UC also diverge with ethnicity and geographic location (Table 3.1). It is currently estimated that there are approximately 800,000 people afflicted with UC in the United States and 1.4 million in Europe (Cosnes et al. 2011). There is a prominent north-south gradient which is in risk. The areas with the maximum rates of prevalence and incidence of UC comprise Northern Europe and North America. In Europe, the incidence rate varies from 1.7 to 13.6 cases per 100,000 individual-years, with a prevalence of nearly 43 to 294 reported cases per 100,000 individuals. In North America, incidence rates varies from 7.6 to 19.5 reported cases per 100,000 individual-years, with the prevalence ranges of 170 to 249 reported cases per 100,000 individuals.

Recent analysis has shown that incidence and prevalence of UC is increasing globally (Molodecky et al. 2012), reasons of which include increased awareness among physicians and patients, improved facilities for diagnostic modalities, improved sanitation, and greater access to health care by the public (Molodecky et al. 2012). In Asia, UC is less frequent than the western countries though the rate of incidence is rising (Molodecky et al. 2012). UC can occur in any age group, but diagnosis prior to age 5 and after 75 is rare. Peak age group with highest incidence is between second and fourth decade, and a second smaller peak has been observed

Table 3.2 Disease presentation of ulcerative colitis and Crohn's disease

Features	UC	CD
Distribution	Continuous, symmetric, and diffuse, with granularity or ulceration found in the entirety of involved segments; however, peri-appendiceal inflammation (cecal patch) is common, even when the cecum is not involved	Often discontinuous and asymmetric, with skipped segments of normal intervening mucosa, especially in early disease
Ileum	Not involved, except as backwash ileitis in panulcerative colitis	Often involved (≈75% of cases)
Rectum	Typically involved with variable proximal distribution	Complete, or more often relative, rectal sparing
Depth of inflammation	Mucosal; transmural only in fulminant disease	Mucosal, submucosal, and transmural
Perianal complications	Not prominent	Often prominent, including large anal skin tags, deep fissures, perianal fistulas that are often complex
Strictures	Rarely present; when present, suggests adenocarcinoma	Often present
Fistulas	Rare	Often present
Inflammation	Diffuse erythema, early loss of vascular markings with mucosal granularity or friability	Focal and asymmetrical, cobblestoning; granularity and friability less commonly seen
Ulceration	Small ulcers in a diffusely inflamed mucosa; deep, ragged ulcers in severe disease	Aphthoid ulcers, linear or serpiginous ulceration; intervening mucosa is often normal
Colonic lumen	Often narrowed in long-standing chronic disease; tubular colon; strictures rare	Strictures are common

between 60 and 70 years. Male:Female ratio is 1:1. It is more common in industrialized areas. Details about causes (features) (Table 3.1) and disease presentation (Table 3.2) of UC and CD were mentioned in Tables 3.1 and 3.2.

3.3 Pathophysiological Consideration

3.3.1 Gut Mucosal Immunity and Inflammation

Human intestine has a surface area of the size of a tennis court, and absorptive surface area of the small intestine is only about 250 sq.m. Billions of plasma cells, dendritic cells, T and memory B cells, and macrophages reside within lamina propria (Mayer et al. 2001; Xiao and Link 1997), instead of a large amount of inflammatory cells present in gut mucosa there is no significant inflammation in healthy individual; this phenomenon is called *physiological/controlled* inflammation. These facts which contradict the law of inflammation in other parts of human body are enough to raise a question that what could be possible mechanisms to dampen the inflammatory response in gut mucosa even for orally given antigenic stimulus. Oral tolerance is the answer (Whitacre et al. 1996; Weiner and Mayer 1996; Xiao and Link 1997; Titus and Chiller 1981).

Oral tolerance is pronouncedly defined as active antigen-specific non-response to antigen which is administered orally (Challacombe and Tomsai 1980; Mattingly and Waksman 1978). Cellular mechanism to suppress mucosal immunity includes physical barrier like mucin, tight junctions preventing invasion of microorganism into cells, local production of defensins maintaining crypt sterility, and mucosal IgA preventing attachment of luminal bacteria thus preventing initiation of inflammation. Even after antigenic challenge to mucosa macrophages, dendrites and lymphocytes are programmed not to respond to them by decreased expression of TLRs (Bullock et al. 2004; Gosselink et al. 2004).

Recent studies show that high antigenic loading was associated with activation of FOXP3 regulatory T-reg cells (Siewert et al. 2008). T-reg cells have a role in maintaining oral tolerance. T_H3 cells have also a role in maintaining oral tolerance by secreting TGF beta which is the most potent suppressor of B and T cells as well as promoting mucosal IgA secretion (Strobel 1996; Kunimoto et al. 1992; Kim and Kagnoff 1990). But paradoxically, numbers of T-reg cells are high in IBD patients and there is peripheral to intestinal shift of T-reg (Wang et al. 2011; Holmén et al. 2006; Hölttä et al. 2008) cells in IBD. Why the T-reg cells fail to protect them from IBD is still under investigation; the reason may be due to an intrinsic, qualitative defect or microenvironmental effects.

3.3.2 Intestinal Microbiota

The intestinal microbiome population comprises the microbes that inhabit in the intestine. These luminal microbiota influences the supplies of nutrients, development of the intestinal immune system and adjusts energy metabolism (Peterson and Gordon 2005). At birth, the gut microbiota is acquired. It changes in due course quite rapidly during the first few years of lifetime. In adult individual, each individual's exclusive population of fecal matter microbiota is mostly stable over time course, but variations occur in response to developmental and environmental factors and in diseases (Eckburg and Relman 2007; Turnbaugh et al. 2009; Wu et al. 2005). Commonly, there are nearly 10^{11}–10^{14} enteric commensal microbiota from approximately 300–500 bacterial species (Quigley and Quera 2006; Pickard et al. 2004). The bulk of commensal bacteria consists of some Gram-positive bacteria like Firmicutes and other Gram-negative bacteria like Bacteroides. Other minor population comprises Actinobacteria, Fusobacteria, Proteobacteria group, and Verrucomicrobia (Eckburg et al. 2005).

There are some reports that commensal bacteria play a noteworthy role in the development process of IBD. First, the empirical antibiotic treatment has been highly effective in some IBD suffering patients (Khan et al. 2011). Second, IBD patients have higher titers against indigenous bacteria (Mitsuyama et al. 2016). Third, different genetic variants that are correlated with bacterial detection, namely

NOD2 (Hugot et al. 2001), and T cell immunity, such as IL23R (Duerr et al. 2006a, b), are implicated in IBD. Fourth, most researches with animal models of colitis need commensal bacteria for the initiation process of intestinal inflammation (Elson et al. 2005). Recent researches have focused on the role of other enteric microorganisms like fungi or viruses for IBD development process. Notably, a research (Yang et al. 2016) showed that compositions and modified amounts of enteric virus were associated to experimental colitis. So intestinal microbiota also deserves a detailed discussion in etiopathogenesis of IBD along with inflammatory cells and gut immunity and genetic contribution.

3.3.3 Genetics and Epigenetics in IBD

Population-based researches have revealed convincing evidences that genetic factors play a significant role in the pathogenesis of IBD, an eight- to ten-fold increased risk of IBD present among relatives of CD and UC probands has been mentioned, most essentially, that there is concordance between the twins (Cho and Brant 2011). Twin studies have revealed the finest evidence for genetic predisposition to IBD, which is effective for CD than UC. Importantly, family and twin studies for IBD have shown that a child has a nine-fold greater risk for developing UC when the other sibling already has it, and the risk is—six-fold increase in the case of CD (Uhlig et al. 2014). Genetic susceptibility is also obvious for patients with IBD as it occurs in several genetic syndromes in early life in setting of a single gene mutation from immune regulation system like in X-linked immune dysregulation (XLID) polyendocrinopathy enteropathy syndrome (PES, defect in FoxP3 for T-reg), Hermansky–Pudlak syndrome, Turner syndrome, Wiskott–Aldrich syndrome (WAS), and glycogen storage disease.

The first susceptibility gene for CD, which was nucleotide-binding oligomerization domain containing 2 (NOD2) was discovered in 2001. NOD2 codes for an intracellular receptor protein for bacterial products present in monocytes causing NFkB activation. The triggering of NOD2 with muramyl dipeptide stimulates autophagy in dendritic cells. DCs isolated from CD patients with susceptibility variants in NOD2 gene showed reduced localization of bacterial population in autophagolysosomes and also are deficient in autophagy induction (Cooney et al. 2010). Genetic analyses have shown, IRGM and ATG16L1, as two autophagy-related genes, both having pivotal role for autophagy in generating immune reactions in IBD. Other genes, associated with autoimmune disease, likely to be PTPN2 and IL23R, are also signifying another facet of Crohn's pathogenesis. Genetic variants that have been reported to bestow a higher risk of CD point the relevance of autophagy, innate immune responses, and phagocytosis in its entire pathogenesis. Other inflammation-related genetic loci are MST1, PTGER4, CCR6, and TNFAIP3. ER stress and metabolism-related genetic loci are SCL22A5, JAK2, AGR2, XBP1, and IRGM. As described earlier, adaptive immunity-related genes implicated in IBD are IL23R, IL-10, IL12B, and PTPN2.

3.3.4 Inflammatory Response in IBD

In the background of genetic susceptibility, various factors come into play to initiate overt inflammation in intestine. A large population of immune cells in the intestinal lamina propria helps in balancing the necessity for immune tolerance of luminal microbes with the requirement to defend against pathogens. The characteristic of active IBD is a significant and marked infiltration into the innate immune cells such as macrophages, neutrophils, dendritic cells, NK-T cells, and B and T cells (adaptive immune cells) present in the lamina propria. Activation of these immune cells in the intestinal mucosa triggers local concentration of cytokines of the interleukin-23–Th17 pathway, interferon-γ, interleukin-1β, and tumor necrosis factor-α (TNF-α), along with other cytokines leading to initiation and propagation of inflammatory cascade. The initial immune reaction to intestinal microbes is highly regulated, and this regulation is the determinant whether defensive inflammatory response or an immune tolerance follows. Commotion of the balance of these responses can guide to IBD in mouse models, and perturbation of the proteins needed for immune function can provoke intestinal inflammation (Elson et al. 2005) (Figs. 3.1 and 3.2).

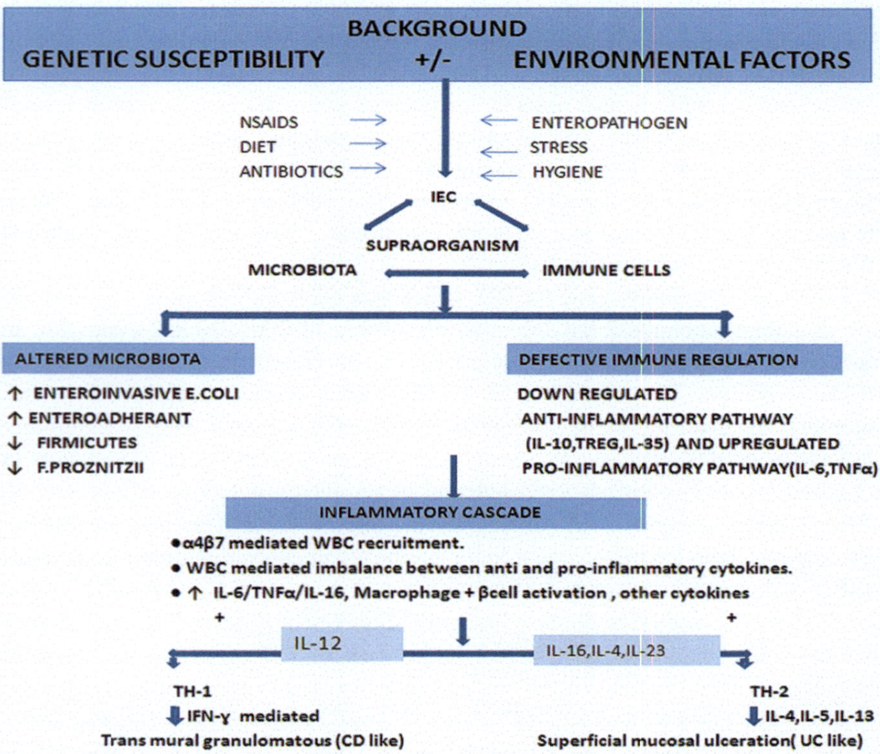

Fig. 3.1 Inflammatory cascade in IBD. The background, regulation, and inflammatory cascade of ulcerative colitis and Crohn's disease was mentioned

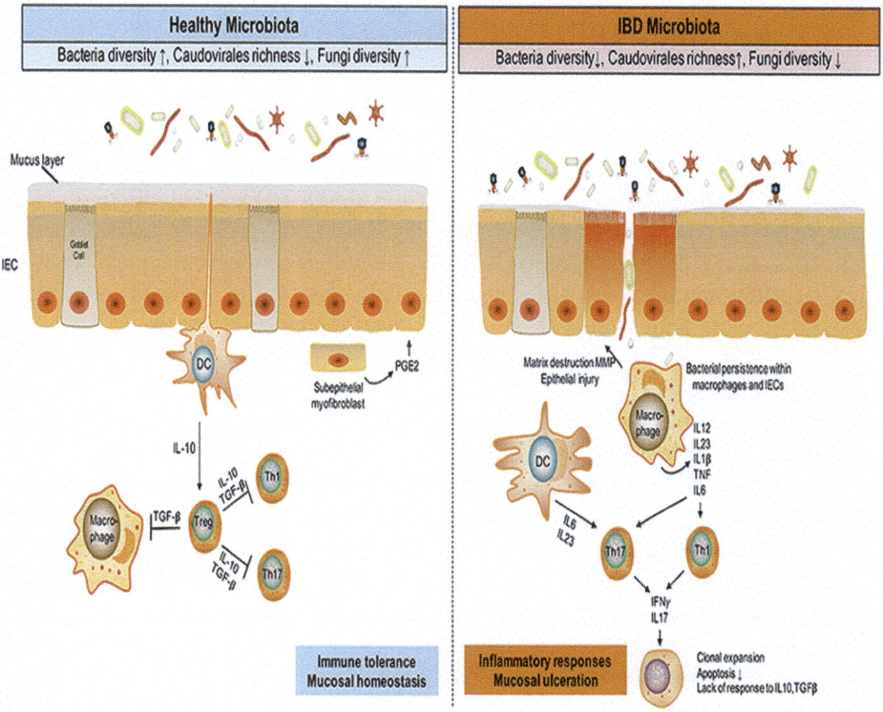

Fig. 3.2 Pathophysiology of IBD: Immune responses and gut microbiota alteration in IBD. The gut microbiota, like viruses,bacteria and fungi, and improper immune responses, involving T-helper 1 (Th1), and Th17, Tregs, are concerned in IBD pathogenesis. During homestasis, gut microbiota trigger an immune tolerance phenotype in the host organism, while in inflammatory IBD like conditions, antigens from dysbiotic microbes trigger Th1 and Th17 cells, causing tissue injury, microbial penetration in decreased mucus layer, and their persistence in the intestinal tissues. This mucosal injury causes subsequent uptake of microbial antigens, viable organisms and TLR ligands, that disseminate the immune responses. TGF, transforming growth factor; MMP, matrix metalloproteinase; DC, dendritic cell. (Image and legend adapted from Zuo T and Ng SC (2018) The gut microbiota in the pathogenesis and therapeutics of inflammatory bowel disease. Front Microbiol. 9:2247. https://doi.org/10.3389/fmicb.2018.02247)

3.3.5 Role of Innate and Adaptive Immune System in Inflammation in IBD

The innate immune system provides an initial, rapid response to microbes in human body. There is a basic difference between innate and adaptive immune system regarding recognition of stimuli for initiating inflammation or killing microbiota. Innate immune cells display receptors that identify general microbial patterns (pattern recognition receptors), in distinction to antigen-specific recognition by receptors of the adaptive immunity. Various types of innate immune receptors in the intestinal epithelial layer not only mediate defense mechanism against luminal microbiota but also balance antigen presenting and epithelial cells for stimulating the tolerance modus that condition immune homeostasis of the intestine (Vijay-Kumar et al. 2007; Lee et al. 2006; Rakoff-Nahoum et al. 2004; Bashir et al. 2004).

Excessive inflow of luminal microbiota and crucial defenses against pathogens require Th1, Th2, and Th17 cells (effector subgroups). But overactivity and expansion of these subgroup cells in relation to regulatory CD4+ T cells can cause intestinal inflammation (Elson et al. 2007; Annunziato et al. 2007; Hue et al. 2006; Fujino et al. 2003; Saruta et al. 2007; Nguyen et al. 2007; Yen et al. 2006).

Regulatory T cells (like forkhead box P3 (Foxp3+) T-reg) and helper T cells (Th1, Th2, and Th17) belonging to CD4+ T cells subgroups secrete typical types of cytokines. Regulation of cytokines secreting cells must be maintained to regulate intestinal immune homeostasis (Makita et al. 2004; Izcue et al. 2006).

The function of B cells in IBD has not been as vastly studied as that of T cells. B cells in intestine secrete IgA antibodies, which in turn provide protection without any scale of inflammation. In presence of flowing antimicrobial antibodies (in patients with IBD) like anti-saccharomyces cerevisiae antibodies and anti-flagellin antibodies (Lodes et al. 2004) but not in healthy subjects indicates B-cell reactivity. In animal models of colitis, both anti-inflammatory (Velázquez et al. 2005) and pro-inflammatory (Olson et al. 2004) roles of B cells have been described.

In Crohn's disease, there is hiked secretion of Th17 cytokine, interleukin-17, Th1 cytokines, interferon-γ, and TNF-α in intestinal lumen (Fujino et al. 2003; Kobayashi et al. 2008). In contrast, in UC, there is usually greater level of Th2 cytokines and interleukin-17 (Fujino et al. 2003; Kobayashi et al. 2008; Fuss et al. 2004). The interleukin-23 pathway is pivotal to the role of Th17 cells. In this pathway, polymorphic variants of multiple genes are engaged in Th17 cell functioning and also associated with both UC and CD (Barrett et al. 2008).

3.3.6 Intestinal Adhesion Molecule and Leukocyte Recruitment

The intestinal epithelium and vasculature have the major role to regulate the entry of leukocytes into the gut. Entry and recruitment of cells into intestinal tissues are adjusted by adhesion molecules like integrins, selectins, and chemokines. Activated T cells in Peyer's patches and mesenteric lymph nodes become "gut-tropic" cells by expressing the chemokine receptor CCR9 and the integrin $\alpha 4\beta 7$ (Jaensson et al. 2008; Mora et al. 2003); this alteration requires retinoic acid. The relative targeting in cellular migration to the intestine is the foundation for specifying these biomolecules for the treatment of IBD.

Various autophagy genes have been implicated in CD including ATG16A1 and IRGM (immune-related GTPase M protein). In mice, ATG16L1 regulates secretion of interleukin-1β and inhibits intestinal inflammation (Saitoh et al. 2008). In mice, the gene expression and morphologic features of Paneth cells are anomalous due to low expression of ATG16L1 (Cadwell et al. 2008). Abnormal Paneth cell morphology also observed in ATG16L1 carriers with CD (Cadwell et al. 2008). NOD2 is an innate immune system-related gene which code for an intracellular sensor peptidoglycan which is a component of bacterial cell wall (Inohara et al. 2003; Girardin et al.

2003). NOD2 are expressed by endothelial cells, epithelial cells, paneth cells, dendritic cells, and macrophages. The NOD2 protein activation by bacterial peptidoglycan triggers mitogen-activated protein (MAP) kinase and NF-κB signaling pathways, which causes secretion of TNF and interleukin-1β cytokines and antimicrobial peptides. Low level of pro-inflammatory cytokines and decreased stimulation of NF-κB on acute stimulation of NOD2 (by bacterial peptidoglycan) have been identified in NOD2 carriers (Abraham and Cho 2006). In NOD2-deficient mice, intestinal inflammation does not develop like most human NOD2 risk allele carriers (Kobayashi et al. 2005). There are three NOD2 gene polymorphisms that are associated with CD resulting in impaired response to peptidoglycan. At least one of the three of the NOD2 gene polymorphisms was expressed in nearly 30% of European ancestry patients. NOD2 carriers are more probable to have an ileal complications and involvement related to fibrostenosis and require intestinal trajection than non-carriers (Lesage et al. 2002). A higher risk of Crohn's disease is conferred by heterozygosity for a polymorphism (by a factor of 1.75–4). A much greater risk was conferred by homozygosity (by a factor of 11–27) (Economou et al. 2004), however, is not sufficient to cause Crohn's disease, indicates a complex multifactorial disorder, where NOD2 polymorphisms alone is not the causative factor. In the intestinal environment, multiple factors contribute toward intestinal antigen-presenting cells in the downregulation of pro-inflammatory cytokines. Factors include interleukin-10 (IL-10) and transforming growth factor β (TGF-β) as inhibitory cytokines. In carriers, during chronic NOD2 stimulation, the NOD2-mediated mechanisms mediated downregulation of pro-inflammatory cytokines are defective and impair NOD2 function (Hedl et al. 2007).

3.3.7 Altered T-Cell Tolerance and IBD

There is a genetic relation/association between ulcerative colitis and the interleukin-10 (inhibitory cytokine) (Franke et al. 2008). In Crohn's disease, the uncommon recessive loss-of-function mutations in either the IL10RB or the IL10RA component of the IL-10 receptor result in the amplified intestinal inflammation caused by defective IL-10 signaling. Sufficiency and importance of IL-10 is highly needed (Glocker et al. 2009).

In Peyer's patch, retinoic acid and TGF-β stimulate proliferation of regulatory T cells (Mucida et al. 2007). In intestine, T-cell tolerance is regulated by local secretion of TGF-β and IL-10 (anti-inflammatory cytokines). Intestinal inflammation in mice is induced by functional or developmental impairment of regulatory T cells. T-cell tolerance is contributed by the autophagy pathway at various stages that elucidate the association of autophagy genes polymorphisms with Crohn's disease which could enhance a patient's susceptibility to intestinal inflammation mediated by T-cell tolerance defects (Nedjic et al. 2008).

IL-23, secreted by dendritic cells and macrophages, may contribute to TR17 survival, proliferation, and also have the function in Th17-independent pathways mediated intestinal inflammation. Levels of Th17 and IL-23 cytokines are high in the colonic mucosa in both ulcerative colitis and Crohn's disease. Heterodimeric receptor (encompassing IL-12RB1 and IL-23R) and heterodimeric IL-23 proteins (constitute p19 and p40 subunits) mediate and engage in interleukin-23 signaling pathways. This engagement stimulates the JAK-STAT (Janus-associated kinase signal transducers and activators of transcription) signaling pathway, which in turn controls the transcription of several genes.

Several researches of genetic correlations between inflammatory bowel disease (Duerr et al. 2006a, b), ankylosing spondylitis (Burton et al. 2007), and Psoriasis (Cargill et al. 2007) with IL23R point that inflammatory bowel disease shares common genetic associations with some other autoimmune diseases.

3.3.8 The Final Inflammatory Cascade in IBD

Once abnormal immune activation by abnormal immune sensing of bacterial pathogen by dendritic cells and parenchymal sets in, stepwise inflammatory response comes into play by sequential activation of T cells and outpouring of local as well as systemic cytokines (Fig. 3.1) leading to intestinal injury. Cytokines like IL-1, IL-6, and TNF have diverse role in intestinal injury. They promote fibrogenesis, collagen deposition activation of tissue metalloproteinase, and production of other inflammatory cytokines. Leukocyte adhesion is mediated by various integrins expressed in intestine. Consistent with these findings, potential therapeutic effects have been noted with inhibiting TNF-alpha and integrins like alpha (α) 4, beta (β) 7. Summarizing those effects of T cells, T_H1 secrets IFN gamma (γ) inducing transmural granulomatous inflammation whereas T_H2 secrets IL-4,5,13 inducing superficial mucosal inflammation as in UC. T_H17 has a role in neutrophilic recruitment. T_H1 cytokine pathway is activated by IL-12, a key cytokine in mucosal ulceration seen in experimental model. IL-4 and IL-23 along with IL-6 and TGF-β induce T_H2 and T_H17, respectively; IL-23 inhibits suppressive function of T-reg cells. Activated macrophage secrets IL-6 and TNF. These inflammatory pathways, cytokines, and integrin molecule are major targets for therapeutics in IBD.

3.4 The Link: Inflammation and CRP

CRP, an acute phase reactant, is the major downstream mediator in an inflammatory event. It is mainly produced by IL-6-triggered hepatic cells (Du Clos 2000; Baumeister et al. 2016). In inflammation, the main function of CRP is the activation of the complement pathway by binding with C1q and final opsonization of pathogens. Thus, CRP can stimulate the fluid phase complement pathways of the host defense mechanism. It can also trigger cell-mediated pathways by initiating complement as well as binding to Fc receptors of IgG. Binding of Fc receptors with CRP leads to the release of pro-inflammatory cytokines. CRP, unlike IgG also has the ability to differentiate/recognize self vs. non-self molecules based on the pattern recognition memory (PRM). IgG, the complement activator identifies pathogenic molecules only by distinct antigenic epitopes. The homopentameric native CRP (nCRP) is characterized by a discoid arrangement of five identical non-covalently bound subunits. Each subunit has a molecular mass of nearly 23 kDa and has 206 amino acids. These subunits arranged in a lectin-fold with a two-layered β-sheet lie in the same orientation around a central pore (Eisenhardt et al. 2009).

CRP is not only secreted by hepatocytes but also locally produced by macrophages, endothelial cells, smooth muscle cells, adipocytes, and lymphocytes. In the elderly, reports show that CRP levels can be influenced by estrogen used in hormone replacement therapy. Traditionally, CRP was used as a marker of infection and biomarker of cardiovascular episodes. Spur of research pinpoint that CRP plays an important function in inflammatory pathways and host responses to infection through phagocytosis, complement activation, apoptosis, nitric oxide (NO) release, and the production of cytokines, like interleukin-6 and tumor necrosis factor-α (Eisenhardt et al. 2009).

CRP binds with phosphorylcholine (PC, as ligand in the presence of calcium ions) present on the surface of microbes. The calcium ions are required for the binding and stability of ligands. Each subunit has a recognition face where the PC-binding site is present in the nCRP molecule (Boncler and Watala 2009). The PC-ligand-binding (recognition) face has two calcium ions per protomer as its characteristic feature. The opposite face binds with the C1q complement protein and also with Fc receptors (Fc gamma (Fcγ)-receptor) (Du Clos and Mold 2004).

The nCRP (pentameric CRP) can irreversibly dissociate at inflammatory sites of infection into five distinct monomers (or free subunits), known as monomeric CRP (mCRP or modified CRP). These mCRP has been observed at either high temperatures in the absence of calcium (Potempa et al. 1983) or high levels of urea (Potempa et al. 1987). The mCRP molecules express different neoepitopes and are differentiated from nCRP by their biological, different antigenic and electrophoretic activities. The two isoforms of CRP, nCRP and mCRP, have been studied with their divergent biological functions in various inflammatory responses. The distinct biological function of the two isoforms can be explained by their binding to varied types of Fc receptors and evoking signaling cascades. The nCRP molecule interacts with low-affinity IgG receptor FcγRIIa (CD32) (Du Clos 2013) while isoform mCRP binds with FcγRIIIa (CD16a) on monocytes and the FcγRIIIb (CD16b) on neutrophils forming low-affinity immune complexes.

Detailed role of CRP in IBD will be discussed in investigation part of this chapter.

3.5 Histopathology

3.5.1 Crohn's Disease

The prevalence of granuloma in CD has been estimated to be nearly 15% in endoscopic series and 70% in surgical specimens. The *aphthous ulcer* is the earliest characteristic lesion of CD. These superficial ulcers are surrounded by a halo of erythema, very minute in size, ranging from 3 mm to a minimum bare visibility. In the small intestine, mostly by destroying the overlying M cells over the lymphoid aggregates, aphthous ulcer arises. In the colon, aphthae may be associated with lymphoepithelial complexes and occur without an endoscopically visible central erosion. In early findings of CD, like aphthous lesions, *non-caseating granulomas* were believed to be also found. Non-caseating granuloma, a high characteristic feature of Crohn's disease, is not universally found and also not unique to Crohn's disease (Yantiss and Odze 2007). There is little or no central necrosis in Crohn's disease in contrast to the granulomas of TB, and mycobacterial cultures and acid-fast stained bacteria are negative. Extra-intestinal granulomas are rare but granulomas may also be found outside the GI tract in the eye, skin, and liver. It is also very crucial to differentiate the granulomas of Crohn's disease from those occurring in association with an injured crypt. The injured crypt afterward represents a response to mucin secreted from injured goblet cells and may be found in UC and different conditions (Yantiss and Odze 2007). The generalized view is that intestinal inflammation is a *transmural process* in Crohn's disease, in contrast to the other superficial inflammation of UC. The transmural process of inflammation, however, cannot be valued on superficial endoscopic biopsy specimens, and it tends to be focal in resected specimens. The highly consistent diagnosis of Crohn's disease involves transmural process. Centralized foci of architectural distortion without chronic inflammation in resected specimens of gut is an observation that pinpoint early superficial lesions like aphthae may be reversible and transient. In chronic disease condition, the aphthae may coalesce into larger ulcers. Serpiginous or linear ulcers can develop when multiple ulcers coalesce in a longitudinal direction. In Crohn's disease, the classic cobblestone appearance results when transverse and linear ulcers traverse and a web of ulcers surround areas of relatively prominent submucosal edema and normal mucosa. Ulcers also descend down to the muscularis propria. Sinus tracts, large ulcers, and strictures are late facets of Crohn's disease. Fistulas and sinuses characterize extensions of fissures; sinus tracts end blindly, and fistulas enter epithelial-lined organs like bowel, bladder, skin, or vagina. Intramural sinus tracts are identified rapidly on barium studies. With infiltration of inflammation to the serosa, serositis can take place, resulting in gripping of bowel to loops of colon, small intestine, or other adjoining organs.

In Crohn's disease, there is always an inclination for proximal colon and distal small intestine involvement. Nearly 30–50% of all patients have disease distressing both colon and ileum. An additional one third have disease restricted to the small intestine, mainly the terminal ileum, and there may be an increased number of people with isolated colonic disease (Jess et al. 2007; Jiang et al. 2006). Gross participation of the stomach, esophagus, or duodenum is rare and nearly always is seen in relation with disease of the colon or distal small intestine. Isolated jejunal involvement is also rare. Focally worsened acute and chronic inflammation may be seen in patients with Crohn's disease in their gastric biopsy specimens reflecting gross involvement (with or without) of the stomach (Annunziata et al. 2012). Thus, the discontinuity nature of the disease makes many variations in different possible location, leading to considerable disparity in clinical presentation.

3.5.2 Ulcerative Colitis

At the time of early clinical presentation, 20% of patients with UC have pancolitis, 35% have disease spreading beyond the sigmoid but approximately 45% patients have disease limited to the rectosigmoid (Langholz et al. 1994). The disease typically is less severe proximally and progressively most severe distally. In contrast to CD, symmetric and continuous involvement is the characteristic of UC with a sharp changeover between uninvolved and diseased segments of the colon. Macroscopically, in mild disease, the mucosa in UC appears edematous, hyperemic, and granular. With progression of disease, punctate ulcers and hemorrhagic mucosa becomes visible. These ulcers can increase and spread into the lamina propria. They often have overhanging edges with irregular shape or may be linear along the line of the teniae coli. Epithelial regeneration alongside multiple attacks leads to formation of pseudopolyps, which is characteristic of acute disease or long-standing UC. Another typical appearance of long-standing disease is featureless and atrophic colonic mucosa, adjunct with narrowing and shortening of the colon. Patients suffering from severe disease can develop acute dilatation of the colon, also featured by grossly ulcerated mucosa where only islands or small fragments of mucosa remaining and thin bowel wall. A perforation in the colon causes fibrinopurulent exudate on the serosal surface of the bowel.

In UC, inflammation is characteristically confined to the mucosa, in distinction/disparity with the transmural involvement of CD. The inflammatory changes typically concentrated at the luminal layer of the muscularis mucosa. Microscopic examination revealed the early stage of UC by congestion of capillaries and venules and edema of the lamina propria, often with extravasation of red blood cells. Increasing inflammatory responses causes the surface epithelial cells to become flattened, finally ulcerate, and it may become undermined in case of deep ulcers. Epithelial regeneration and the subside of inflammatory infiltrate represent the healing phase of UC. Cuboidal epithelial cells with large, eccentric nuclei and well-formed nucleoli are developed during regenerative phases. These features are also common in dysplasia. Thus, during acute inflammation, distinction of dysplasia in UC should be made very cautiously. Actual dropout of glands or crypt architectural distortion are a classic histologic feature of chronic quiescent UC.

3.6 Clinical Features

Symptoms likely to differ according to the extent of disease, site of involvement and absence or presence of extra-intestinal manifestation (Both et al. 1983). The major symptoms of UC are diarrhea with rectal bleeding, tenesmus, passage of mucous and crampy abdominal pain; patient can either present acutely or may present with chronic diarrhea; rarely patient may present with proctitis or proctosigmoiditis. The complaint of patients with proctitis is usually the fresh blood passing, either separately from the streaked or stool on the surface of a hard or normal stool (Rao et al. 1988). In UC, the pathophysiology of diarrhea engross several processes, but

Table 3.3 Diagnosis, clinical features, radiology and colonoscopy, and histology of UC

Diagnosis	Clinical features	Radiology and colonoscopy	Histology
Pseudomembranous colitis	History of hospitalization or prior antibiotic use, presents with loose stool	Edematous shaggy outline of colon, pseudomembrane may be seen in colonoscopy	Mimics ischemic colitis. Fibropurulent exudation
Infectious colitis	Sudden onset, often food- or water-borne, may present with fever, pain in abdomen, toxin can be demonstrated in stool	Nonspecific findings	Superficial neutrophilic infiltrate, crypt architecture is normal, crypt abscess may be seen
Amoebic colitis	History of travel to endemic area. ELISA for amebic lectin antigen may be helpful	Discrete ulcers, ameboma, strictures	Similar to UC, flask-shaped ulcer. Amoeba in Lamina propria
Microscopic colitis	*Normal colonoscopy* with watery diarrhea	Normal	Chronic inflammatory infiltrate, thickened subepithelial collagen band, increased intra-epithelial lymphocytes
Ischemic colitis	Occurs in older patients with concomitant atherosclerotic disease. Sudden pain in abdomen may be there	Segmental spenic flexure is most commonly involved. Ulcer after 24–72 h. Thumbprinting is early finding	Mucosal necrosis with ghost cell

failure to absorb water and salt is the predominant factor (Sandle et al. 1990) (Table 3.3). The differential diagnosis of UC includes infective etiology like *Campylobacter jejuni*, *Aeromonas hydrophila*, Chlamydia, *Clostridium difficile*, CMV, *Salmonella*, and Shigella. Other non-infectious causes include CD, diversion colitis, diverticulitis, eosinophilic colitis, microscopic colitis, ischemic colitis, and radiation colitis.

Toxic megacolon, a catastrophic complication of UC may present as acute abdomen. Per digital examination may reveal signs of proctitis which includes tender anal canal and blood. In toxic megacolon, hepatic dullness is lost. Only 15% patients with UC present with acute catastrophic illness. Severe attack is characterized by massive bleeding. Patients requiring 6–8 units of blood usually get benefit from colectomy. Toxic megacolon is defined as right or transverse colon diameter of more than 6 cm with loss of haustration. It is present only in 5% of the patients, and it can be triggered by dyselectrolytemia or narcotics. 50% of the patients with dilation will respond to medical therapy only; medically refractory dilation requires surgery. Though rare perforation is the most dangerous complication. Strictures occur in 5–10% of the patients.

3.6.1 Severity

Truelove and Witts criteria is used to define severity in acute severe UC.

Crohn's disease usually present with acute or chronic bowel symptoms. Patients with Crohn's disease have a longer mean time to diagnose than in patients with UC, and as many as 25% of the patients have a delay in diagnosis of over 2 years from the onset of symptoms (Vavricka et al. 2012). As we have previously discussed, Crohn's disease is a transmural process; there are two main varieties of evolving Crohn's disease, i.e., fibrostenotic obstructing pattern or a penetrating fitulizing pattern. Clinical manifestation mainly depends on the site of involvement.

Terminal ileum is the most common site of inflammation so they present as pain in the right lower quadrant of the abdomen with or without lump and diarrhea. They may mimic acute appendicitis. Low-grade fever with colicky pain may be associated findings. Weight loss is also common in CD, patient may lose around 10–20% of body weight. Jejunoileitis may present with features of malabsorption. Nutritional deficiencies are also part of low oral intake. Diarrhea is a part of early disease, and it can be due to bacterial overgrowth in pre-stenotic stasis, bile acid malabsorption in dysfunctional ileal segment, decreased water absorption, or increased electrolyte secretion. Colitis may present with low-grade fever, malaise, diarrhea, crampy abdominal pain, and sometimes with hematochezia. Frank bleeding in CD is much more less common than UC. Perianal disease present in one third patients with CD. Perianal disease may present with perianal abscess, large hemorrhoides, anal stricture, and ano-rectal fistula. Patients with perianal disease may have normal colonoscopy. Upper GI tract disease presents as *H. Pylori*-negative gastritis with epigastric pain, nausea, and vomiting. The second part of duodenum is mostly affected.

3.6.2 Extra-Intestinal Manifestation

Patient with IBD may come with extra-intestinal manifestations involving joints, skin, hepatobiliary, or eyes. Dermatological diseases include erythema nodosum, pyoderma gangrenosum, Sweet's syndrome, pyoderma vegetans, pyostomatitis vegetans, and metastatic CD. Perianal skin tags are found in 75% patients with CD. Oral mucosal lesion is more commonly seen in CD than UC, which includes aphthous ulcer and cobblestone lesion. Erythema nodosum is more commonly seen in CD than UC (15% vs. 10%). Attacks correlate with bowel activity. Pyoderma gangrenosum is seen up to 12% of the patients. Peripheral arthritis develops in 15–20% of the patients of IBD and worsens with bowel activity. It is asymmetric, polyarticular, and migratory. Frequently involved joints are large joints of upper and lower extremities. Peripheral arthropathy can be classified into two distinct types (Orchard et al. 1998). Type 1 is pauciarticular and asymmetrical, typically involving the large joints (elbows, knees, and ankles) and affecting less than five joints. It usually expresses with self-limited, acute episodes that matches the underlying bowel disease activity. Type 2 arthropathy is polyarticular and symmetrical, typically involving the small joints and affecting five or more joints. This type shows with persistent symptoms independent of the colitis process.

Both forms are seronegative and non-deforming. The involved joints are erythematous, hot, and swollen. In treatment of colitis, the peripheral arthropathy usually responds. Rest, intra-articular glucocorticoid injection, physical therapy, and therapeutic arthrocentesis also can assist to manage symptoms. Ankylosing spondylitis occurs in 10% patients with IBD. Sacroiliits occurs early in IBD and symmetric and mostly asymptomatic. Eye manifestation includes conjunctivitis and uveitis. Other extra-intestinal manifestations are listed below:

3.7 Extra-Intestinal Manifestation of UC

3.7.1 Musculo-Skeletal

Osteoporosis, Ankylosing spondylitis, Osteopenia, Osteomalacia, Osteonecrosis, Sacroiliitis, Peripheral arthropathy.

3.7.2 Hepatobiliary

Autoimmune hepatitis, Pericholangitis, Cholangiocarcinoma, Hepatic steatosis, PSC.

3.7.3 Cutaneous/Oral

Aphthous stomatitis, Angular stomatitis, Oral ulcerations, Erythema nodosum, Pyoderma gangrenosum, Psoriasis, Sweet's syndrome (acute febrile neutrophilic dermatosis), Pyostomatitis vegetans.

3.7.4 Ophthalmologic

Retinal vascular disease, Conjunctivitis, Uveitis, Episcleritis, Scleritis, iritis.

3.7.5 Hematologic

Autoimmune hemolytic anemia, Anemia of chronic disease, Iron deficiency anemia, Hypercoagulable state, Leukopenia or thrombocytopenia, Leukocytosis or thrombocytosis.

Depending on the definition, nearly 6–25% of Crohn's disease patients have an extra-intestinal manifestation (EIM) of IBD (Peyrin-Biroulet et al. 2011). EIMs are more frequent among patients with colonic involvement than in patients with no colonic inflammation. EIMs are found to arise more often in Crohn's disease than in UC. One fourth of those affected have more than 1 EIM (Farmer et al. 1975). Many of these EIMs are common to UC, Crohn's disease, and to other non-idiopathic inflammatory conditions. For example, patients with extensive UC are at risk for cholelithiasis whereas patients with ileal Crohn's disease are at increased risk for cholelithiasis. The extent of ileal resections appears to be the major risk factor for this complication. Some EIMs take place as a direct result of the bowel disease (like resultant nephrolithiasis from oxalate malabsorption). In the case of inflammatory joint, mucocutaneous, and ocular EIMs, the pathophysiology in the intestine is an activated inflow of mononuclear cells but homing deviantly to the involved extra-intestinal organs.

3.8 Diagnosis

Active disease is related with increased inflammatory markers like ESR, CRP, and increased platelet count and low albumin and hemoglobin. Fecal lactoferrin is a useful marker for intestinal inflammation. Fecal calprotectin also correlates with histologic inflammation, pouchitis, and relapse. Fecal lactoferrin and fecal calprotectin are useful to rule out intestinal inflammation and to distinguish between IBS and IBD. Serological tests include pANCA (found in 60–65% patients in UC, and only one fourth in CD) and anti-saccharomyces cerevisiae is mostly associated with CD (40–75% vs. 5% in UC).

Plain film of the abdomen was observed in X-ray of patients with a severe attack of UC with an acute abdomen. On plain abdominal films, the presence of intraperitoneal air may be missed, however, in CT (with better diagnostic analysis) showed better detection of complications than a plain abdominal radiology in disease and its extent.

Earliest radiological change in barium enema is fine mucosal granularity. Severe disease may be associated with thickened superficial ulcers, whereas deep ulcers are seen as collar button ulcers which denote mucosa-penetrating ulcers. Haustral folds may become thickened and edematous in advanced disease but are normal in mild disease. Loss of haustral folds occurs in long-standing disease. Endoscopically mild disease is characterized by absent vascular pattern friability, erythema, and erosions and in case of severe disease it is identified by ulceration and bleeding. In long-standing disease, colon becomes shortened, narrowed with polyps. Polyps can be of post-inflammatory variety, pseudopolyps, adenomatous polyps, or carcinoma. CT or

MRI are not as helpful as endoscopy though they may show mural thickening, adenopathy, inhomogeneous wall density.

Radiological findings in early CD are thickened folds and aphthous ulceration. Longitudinal and transverse ulceration may give rise to cobblestone appearance. Earliest finding in CD is aphthous ulcer. Small ulcers are intervened by normal mucosa in between. Due to transmural inflammation, intestine has decreased distensibility and narrow lumen. Radiological string sign represents long areas of circumferencial inflammation and fibrosis.

In case of CD, CT or MR enteroclysis has been shown to be superior than small bowel follow through for detecting extraluminal complications including fistula and sinus tracts, and they are as efficient as SBFT to detect mucosal inflammation. So CTE or MRE is the first modality of investigation in complicated CD; MRE is preferred over CTE in pelvic lesions such as ischiorectal abscess and perianal diseases.

Endoscopically, CD is characterized by rectal sparing, aphthous ulceration, skip lesions, and fistula. Upper GI endoscopy may allow to detect gastro-duodenal CD.

As ileal involvement is predominant feature in CD, most of the differential diagnosis includes differential diagnosis of right iliac fossa pain or lump. These include ileocecal tuberculosis, ectopic pregnancy, appendicitis, backwash ileitis in UC, endometriosis, tubo-ovarian pathology, and ileitis with spondyloarthropathy. Whereas colonic involvement has wide differential diagnosis of UC (Table 3.3), diversion colitis, eosinophilic colitis, and infective colitis as described in differentials in UC.

3.9 CRP in IBD

In 1930, C-reactive protein was first described by Tillet and Francis (1930) when serum of pneumonia patients precipitated in contact with bacterial "C Fraction." This fraction no longer persisted in serum after the resolution of pneumonia. Moreover, CRP was persistently present in severe pneumonia cases. CRP, an acute phase protein, belong to conserved pentraxin family. CRP, a pentameric protein, is synthesized by hepatocytes in acute phase conditions. The half-life of CRP is approximately 19 h. Its level rapidly rises and falls with the onset of and resolution of inflammatory condition, respectively. Healthy subjects have low levels of circulatory CRP, usually less than 1 mg/L, but may rise to nearly 100-fold or more during acute inflammation. However, CRP is not a specific indicator for intestinal inflammation, but it is a prognostic or diagnostic marker in many diseases. Quantification of CRP may be increased in conditions like extra-intestinal inflammation or infection. CRP has been studied extensively in diseases to predict disease outcomes after myocardial infarction or diagnosis of multiple myeloma. It is less elaborately studied in gastroenterology. But, in IBD, CRP has been significantly correlated with other inflammatory biomarkers such as ESR, anemia, thrombocytosis, and hypoalbuminemia. As a biomarker of inflammation, CRP is prospective because it is cost-effective, minimally invasive, and evaluative in fast results.

3.9.1 CRP and Endoscopic Correlation

CRP is often used to check for occult internal inflammation in case of clinically asymptomatic patients. CRP is more frequently increased in active transmural CD than in moderate to mild mucosal inflammation with UC (Fagan et al. 1982a, b; Vermeire et al. 2005; Saverymuttu et al. 1986). Though not always specific or accurate, clinical disease activity in children and adults with CD has been shown to associate with CRP level (Karoui et al. 2007; Tilakaratne et al. 2010). But, 20–25% of the patients with CD having flares do not exhibit higher CRP levels due to genetic single nucleotide polymorphisms in the CRP gene, which hamper CRP synthesis (Jones et al. 2008). In multiple studies, CRP has been associated with significant intestinal lesion on endoscopy.

Solem et al. (2005) showed in a retrospective cohort of 104 CD patients of which 75% CD patients had normal CRP level and with normal ileocolonoscopy. On the other hand, CRP increased level were significantly correlated with active mucosal inflammation on colonoscopy (OR = 3.5, 95%CI: 1.4–8.9) usually defined as ulcerations, erosions, exudate, spontaneous bleeding, granularity, friability, extensive erythema, inflammatory-appearing nodularity, cobblestoning, and masses. CRP elevations (>0.8 mg/dL) were significantly related with mild to severe clinical activity (OR = 4.5, 95% CI: 1.1–18.3) as mentioned in ACG clinical practice guidelines (Hanauer et al. 2001). Henriksen et al. (2008a, b) showed that CRP levels varies according to disease types in 371 UC patients and 176 CD patients. For CD, there were no significant variation in disease localization (colitis, ileitis, or ileocolitis) with CRP levels thus revealing that isolated ileal disease also result in rise of CRP. For both CD and UC, CRP responses rise based on the degree of disease. However, the mean and median values of CRP in UC were within the normal range of CRP (<10 mg/L) for all kinds of disease subgroups, making CRP less enlightening.

Mosli et al. (2015) conducted a meta-analysis of 19 research (n = 2499 IBD patients) to exemplify CRP correlation with endoscopic disease activity. For IBD, CRP concentration had a pooled specificity and sensitivity of 92% and 49%, respectively. There were an insufficient number of research to estimate separate CRP performance metrics for CD and UC. The suggested CRP cut-off of more than 5 mg/dL is used to monitor active endoscopic diseases.

3.9.2 IBD Relapse and CRP: Any Link?

Increased CRP concentration is associated with clinical relapse in both long-term and short-term follow-up (Boirivant et al. 1988; Koelewijn et al. 2008; Bitton et al. 2008; Consigny et al. 2006). Various research have showed a higher risk of relapse with the relative risk of 3–58 (Travis et al. 1996; Gelbmann 2000). In severe UC flares, increased CRP, collective with low serum albumin and high stool frequency, has been related with higher possibility of failure to counter medical therapy (Vargas et al. 2013). There is also a 6-times increased risk of hospitalization (OR = 6.82, 95% CI: 2.5–18.58; $p < 0.0001$) with high CRP in patients with CD (Vargas et al. 2013). In a GETAID trial, analysis of nearly 71 CD patients in medically induced clinical remission had CRP, erythrocyte sedimentation rate (ESR), complete blood count, and α-1 antitrypsin checked after every 6 weeks (Consigny et al. 2006). Nearly 38 clinically relapsed patients described as a CDAI higher than 150 or height of at least 100 points from baseline, after a median of 31 weeks. CRP higher than 20 mg/L and ESR more than 15 mm revealed clinical relapse. Concentration of CRP were marked to increase in 4–6 months before clinical relapse, proposing that regular measurement of every biomarkers after 3–4 months could alert the physicians that a change in therapy is highly needed. Attaining not only clinical remission but also mucosal healing may cause greater rates of remission or long-term response. The ACCENT-1 trial and its post hoc analysis showed in clinical remission of 137 patients with CD had CRP concentrations computed after induction with infliximab. At 14 weeks, CRP less than 0.5 mg/dL was shown in 56.6% of the patients whereas CRP greater than 0.5 mg/dL in 37.2% of the patients revealed a maintained response to infliximab through 54 weeks ($p = 0.005$) (Vargas et al. 2013).

Rapid and early restoration of CRP levels connects with prolonged long-term response to adalimumab and infliximab (Vargas et al. 2013). Jürgens et al. (2011) examined 268 CD patients who had responded to infliximab induction. Among these patients, 73.5% (197 patients) had increased CRP levels at baseline. 46.7% (92 patients) had normal CRP level (<3 mg/L) at week 4, and another 14.7% (29 patients) attain normal CRP level after 10 weeks. Long-term benefit ($p < 0.001$) of Kaplan–Meier curves includes CRP normalization after 4 weeks of therapy within 5 years with a PPV of 63%. Karmiris et al. 2009 (Tibble et al. 2000) reported alike findings for CD patients with baseline increased CRP and normalization of CRP (<3 mg/L) at both 4 weeks and 12 weeks envisaging less frequent discontinuation of adalimumab and longer prolonged clinical benefit up to a follow-up of 2 years. Kim and Kagnoff (1990) showed low CRP at 12 weeks (<10 mg/L) as being an indicator of clinical remission at 52 weeks (OR = 4.61, $p < 0.001$) during the first year period of adalimumab therapy. Conversely, CRP levels are mostly increased in patients who had lost reactions to biologics (Jürgens et al. 2011). Increased CRP may be a symbol of low drug level and an indication of ensuing loss of response and clinical relapse. In a study by Jürgens et al. (2011), 57 patients with CD who were responders to induction with infliximab had infliximab and CRP levels estimated at 14 weeks. Clinical response after induction in 75% of the patients showed a decrease in infliximab levels before loss of response by 54 weeks. In 60–80% of the patients with higher CRP (>5 mg/L), the infliximab level was less than 1 μg/mL. CRP has also been exposed to correlate superior with low infliximab levels (<1 μg/mL) than with clinical assessment via CDAI. Higher CRP levels are also related with a higher risk of surgery. In a Norwegian study, UC patients with a CRP higher than 23 mg/L at diagnosis were 4.8 times more possibly will have surgery in the future (95% CI: 1.5–15.1, $p = 0.02$). At 1 year, UC patients with a CRP concentration higher than 10 mg/L were three times more possibly require surgery in the upcoming 4 years (95% CI: 1.1–7.8, $p = 0.02$). CD patients having terminal ileitis were six times more likely to require future surgery if CRP concentrations at diagnosis were above 53 mg/L (95% CI: 1.1–31.9, $p = 0.03$).

3.9.3 CRP to Differentiate Between IBD and IBS

The general symptoms of IBS like altered bowel habits, abdominal pain, or discomfort overlaps with a number of other diseases. The specialists and the primary care providers need noninvasive identification tests which will be cost-effective and confident for diagnosis of IBS. Stacy B et al. performed a meta-analysis for evaluating the role of CRP to rule out IBD in patients with bowel symptoms. In 1252 identified systematic literature review after thorough cross-referencing search of medical subject headings, with elaborate evaluation, recognized 140 potentially pertinent journal abstracts/papers for CRP, ESR, fecal lactoferrin, and fecal calprotectin. Ultimately, 67 abstracts and full manuscripts were reviewed in elaboration of which 12 studies fulfilled inclusion criteria. For the CRP analysis, four studies (Tibble et al. 2000; Oikonomou et al. 2012; Schoepfer et al. 2013; Dolwani et al. 2004) had suitable data. High CRP levels were quite predicted in IBD. Levels >2.7 mg/dL have more than a 90% likelihood of IBD, but the opposite is also true. A level of ≥1.7 mg/dL has a >52% predictive probability of IBD. The presence of IBD is strongly argued by low levels of CRP. For IBS, the predictive probability of CRP increased constantly to a 45.7% peak likelihood of IBS at 1.3 mg/dL concentration and gradually declines to <10% predictive probability at a CRP level of ≥2.7 mg/dL.

Among the different serum biomarkers assessed, evaluation showed that CRP at a concentration of <0.5 mg/dL bestows a 1% or lower probability of IBD in patients. The reference range values differ among various laboratories, but a CRP concentration >1 mg/dL is generally considered abnormal. It is imperative to admit that IBD patients can have normal CRP levels (Shine et al. 1985; Henriksen et al. 2008a, b). An increased CRP is more prominent in patients with Crohn's disease than ulcerative colitis. Henriksen et al. (2008a, b) in a prospective population-based study found a CRP level of ≤10 mg/L in 71% of ulcerative colitis patients and in a 25% of Crohn's disease patients. Unfortunately, no data is recorded regarding the absolute CRP values among patients with a CRP level of <10 mg/L. The range for both ulcerative colitis and Crohn's disease patients initiated at 0 and expanded up to 277 and 266, respectively. Thus, the findings are still likely valid as the normal reference range fixed at <0.5 mg/dL.

3.10 Malignancy in IBD

Higher risk of colorectal cancer (CRC) associated with patients with UC. CRC risk factors include family history of CRC, extent and duration of the disease, PSC, severity of inflammation, age at diagnosis of disease, the presence of pseudopolyps, and sometimes backwash ileitis (Karmiris et al. 2009; Rutter et al. 2004; Brentnall et al. 1996). Although the risk of CRC can be virtually eliminated by prophylactic colectomy. However, without any indication for colectomy, patients particularly are unwilling to undergo surgery. Thus, to reduce the risk of CRC associated with UC, effort on colonoscopic surveillance programs was initiated. When doing surveillance colonoscopy, most authorities recommend colonoscopy with biopsies every 1–3 years in UC patients' which extend beyond the rectum and who are suffering for 8–10 years (Askling et al. 2001).

In Crohn's disease, the approximate risk of CRC has differed widely, spanning from no more risk than that of the general study group to an estimated normalized incidence ratio as elevated as 26.6 (Lewis et al. 1999). One meta-analysis estimated the absolute risk at 10, 20, and 30 years to be 2.9%, 5.6%, and 8.3%, respectively. For patients with Crohn's colitis, this was higher than expected in the general population, but not different from what was reported for patients with UC. For patients with isolated ileal disease, the relative risk of colon cancer was not statistically higher than those without IBD. The overall rate of colon cancer in Crohn's disease over the past 30 years may have decreased (Canavan et al. 2006). The general recommendation for CRC surveillance for patients with Crohn's disease that involves the colon is to follow the same guidelines as for UC.

3.11 IBD and Pregnancy

Patients with mild IBD have normal fertility though fallopian tube can be scarred due to inflammatory process mostly right sided due to proximity of terminal ileum. Perianal disease may be problematic in vaginal delivery; otherwise, vaginal mode of delivery is preferred. Glucocorticoids and sulfasalazine are safe during pregnancy. Male patient may experience drug-induced infertility. The majority of the relevant literature demonstrates a negative effect of IPAA on fertility (Cornish et al. 2007; Rajaratnam et al. 2011) and pregnancy may be at least 5 times less likely to occur compared with the general population. In a meta-analysis by Rajaratnam and researchers, a statistically significant rise of 3.91 in the relative risk of infertility was shown after IPAA surgery (Cornish et al. 2007). Methotrexate and anti-TNF are contraindicated on pregnancy.

3.12 Management

Because neither surgical nor medical therapy offers a cure for Crohn's disease, the primary ambitions of therapy are to persuade and maintain remission. The definition of remission is evolving everyday and has moved beyond the subjective disease activity indices to mucosal finding on endoscopic images and histological evidences on biopsy. So remission of disease activity is now closely associated with improved long term treatment goals.

3.12.1 Drugs Used in Management of IBD

5 ASA Agents

Sulfasalazine, olsalazine, and balsalazide. Mainstay of treatment in mild to moderate UC. Controlled release formulations are mesalamine. Sulfasalazine is composed of sulfapyridine and 5 ASA agent. Dose ranges from 3 to 6 g/dL in acute attack and for maintenance it is 2–4 g/d. Side effects include rashes, nausea, vomiting, fever, hepatitis, hypersensitivity reaction, hypersensitivity pneumonitis, pancreatitis, and sometimes paradoxical worsening of colitis. It can cause folate deficiency if patient is not prescribed folate along with it.

Glucocorticoids

IV hydrocortisone (300md/d) or oral prednisolone are the mostly used regimens mainly indicated in moderate to severe UC. Active UC unresponsive to 5 ASA agents can be treated with 40–60 mg of prednisolone. Topically applied steroid is helpful in distal colitis though topical ASA agents are more effective than topical GC in distal colitis. New steroid budesonide has been approved orally for UC, and it is devoid of chronic GC toxicity (dose 9 mg/d for 8 weeks without tapering). Controlled release of budesonide is as effective as prednisolone for ileocolonic CD.

GC has no role in maintenance of remission. Once remission is induced it can be tapered off as no more than 5 mg/week according to clinical response. Side effects profile is well known and includes weight gain, insulin resistance, secondary DM osteoporosis, and increased susceptibility to infections and cataract.

Antibiotics

Antibiotics have no role in the treatment of active or quiescent UC. Pouchitis, complicating postoperative patients, usually respond to metronidazole with or without ciprofloxacin. Tablet metronidazole (15–20 mg/d) is indicated in active fistulous and perianal CD. Metronidazole is associated with metallic taste, peripheral neuropathy, and GI upset.

Azathioprine and 6 MP

They are indicated in glucocorticoid-dependent IBD. Azathioprine (2–3 mg/d) and 6 MP (1–1.5 mg/d) have been successfully used in one third of patients with UC or CD who were steroid dependent. They are also used in maintenance therapy in IBD with active perianal disease and fistula. They both are used in postoperative prophylaxis. Side effects include nausea, rashes, fever, and hepatitis.

Methotrexate

Intramuscular methotrexate (25 mg/week) is useful in inducing remission and reducing GC dosage. For maintenance, 15 mg per week IM is used in CD. Side effects are megaloblastic anemia, hepatic and pulmonary fibrosis, and leukopenia.

Cyclosporin

IV CSA (2–4 mg/kg/d) is used in active severe UC not responding to IV GC. Side effects include hypertension, seizure, gingival hypertrophy, rashes, tremor, and parasthesia.

Tacrolimus

It has been used in refractory IBD in children.

Anti-TNF Therapy

The critical role of TNF in granuloma formation postulated by the Dutch investigator van Deventer paved the way for anti-TNF agent which might prove effective for granulomatous bowel disease. However, contradictory reports about the importance of TNF in IBD was also noted. Prolonged and rapid improvement in disease activity, along with mucosal healing in many cases was subsequently demonstrated by open-label trials (Derkx et al. 1993; van Dullemen et al. 1995; Baert et al. 1999). An early short-term (12-week) RCT provided strong confirmation of the initial impression of efficacy (Targan et al. 1997).

First biologic to be approved in CD was infliximab. It is used in moderate to severe active UC. It is also effective for refractory perianal and enterocutaneous fistulous CD. Active CD refractory to GC, 6 MP ASA may respond to infliximab. Other anti-TNF molecule includes adalimumab, which has been approved for moderate to severe CD. Certolizumab is helpful in induction of remission in active inflammatory CD. Side effects of anti-TNF therapy include non-Hodgkin lymphoma, hepatosplenic T cell leukemia, and reactivation of old hepatitis B to tuberculosis.

Other potential new therapies are anti-integrins like natalizumab and vedolizumab.

3.12.2 Treatment Algorithm: Choosing Drugs in Sequential Order Depending on Severity of IBD

Mild to Moderate UC

5 ASA > rectal GC > oral GC > 6MP/azathioprine> anti-TNF.

Moderate to Severe UC

Oral GC > IV GC > 6MP/azathioprine + infliximab > adalimumab/golimumab > IV CSA/vedolizumab.

Mild to Moderate CD

Budesonide > 5 ASA > prednisolone > 6 MP/azathioprine/MTX > anti-TNF.

Moderate to Severe CD

6 MP/azathioprine/MTX + anti-TNF > anti-integrins > IV GC > TPN.

3.12.3 Nutritional Therapy

As dietary antigen may stimulate enteral immune system, gut rest is important in active severe CD. Bowel rest and TPN is as effective as steroid in inducing remission but not in maintenance. Micronutrient deficiencies are to be identified and treated.

3.12.4 Surgery

Nearly 50% of the patients in UC may require surgery in 10 years. The choice of operation is ileoanal J pouch anastomosis. Most patients with CD also require surgery in their lifetime. Surgery is related to duration of disease and site of involvement.

3.13 Indications of Surgery in UC and CD is Outlined

Indication of surgery in UC: Intractable, fulminant disease, colonic perforation, toxic megacolon, colonic dysplasia, colon CA, prophylaxis of CRC, and massive colonic hemorrhage. In CD, indications are stricture, obstruction and fistula, medically refractory disease, and abscess involving small intestine. For colonic disease, they are same as UC.

3.14 Morbidity and Mortality

Course of disease in CD is highly variable with high rate of relapse. Mortality in CD is slightly higher than general population mostly due to pulmonary disease, steatohepatitis, and CRC. In patients with UC, the mortality is same as general population.

References

Abraham C, Cho JH (2006) Functional consequences of NOD2 (CARD15) mutations. Inflamm Bowel Dis 12:641–650

Annunziata ML, Caviglia R, Papparella LG et al (2012) Upper gastrointestinal involvement of Crohn's disease: a prospective study on the role of upper endoscopy in the diagnostic work-up. Dig Dis Sci 57:1618–1623

Annunziato F, Cosmi L, Santarlasci V et al (2007) Phenotypic and functional features of human Th17 cells. J Exp Med 204:1849–1861

Askling J, Dickman PW, Karlen P et al (2001) Family history as a risk factor for colorectal cancer in inflammatory bowel disease. Gastroenterology 120:1356–1362

Baert F, D'Haens G, Peeters M et al (1999) Tumor necrosis factor alpha antibody (infliximab) therapy profoundly downregulates the inflammation in Crohn's ileocolitis. Gastroenterology 116:22–28

Barrett JC, Hansoul S, Nicolae DL et al (2008) Genome-wide association defines more than 30 distinct susceptibility loci for Crohn's disease. Nat Genet 40:955–962
Bashir ME, Louie S, Shi HN, Nagler C (2004) Toll-like receptor 4 signaling by intestinal microbes influences susceptibility to food allergy. J Immunol 172:6978–6987
Baumeister D, Akhtar R, Ciufolini S, Pariante CM, Mondelli V (2016) Childhood trauma and adulthood inflammation: a meta-analysis of peripheral C-reactive protein, inetleukin-6 and tumour necrosis factor-α. Mol Psychiatry 21:642–649
Bernstein C, Wajda A, Svenson L et al (2006) The epidemiology of inflammatory bowel disease in Canada: a population based study. Am J Gastroenterol 101:1559–1568
Bitton A, Dobkin PL, Edwardes MD, Sewitch MJ, Meddings JB, Rawal S, Cohen A, Vermeire S, Dufresne L, Franchimont D, Wild GE (2008) Predicting relapse in Crohn's disease: a biopsychosocial model. Gut 57:1386–1392. https://doi.org/10.1136/gut.2007.134817
Boirivant M, Leoni M, Tariciotti D, Fais S, Squarcia O, Pallone F (1988) The clinical significance of serum C reactive protein levels in Crohn's disease. Results of a prospective longitudinal study. J Clin Gastroenterol 10:401–405
Boncler M, Watala C (2009) Regulation of cell function by isoforms of C-reactive protein: a comparative analysis. Acta Biochim Pol 56(1):17–31
Both H, Torp-Pedersen K, Kreiner S et al (1983) Clinical appearance at diagnosis of ulcerative colitis and Crohn's disease in a regional patient group. Scand J Gastroenterol 18:987–991
Brentnall TA, Haggitt RC, Rabinovitch PS et al (1996) Risk and natural history of colonic neoplasia in patients with primary sclerosing cholangitis and ulcerative colitis. Gastroenterology 110:331–338
Bullock NR, Booth JC, Gibson GR (2004) Comparative composition of bacteria in the human intestinal microflora during remission and active ulcerative colitis. Curr Issues Intest Microbiol 5:59–64
Burton PR, Clayton DG, Cardon LR et al (2007) Association scan of 14,500 nonsynonymous SNPs in four diseases identifies auto immunity variants. Nat Genet 39:1329–1337
Cadwell K, Liu JY, Brown SL et al (2008) A key role for autophagy and the autophagy gene Atg16L1 in mouse and human intestinal Paneth cells. Nature 456:259–263
Canavan C, Abrams KR, Mayberry J (2006) Meta-analysis: colorectal and small bowel cancer risk in patients with Crohn's disease. Aliment Pharmacol Ther 23:1097–1104
Cargill M, Schrodi SJ, Chang M et al (2007) A large-scale genetic association study confirms IL12B and leads to the identification of IL23R as psoriasis-risk genes. Am J Hum Genet 80:273–290
Castro M, Papadatou B, Baldassare M et al (2008) Inflammatory bowel disease in children and adolescents in Italy: data from the pediatric national IBD register (1996–2003). Inflamm Bowel Dis 14:1246–1252
Challacombe SJ, Tomsai TB (1980) Systemic tolerance and secretory immunity after oral imunization. J Exp Med 152:1459–1472
Charpentier C, Salleron J, Savoye G et al (2014) Natural history of elderly-onset inflammatory bowel disease: a population-based cohort study. Gut 63(3):423–432
Cho JH, Brant SR (2011) Recent insights into the genetics of inflammatory bowel disease. Gastroenterology 140:1704–1712. https://doi.org/10.1053/j.gastro.2011.02.046
Consigny Y, Modigliani R, Colombel JF, Dupas JL, Lémann M, Mary JY (2006) A simple biological score for predicting low risk of short term relapse in Crohn's disease. Inflamm Bowel Dis 12:551–557
Cooney R, Baker J, Brain O, Danis B, Pichulik T, Allan P et al (2010) NOD2 stimulation induces autophagy in dendritic cells influencing bacterial handling and antigen presentation. Nat Med 16:90–97. https://doi.org/10.1038/nm.2069
Cornish JA, Tan E, Teare J et al (2007) The effect of restorative proctocolectomy on sexual function, urinary function, fertility, pregnancy and delivery: a systematic review. Dis Colon Rectum 50:1128–1138
Cosnes J, Gower-Rousseau C, Seksik P et al (2011) Epidemiology and natural history of inflammatory bowel diseases. Gastroenterology 140:1785–1794

Derkx B, Taminiau J, Radema S et al (1993) Tumour-necrosis factor antibody treatment in Crohn's disease. Lancet 342:173–174

Dolwani S, Metzner M, Wassell JJ et al (2004) Diagnostic accuracy of faecal calprotectin estimation in prediction of abnormal small bowel radiology. Aliment Pharmacol Ther 20:615–621

Du Clos TW (2000) Function of C-reactive protein. Ann Med 32(4):274–278

Du Clos TW (2013) Pentraxins: structure, function and role in inflammation. ISRN Inflamm 2013:379040

Du Clos TW, Mold C (2004) C-reactive protein: an activator of innate immunity and a modulator of adaptive immunity. Immunol Res 30(3):261–277

Duerr RH, Taylor KD, Brant SR et al (2006a) A genome-wide association study identifies IL23r as an inflammatory bowel disease gene. Science 314:1461–1463

Duerr RH, Taylor KD, Brant SR, Rioux JD, Silverberg MS, Daly MJ, Steinhart AH, Abraham C, Regueiro M, Griffiths A, Dassopoulos T, Bitton A, Yang H, Targan S, Datta LW, Kistner EO, Schumm LP, Lee AT, Gregersen PK, Barmada MM, Rotter JI, Nicolae DL, Cho JH (2006b) A genome-wide association study identifies IL23R as an inflammatory bowel disease gene. Science 314:1461–1463

van Dullemen H, van Deventer S, Hommes D et al (1995) Treatment of Crohn's disease with anti–tumor necrosis factor chimeric monoclonal antibody (cA2). Gastroenterology 109:129–135

Eckburg PB, Relman DA (2007) The role of microbes in Crohn's disease. Clin Infect Dis 44:256–262

Eckburg PB, Bik EM, Bernstein CN, Purdom E, Dethlefsen L, Sargent M, Gill SR, Nelson KE, Relman DA (2005) Diversity of the human intestinal microbial flora. Science 308:1635–1638

Economou M, Trikalinos TA, Loizou KT, Tsianos EV, Ioannidis JP (2004) Differential effects of NOD2 variants on Crohn's disease risk and phenotype in diverse populations: a metaanalysis. Am J Gastroenterol 99:2393–2404

Eisenhardt SU, Thiele JR, Bannasch H, Stark GB, Peter K (2009) C-reactive protein: how conformational changes influence inflammatory properties. Cell Cycle 8(23):3885–3892

Elson CO, Cong Y, McCracken VJ, Dimmitt RA, Lorenz RG, Weaver CT (2005) Experimental models of inflammatory bowel disease reveal innate, adaptive, and regulatory mechanisms of host dialogue with the microbiota. Immunol Rev 206:260–276

Elson CO, Cong Y, Weaver CT et al (2007) Monoclonal anti-interleukin 23 reverses active colitis in a T cell-mediated model in mice. Gastroenterology 132:2359–2370

Fagan EA, Dyck RF, Maton PN et al (1982a) Serum levels of C-reactive protein in Crohn's disease and ulcerative colitis. Eur J Clin Invest 12:351–359

Fagan EA, Dyck RF, Maton PN, Hodgson HJ, Chadwick VS, Petrie A, Pepys MB (1982b) Serum levels of C-reactive protein in Crohn's disease and ulcerative colitis. Eur J Clin Invest 12:351–359

Farmer R, Hawk W, Turnbull RBJ (1975) Clinical patterns in Crohn's disease: a statistical study of 615 cases. Gastroenterology 68:627–635

Franke A, Balschun T, Karlsen TH et al (2008) Sequence variants in IL10, ARPC2 and multiple other loci contribute to ulcerative colitis susceptibility. Nat Genet 40:1319–1323

Fujino S, Andoh A, Bamba S et al (2003) Increased expression of interleukin 17 in inflammatory bowel disease. Gut 52:65–70

Fuss IJ, Heller F, Boirivant M et al (2004) Nonclassical CD1d-restricted NK T cells that produce IL-13 characterize an atypical Th2 response in ulcerative colitis. J Clin Invest 113:1490–1497

Gelbmann CM (2000) Prediction of treatment refractoriness in ulcerative colitis and Crohn's disease—do we have reliable markers? Inflamm Bowel Dis 6:123–131

Girardin SE, Boneca IG, Viala J et al (2003) Nod2 is a general sensor of peptidoglycan through muramyl dipeptide (MDP) detection. J Biol Chem 278:8869–8872

Glocker E-O, Kotlarz D, Boztug K et al (2009) Inflammatory bowel disease and mutations affecting the interleukin-10 receptor. N Engl J Med 361:2033–2045

Gosselink MP, Schouten WR, van Lieshout LM et al (2004) Delay of the first onset of pouchitis by oral intake of the probiotic strain Lactobacillus rhamnosus GG. Dis Colon Rectum 47:876–884

Hanauer SB, Sandborn W, Practice Parameters Committee of the American College of Gastroenterology (2001) Management of Crohn's disease in adults. Am J Gastroenterol 96:635–643
Hedl M, Li J, Cho JH, Abraham C (2007) Chronic stimulation of Nod2 mediates tolerance to bacterial products. Proc Natl Acad Sci U S A 104:19440–19445
Henriksen M, Jahnsen J, Lygren I et al (2008a) C-reactive protein: a predictive factor and marker of infl ammation in infl ammatory bowel disease. Results from a prospective population-based study. Gut 57:1518–1523
Henriksen M, Jahnsen J, Lygren I, Stray N, Sauar J, Vatn MH, Moum B (2008b) C-reactive protein: a predictive factor and marker of inflammation in inflammatory bowel disease. Results from a prospective population-based study. Gut 57:1518–1523
Holmén N, Lundgren A, Lundin S et al (2006) Functional CD4+CD25 high regulatory T cells are enriched in the colonic mucosa of patients with active ulcerative colitis and increase with disease activity. Inflamm Bowel Dis 12:447–456
Hölttä V, Klemetti P, Sipponen T et al (2008) IL-23/IL-17 immunity as a hallmark of Crohn's disease. Inflamm Bowel Dis 14:1175–1184
Hue S, Ahern P, Buonocore S et al (2006) 61. Interleukin-23 drives innate and T cellmediated intestinal inflammation. J Exp Med 203:2473–2483
Hugot JP, Chamaillard M, Zouali H, Lesage S, Cezard JP, Belaiche J, Almer S, Tysk C, O'Morain CA, Gassull M, Binder V, Finkel Y, Cortot A, Modigliani R, Laurent-Puig P, Gower-Rousseau-C, Macry J, Colombel JF, Sahbatou M, Thomas G (2001) Association of NOD2 leucine-rich repeat variants with susceptibility to Crohn's disease. Nature 411:599–603
Inohara N, Ogura Y, Fontalba A et al (2003) Host recognition of bacterial muramyl dipeptide mediated through NOD2: implications for Crohn's disease. J Biol Chem 278:5509–5512
Izcue A, Coombes JL, Powrie F (2006) Regulatory T cells suppress systemic and mucosal immune activation to control intestinal inflammation. Immunol Rev 212:256–271
Jacobsen B, Fallingborg J, Rasmussen H et al (2006) Increase in incidence and prevalence of inflammatory bowel disease in northern Denmark: a population-based study, 1978–2002. Gastroenterol Hepatol 18:601–606
Jaensson E, Uronen-Hansson H, Pabst O et al (2008) Small intestinal CD103+ dendritic cells display unique functional properties that are conserved between mice and humans. J Exp Med 205:2139–2149
Jess T, Riis L, Vind I et al (2007) Changes in clinical characteristics, course, and prognosis of inflammatory bowel disease during the last 5 decades: a population based study from Copenhagen, Denmark. Inflamm Bowel Dis 13:481–489
Jiang L, Xia B, Li J et al (2006) Retrospective survey of 452 patients with inflammatory bowel disease in Wuhan city, central China. Inflamm Bowel Dis 12:212–217
Jones J, Loftus EV, Panaccione R, Chen LS, Peterson S, McConnell J, Baudhuin L, Hanson K, Feagan BG, Harmsen SW, Zinsmeister AR, Helou E, Sandborn WJ (2008) Relationships between disease activity and serum and fecal biomarkers in patients with Crohn's disease. Clin Gastroenterol Hepatol 6:1218–1224
Jürgens M, Mahachie John JM, Cleynen I, Schnitzler F, Fidder H, van Moerkercke W, Ballet V, Noman M, Hoffman I, van Assche G, Rutgeerts PJ, van Steen K, Vermeire S (2011) Levels of C-reactive protein are associated with response to infliximab therapy in patients with Crohn's disease. Clin Gastroenterol Hepatol 9:421–427.e1
Kappelman M, Rifas-Shiman S, Kleinman K et al (2007) The prevalence and geographic distribution of Crohn's disease and ulcerative colitis in the United States. Clin Gastroenterol Hepatol 5:1424–1429
Karmiris K, Paintaud G, Noman M, Magdelaine-Beuzelin C, Ferrante M, Degenne D, Claes K, Coopman T, Van Schuerbeek N, Van Assche G, Vermeire S, Rutgeerts P (2009) Influence of trough serum levels and immunogenicity on long-term outcome of adalimumab therapy in Crohn's disease. Gastroenterology 137:1628–1640
Karoui S, Ouerdiane S, Serghini M, Jomni T, Kallel L, Fekih M, Boubaker J, Filali A (2007) Correlation between levels of C-reactive protein and clinical activity in Crohn's disease. Dig Liver Dis 39:1006–1010

Khan KJ, Ullman TA, Ford AC, Abreu MT, Abadir A, Marshall JK, Talley NJ, Moayyedi P (2011) Antibiotic therapy in inflammatory bowel disease: a systematic review and meta-analysis. Am J Gastroenterol 106:661–673

Kim PH, Kagnoff MF (1990) Transforming growth factor beta 1 increases IgA isotype switching at the clonal level. J Immunol 145:3773–3778

Kobayashi KS, Chamaillard M, Ogura Y et al (2005) Nod2-dependent regulation of innate and adaptive immunity in the intestinal tract. Science 307:731–734

Kobayashi T, Okamoto S, Hisamatsu T et al (2008) IL-23 differentially regulates the Th1/Th17 balance in ulcerative colitis and Crohn's disease. Gut 57:1682–1689

Koelewijn CL, Schwartz MP, Samsom M, Oldenburg B (2008) C-reactive protein levels during a relapse of Crohn's disease are associated with the clinical course of the disease. World J Gastroenterol 14:85–89

Kunimoto DY, Ritzel M, Tsang M (1992) The roles of IL-4, TGF-beta and LPS in IgA switching. Eur Cytokine Netw 3:407–415

Langholz E, Munkholm P, Davidsen M et al (1994) Course of ulcerative colitis: analysis of changes in disease activity over years. Gastroenterology 107:3–11

Lee J, Mo JH, Katakura K et al (2006) Maintenance of colonic homeostasis by distinctive apical TLR9 signalling in intestinal epithelial cells. Nat Cell Biol 8:1327–1336

Lesage S, Zouali H, Cézard JP et al (2002) CARD15/NOD2 mutational analysis and genotype-phenotype correlation in 612 patients with inflammatory bowel disease. Am J Hum Genet 70:845–857

Lewis J, Deren J, Lichtenstein G (1999) Cancer risk in patients with inflammatory bowel disease. Gastroenterol Clin North Am 28:459–477. x

Lodes MJ, Cong Y, Elson CO, Mohamath R, Landers CJ, Targan SR, Fort M, Hershberg RM (2004) Bacterial flagellin is a dominant antigen in Crohn disease. J Clin Invest 113:1296–1306

Loftus C, Loftus EVJ, Harmsen W et al (2007) Update on the incidence and prevalence of Crohn's disease and ulcerative colitis in Olmsted County, Minnesota, 1940-2000. Inflamm Bowel Dis 13:254–261

Makita S, Kanai T, Oshima S et al (2004) CD4+CD25bright T cells in human intestinal lamina propria as regulatory cells. J Immunol 173:3119–3130

Mattingly JA, Waksman BH (1978) Immunologic suppression after oral administration of antigen. I. Specific suppressor cells formed rat Peyer's patches after oral administration. J Immunol 121:1878–1883

Mayer L, Sperber K, Chan L et al (2001) Oral tolerance to protein antigens. Allergy 56(Suppl 67):12–15

Mitsuyama K, Niwa M, Takedatsu H, Yamasaki H, Kuwaki K, Yoshioka S, Yamauchi R, Fukunaga S, Torimura T (2016) Antibody markers in the diagnosis of inflammatory bowel disease. World J Gastroenterol 22:1304–1310

Molodecky NA, Soon IS, Rabi DM, Ghali WA, Ferris M, Chernoff G, Benchimol EI, Panaccione R, Ghosh S, Barkema HW, Kaplan GG (2012) Increasing incidence and prevalence of the inflammatory bowel disease with time, based on systematic review. Gastroenterology 142 (1):46–54. https://doi.org/10.1053/j.gastro.2011.10.001

Mora JR, Bono MR, Manjunath N et al (2003) Selective imprinting of gut-homing T cells by Peyer's patch dendritic cells. Nature 424:88–93

Mosli MH, Zou G, Garg SK, Feagan SG, MacDonald JK, Chande N, Sandborn WJ, Feagan BG (2015) C-reactive protein, fecal calprotectin, and stool lactoferrin for detection of endoscopic activity in symptomatic inflammatory bowel disease patients: a systematic review and meta-analysis. Am J Gastroenterol 110:802–819. quiz 820

Mucida D, Park Y, Kim G et al (2007) Reciprocal TH17 and regulatory T cell differentiation mediated by retinoic acid. Science 317:256–256

Nedjic J, Aichinger M, Emmerich JMizushima N, Klein L (2008) Autophagy in thymic epithelium shapes the T-cell repertoire and is essential for tolerance. Nature 455:396–400

Nerich V, Monnet E, Etienne A et al (2006) Geographical variations of inflammatory bowel disease in France: a study based on national health insurance data. Inflamm Bowel Dis 12:218–226
Nerich V, Jantchou P, Boutron-Ruault MC et al (2011) Low exposure to sunlight is a risk factor for Crohn's disease. Aliment Pharmacol Ther 33:940–945
Nguyen DD, Maillard MH, Cotta-de-Almeida V et al (2007) Lymphocyte-dependent and Th2 cytokine-associated colitis in mice deficient in Wiskott-Aldrich syndrome protein. Gastroenterology 133:1188–1197
Oikonomou KA, Kapsoritakis AN, Theodoridou C et al (2012) Neutrophil gelatinase-associated lipocalin (NGAL) in infl ammatory bowel disease: association with pathophysiology of inflammation, established markers, and disease activity. J Gastroenterol 47:519–530
Olson TS, Bamias G, Naganuma M et al (2004) Expanded B cell population blocks regulatory T cells and exacerbates ileitis in a murine model of Crohn disease. J Clin Invest 114:389–398
Orchard TR, Wordsworth BP, Jewell DP (1998) Peripheral arthropathies in inflammatory bowel disease: their articular distribution and natural history. Gut 42:387–391
Peterson DA, Gordon JI (2005) Host-bacterial mutualism in the human intestine. Science 307:1915–1920
Peyrin-Biroulet L, Loftus EV Jr, Colombel JF et al (2011) Longterm complications, extraintestinal manifestations, and mortality in adult Crohn's disease in population-based cohorts. Inflamm Bowel Dis 17:471–478
Pickard KM, Bremner AR, Gordon JN, MacDonald TT (2004) Microbial-gut interactions in health and disease. Immune Respon Best Pract Res Clin Gastroenterol 18:271–285
Potempa LA, Maldonado BA, Laurent P, Zemel ES, Gewurz H (1983) Antigenic, electrophoretic and binding alterations of human C-reactive protein modified selectively in the absence of calcium. Mol Immunol 20(11):1165–1175
Potempa LA, Siegel JN, Fedel BA, Potempa RT, Gewurz H (1987) Expression, detection and assay of a neoantigen (Neo-CRP) associated with a free, human C-reactive protein subunit. Mol Immunol 24(5):531–541
Quigley EM, Quera R (2006) Small intestinal bacterial overgrowth: roles of antibiotics, prebiotics, and probiotics. Gastroenterology 130:S78–S90
Rajaratnam SG, Eglinton TW, Hider P et al (2011) Impact of ileal pouch-anal anastomosis on female fertility: Meta-analysis and systematic review. Int J Colorectal Dis 26:1365–1374
Rakoff-Nahoum S, Paglino J, Eslami-Varzaneh F, Edberg S, Medzhitov R (2004) Recognition of commensal microflora by tolllike receptors is required for intestinal homeostasis. Cell 118:229–241
Rao SS, Holdsworth CD, Read NW (1988) Symptoms and stool patterns in patients with ulcerative colitis. Gut 29:342–345
Rutter M, Saunders B, Wilkinson K et al (2004) Severity of inflammation is a risk factor for colorectal neoplasia in ulcerative colitis. Gastroenterology 126:451–459
Saitoh T, Fujita N, Jang MH et al (2008) Loss of the autophagy protein Atg16L1 enhances endotoxin-induced IL-1beta production. Nature 456:264–268
Sandle GI, Higgs N, Crowe P et al (1990) Cellular basis for defective electrolyte transport in inflamed human colon. Gastroenterology 99:97–105
Saruta M, Yu QT, Avanesyan A, Fleshner PR, Targan SR, Papadakis KA (2007) Phenotype and effector function of CC chemokine receptor 9-expressing lymphocytes in small intestinal Crohn's disease. J Immunol 178:3293–3300
Saverymuttu SH, Hodgson HJ, Chadwick VS, Pepys MB (1986) Differing acute phase responses in Crohn's disease and ulcerative colitis. Gut 27:809–813
Schoepfer AM, Beglinger C, Straumann A et al (2013) Fecal calprotectin more accurately refl ects endoscopic activity of ulcerative colitis than the Lichtiger Index, C-reactive protein, platelets, hemoglobin, and blood leukocytes. Inflamm Bowel Dis 19:332–341
Shine B, Berghouse L, Jones JE et al (1985) C-reactive protein as an aid in the diff erentiation of functional and infl ammatory bowel disorders. Clin Chim Acta 148:105–109
Siewert C, Lauer U, Cording S et al (2008) Experience-driven development: effector/memory-like alphaE+Foxp3+ regulatory T cells originate from both naive T cells and naturally occurring naive-like regulatory T cells. J Immunol 180:146–155

Solem CA, Loftus EV, Tremaine WJ, Harmsen WS, Zinsmeister AR, Sandborn WJ (2005) Correlation of C-reactive protein with clinical, endoscopic, histologic, and radiographic activity in inflammatory bowel disease. Inflamm Bowel Dis 11:707–712
Strobel SS (1996) Neonatal oral tolerance. Ann N Y Acad Sci 778:88–102
Targan S, Hanauer S, van Deventer S et al (1997) A short-term study of chimeric monoclonal antibody cA2 to tumor necrosis factor alpha for Crohn's disease. Crohn's Disease cA2 Study Group. N Engl J Med 337:1029–1035
Tibble J, Teahon K, Thjodleifsson B et al (2000) A simple method for assessing intestinal inflammation in Crohn's disease. Gut 47:506–513
Tilakaratne S, Lemberg DA, Leach ST, Day AS (2010) C-reactive protein and disease activity in children with Crohn's disease. Dig Dis Sci 55:131–136
Tillett WS, Francis T (1930) Serological reactions in pneumonia with a non-protein somatic fraction of pneumococcus. J Exp Med 52:561–571
Titus RG, Chiller JM (1981) Orally induced tolerance. Definition at the cellular level. Int Arch Allergy Appl Immunol 65:323–338
Travis SP, Farrant JM, Ricketts C, Nolan DJ, Mortensen NM, Kettlewell MG, Jewell DP (1996) Predicting outcome in severe ulcerative colitis. Gut 38:905–910
Turnbaugh PJ, Hamady M, Yatsunenko T et al (2009) A core gut microbiome in obese and lean twins. Nature 457:480–484
Uhlig HH, Schwerd T, Koletzko S, Shah N, Kammermeier J, Elkadri A et al (2014) The diagnostic approach to monogenic very early onset inflammatory bowel disease. Gastroenterology 147 (5):990–1007. https://doi.org/10.1053/j.gastro.2014.07.023
Vargas EJ, Ramos Rivers CM, Regueiro M, Baidoo L, Barrie A, Schwartz M, Swoger JM, Coates M, Dunn MA, Dudekula A, Binion DG (2013) Silent Crohn's disease: elevated c reactive protein in asymptomatic patients and risk of subsequent hospitalization. Gastroenterology 144:S-102
Vavricka SR, Spigaglia SM, Rogler G et al (2012) Systematic evaluation of risk factors for diagnostic delay in inflammatory bowel disease. Inflamm Bowel Dis 18:496–505
Velázquez P, Wei B, Braun J (2005) Surveillance B lymphocytes and mucosal immuno regulation. Springer Semin Immunopathol 26:453–462
Vermeire S, Van Assche G, Rutgeerts P (2005) The role of C-reactive protein as an inflammatory marker in gastrointestinal diseases. Nat Clin Pract Gastroenterol Hepatol 2:580–586
Vijay-Kumar M, Sanders CJ, Taylor RT et al (2007) Deletion of TLR5 results in spontaneous colitis in mice. J Clin Invest 117:3909–3921
Wang Y, Liu XP, Zhao ZB et al (2011) Expression of CD4+ forkhead box P3 (FOXP3)+ regulatory T cells in inflammatory bowel disease. Dig Dis Sci 12:286–294
Weiner HL, Mayer LF (1996) Oral tolerance: Mechanisms and applications. Introduction. Ann N Y Acad Sci 778:xiii–xviii
Whitacre CC, Gienapp IE, Meyer A et al (1996) Treatment of autoimmune disease by oral tolerance to autoantigens. Clin Immunol Immunopathol 80:S31–S39
Wu L, Estrada O, Zaborina O et al (2005) Recognition of host immune activation by Pseudomonas aeruginosa. Science 309:774–777
Xiao BG, Link H (1997) Mucosal tolerance: a two-edged sword to prevent and treat autoimmune diseases. Clin Immunol Immunopathol 85:119–128
Yang JY, Kim MS, Kim E, Cheon JH, Lee YS, Kim Y, Lee SH, Seo SU, Shin SH, Choi SS, Kim B, Chang SY, Ko HJ, Bae JW, Kweon MN (2016) Enteric viruses ameliorate gut inflammation via Toll-like receptor 3 and Toll-like receptor 7-mediated interferon-beta production. Immunity 44:889–900
Yantiss RK, Odze RD (2007) Pitfalls in the interpretation of nonneoplastic mucosal biopsies in inflammatory bowel disease. Am J Gastroenterol 102:890–904
Yen D, Cheung J, Scheerens H et al (2006) IL-23 is essential for T cell-mediated colitis and promotes inflammation via IL-17 and IL-6. J Clin Invest 116:1310–1316

C-Reactive Protein (CRP) and Markers of Oxidative Stress in Acute Myocardial Infarction

4

Mriganka Baruah

Contents

4.1 Introduction 97
4.1.1 CRP: History and Nomenclature 97
4.1.2 CRP: Biochemistry and Genetics 98
4.2 CRP: Diagnostic Use 99
4.3 CRP and Its Diagnostic Use 99
4.4 Acute Myocardial Infarction (AMI) 100
4.4.1 Pathogenesis of Acute Myocardial Infarction (AMI) 101
4.4.2 Infective Theory of Acute Myocardial Infarction (AMI) 104
4.4.3 Risk Factors of Acute Myocardial Infarction (AMI) 104
4.4.4 Diagnostic Test 105
4.4.5 Role of C-Reactive Protein (CRP) in Acute Myocardial Infarction 105
4.4.6 Oxidative Stress in Acute Myocardial Infarction 106
4.4.7 Malondialdehyde (MDA) 108
4.4.8 Uric Acid 108
4.5 Discussion 109
4.6 Conclusion 112
References 113

Abstract

Acute myocardial infarction (MI or AMI) is a medical emergency and an important health problem which requires urgent medical attention in terms of early diagnosis and management. AMI stimulates inflammatory reaction which induces myocardial injury. This inflammatory response is depicted by a rise in C-reactive protein (CRP) which is a marker of chronic vascular inflammation, and it also correlates with cardiac outcomes following AMI. Oxidative stress which is balance between the prooxidants and the antioxidants of our body plays a critical

M. Baruah (✉)
ESIC Medical College, Kolkata, India

W. Ansar, S. Ghosh (eds.), *Clinical Significance of C-reactive Protein*,
https://doi.org/10.1007/978-981-15-6787-2_4

role in the pathogenesis of AMI. An index of lipid peroxidation marker serum malondialdehyde (MDA) and uric acid represent the production of free radical and thereby oxidative stress. The results of blood estimation of CRP, serum uric acid, and MDA from patients of AMI were found to be statistically significant ($p < 0.01$) in comparison to the control individuals. There is also a significant positive correlation between CRP and MDA on AMI. Thus, inflammation and oxidative stress may play an important role in AMI. Very high CRP, MDA, and uric acid at presentation of AMI also correlates with adverse outcome in AMI. Hence, these parameters may be considered markers in predicting outcome including death in patients of AMI. Thus, anti-inflammatory and antioxidant medication may be beneficial in AMI and gradual monitoring of these parameters may improve patient outcome and is highly recommended.

Keywords

C-reactive protein (CRP) · Oxidative stress · Acute myocardial infarction · Malondialdehyde (MDA) · Uric acid

Abbreviations

AMI	Acute myocardial infarction
AMP	Adenosine monophosphate
AP	Angina pectoris
ATP	Adenosine triphosphate
CK	Creatine phosphokinase (creatine kinase)
CK	Creatinine kinase
CK-MB	Creatinine kinase-MB fraction (creatine kinase isoenzyme)
CL	Confidence intervals
CRP	C-reactive protein
CVS	Cardiovascular
DNA	Deoxyribonucleic acid
ELISA	Enzyme-linked immunosorbent assay
ESR	Erythrocyte sedimentation rate
ETC	Electron transport chain
HF	Heart failure
HR	Hazard ratio
IL-1	Interleukin-1
IL-6	Interleukin-6
LDH	Lactate dehydrogenase
LDL	Low-density lipoprotein
MDA	Malondialdehyde
MDA-LM	Malondialdehyde-like material
MI	Myocardial infarction
NF-κB	Nuclear factor kappa-light-chain-enhancer of activated B cells
NO	Nitric oxide

OR	Odds ratio
OS	Oxidative stress
PAA	Plasminogen activator inhibitor-1
PC	Phosphatidylcholine
PC	Phosphocholine
RA	Rheumatoid arthritis
ROS	Reactive oxygen species
SAA	Serum amyloid A protein
SD	Standard deviation
SGOT	Serum glutamic oxaloacetic transaminase
STEMI	ST-segment elevation myocardial infarction
TF	Tissue factor
TNF	Tumor necrosis factor
UC	Ulcerative colitis

Highlights

1. Inflammatory stimulus like infection or oxidized LDL alters the arterial endothelial lining and makes it "sticky" by the appearance of adhesion molecules and the secretion of chemokines on the luminal wall surface. The sticky endothelium captures or attracts circulating inflammatory cells or monocytes on the lining of arterial wall.
2. Some monocytes adhere to the surface of the endothelium and gradually transform into tissue macrophages. Tissue macrophages ingest lipid deposits in the tissue to form foam cells. The continuous stream of monocytes into the wall of the artery causes the development of lesion. Lesion is formed of advanced fibrous plaque from the initial fatty streak. The formation of plaque causes the development of atherosclerosis.
3. As inflammation is the major contributor in the pathogenesis of atherosclerosis, inflammatory markers could predict the clinical outcome of cardiac patients. The indicators of inflammation are ICAM-1, VCAM-1, IL-1, IL-6, tumor necrosis factors (TNF), and C-reactive protein (CRP).
4. An index of lipid peroxidation marker serum malondialdehyde (MDA) and uric acid represents the production of free radical and thereby oxidative stress. The role of CRP, MDA, and uric acid in AMI patients was also investigated.

4.1 Introduction

4.1.1 CRP: History and Nomenclature

In serum of patients with acute inflammation, a substance was found that reacted with the *C*-polysaccharide of *pneumococcus,* which was named by Tillett and Francis in 1930, as C-reactive protein (CRP) (Tillett and Francis Jr 1930). CRP was elevated in people with a variety of diseases/disease state, including carcinomas; hence, it was previously thought to be a pathogenic secretion but the debate gets closed by the discovery of hepatic synthesis and secretion of CRP. CRP initiating recognition and phagocytosis of damaged cells by binding to ligands like phosphocholine (PC) (Pepys and Hirshfield 2003).

4.1.2 CRP: Biochemistry and Genetics

The CRP gene is located on the chromosome (1q21–q23). CRP is a 224 residue protein (NCBI Entrez Protein #CAA39671, 2003) is an annular pentameric disc shaped with a monomer molar mass of 25,106 Da. Proteins with this sort of configuration are referred to as pentraxins. Native CRP is somewhat different because it has 10-subunits forming two pentameric discs, with a molecular mass of nearly 251,060 Da (https://www.ncbi.nlm.nih.gov/gene/1401, updated February, 2020).

CRP levels rise dramatically during inflammatory processes within the body; hence, it is categorized as a member of a class of acute phase reactants. Various cytokine mediators like interleukin-6, interleukin-1, and tumor necrosis factor-α (TNF-α) is synthesized mainly by cells like macrophages within hours of injury or the onset of inflammation (Pepys and Hirschfield 2003) and also by adipocytes (Lau et al. 2005) which stimulate the liver cells to synthesize CRP and the plasma levels of CRP rises and peaks within 24–48 h. Moreover, local inflammatory cells in the areas of tissue damage, infection, etc. also secrete CRP.

CRP is remarkably consistent regardless of the underlying inflammatory condition and the plasma half-life of CRP is approximately 20 h. The plasma levels of CRP and its rate of synthesis have a direct correlation, making it unique in an acute phase reactant category. The exact physiologic role of this acute phase protein is yet uncertain to some extent though it is one of the most studied among the acute phase proteins. CRP exhibits a variety of pro-inflammatory and pro-coagulant characteristics (Fusman et al. 2002) thought to be associated with cardiovascular disease. It enhances phagocytosis by macrophages, by complement binding (Nijmeijer et al. 2003) to damaged and foreign cells, macrophages, express a receptor for CRP. It binds to many pneumococcal polysaccharides (Volanakis and Kaplan 1971) to phosphatidylcholine (PC) (Tsujimoto et al. 1980) and to damaged cell membranes (Kushner and Kaplam. 1961; Volanakis and Wirtz 1979; Volanakis and Narkates 1981). It also enhances the activity of phagocytic cells (Pepys et al. 1981), modifies T-cell lymphocyte function (Mortensen et al. 1975; Volanakis and Narkates 1981), destroys leukocytes (Wood et al. 1951; Li et al. 1994), activates the complement system (Li et al. 1994), which acts as an early defense protector against various infections and plays an important role in innate immunity (Fig. 4.1). It produces tissue factor, the primary initiator of coagulation by stimulating mononuclear cells. Although there is no in vivo demonstration but aggregated CRP binds LDL cholesterol in vitro, a characteristic that has been proposed to implicate CRP in the progression of atherosclerosis (Zhang et al. 1999).

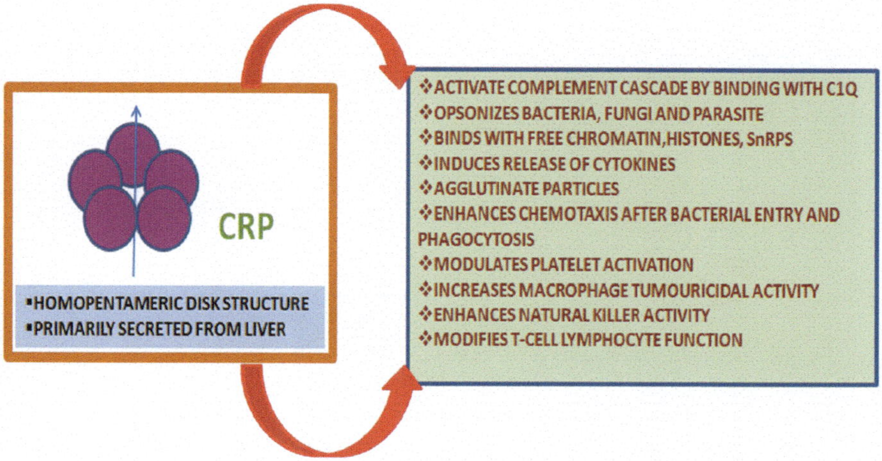

Fig. 4.1 Functional role of CRP. The biological function of CRP in host defense includes phagocytosis, complement activation and clearance of nuclear debris. Other functions also noted (Du Clos 2000; Wu et al. 2015; Marnell et al. 2005; Du Clos and Mold 2004)

4.2 CRP: Diagnostic Use

CRP rises up to 50,000 fold in acute inflammation, such as infection. It rises above normal limits within 6 h and peaks at 48 h. Its half-life is constant, and therefore its level is mainly determined by the rate of production. Serum amyloid A (SAA) is a related acute phase marker that responds rapidly in similar circumstances (Pepys and Hirschfield 2003).

4.3 CRP and Its Diagnostic Use

CRP is used diagnostically mainly as a marker of inflammation. CRP production may be interfered by other known factors apart from liver failure (Pepys and Hirschfield 2003). Gradual monitoring and measuring of CRP values can prove useful in determining disease prognosis or effectiveness of treatments (Fig. 4.2). Bacterial infection tends to have a higher CRP level than viral infections. Determination of CRP can be done by various analytical methods such as ELISA (Enzyme-Linked Immunosorbent Assay), rapid immunodiffusion, immunoturbidimetry, and visual agglutination.

Various chemical insults result in inflammation leading to arterial damage. CRP being a biomarker for inflammation and infection can be used as a clinical indicator for risk related to heart diseases. Since elevated CRP can be caused by many things, this is not a very critical prognostic indicator (Lloyd-Jones et al. 2003). Nevertheless,

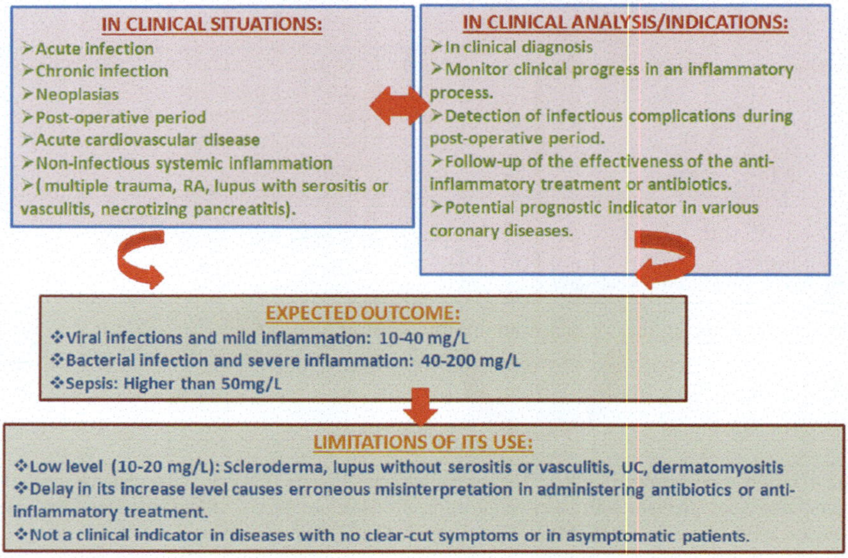

Fig. 4.2 Diagnostic relevance and limitations of CRP in clinics. In cases of a acute or chronic infection or systemic inflammatory disease, high CRP level is an indicator of the disease. CRP is incapable of specifically differentiating/defining between infectious and inflammatory diseases, nor between viral and bacterial diseases. In already diagnosed inflammatory or chronic infectious disease, CRP level acts as a predictor in treatment modalities with antibiotics or anti-inflammatory drugs. *RA* rheumatoid arthritis, *UC* ulcerative colitis (Aguiar et al. 2013)

CRP level more than 2.4 mg/L has been correlated with twice risk of a coronary event in comparison to levels below 1 mg/L (Pepys and Hirschfield 2003).

4.4 Acute Myocardial Infarction (AMI)

Acute myocardial infarction (MI or AMI), more frequently referred to as a heart attack, is a life-threatening state, a medical emergency caused by an abrupt disruption of blood supply to a part of the heart leading to damage, and potential death of heart tissue from the unexpected obstruction of a coronary artery by a blood clot or coronary thrombosis (Boersma et al. 2003) previously named by atherosclerosis (Harrison's 17th edition 2008) or following the rupture of a vulnerable atherosclerotic plaque which is an unstable collection of white blood cells and various lipids within the wall of the artery. If blood flow is not restored within 20–40 min, irreversible death of the heart muscle will begin to occur. Injury to the heart muscle causes chest pain and pressure. The dead heart muscle is replaced by scar tissue (Boersma et al. 2003), through the ages it has astonished all with its abrupt and disastrous outcome. It has always been a concern to every physician. Its presentation is often serious with increased morbidity and mortality irrespective of age. In many occasions, this disease is seen to have premature onset with serious complication. Although myocardial infarction is often depicted as a modern disease, it was clearly recognized before the modern era by the ancient Egyptians. The earliest record of sudden death, probably due to atherosclerotic occlusion of the coronary arteries, is portrayed in the tomb of Egyptians nobleman belonging to the 6th dynasty (Bruetsch 1959). AMI once supposed to be a medical curiosity has now become a challenge and an health problem deserving serious concern in spite of impressive advancement in diagnosis and management over the last three decades. The principal major risk factor for myocardial infarction, identified in several Indian population are smoking, hypertension, obesity, and diabetes (Gupta et al. 1995; Bhatia 1995).

4.4.1 Pathogenesis of Acute Myocardial Infarction (AMI)

Coronary atherosclerosis with superimposed coronary thrombosis is the cause of almost all myocardial infarction (Fallon 1996). The intraluminal thrombus demonstration consequent at the time of emergency coronary surgery and the display of recanalization with intracoronary thrombolytic therapy (Rentrop et al. 1998) paved the way for intracoronary thrombus in acute coronary occlusion. Subsequent clarification led to the perception that the initiating mechanism of coronary occlusions which are the result of coronary spasm, intraplaque hemorrhage, and luminal thrombosis may be due to fissuring or rupture of a vulnerable atherosclerotic plaque (Falk E. 1999) (Fig. 4.3, Box 4.1).

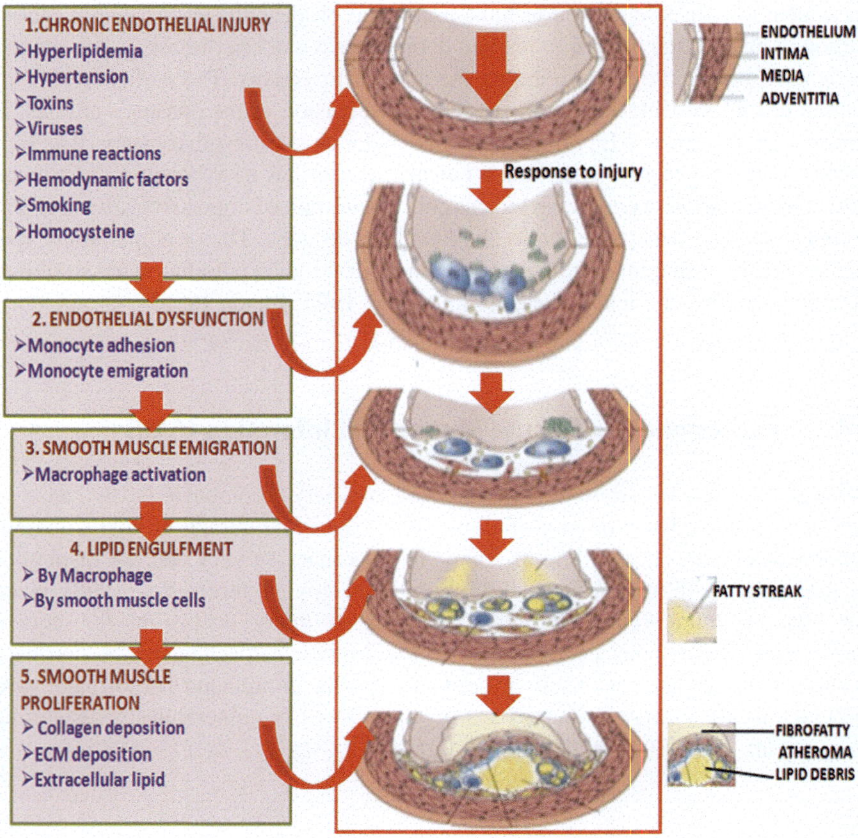

Fig. 4.3 Thrombus formation and steps of atherogenesis. The steps of atherogenesis includes chronic endothelial injury, endothelial dysfunction (like increase in leukocyte adhesion, vascular permeability, thrombosis), accretion of lipoprotein in the vessel wall (LDL and its oxidized form), monocyte adhesion to the endothelium (adhesion and then migration to intima, transformation of monocyte to macrophages and foam cells),platelet adhesion, release of factors from activated platelets (vascular wall cells and macrophages inducing smooth muscle cell recruitment) smooth muscle emigration(from media to intima), smooth muscle proliferation (ECM and collagen deposition, extracellular lipid), accumulation of lipid (extracellular and intracellular in macrophages and smooth muscle cells) leading to formation of well-developed plaque(Robbins Basic Pathology**,** 10th Edition**,** Editors: Vinay Kumar Abul Abbas Jon Aster**,** Published Date: 28th March 2017, eBook ISBN: 9780323394147, Imprint: Elsevier)

Box 4.1: The Pathogenesis of Myocardial Infarction (MI)

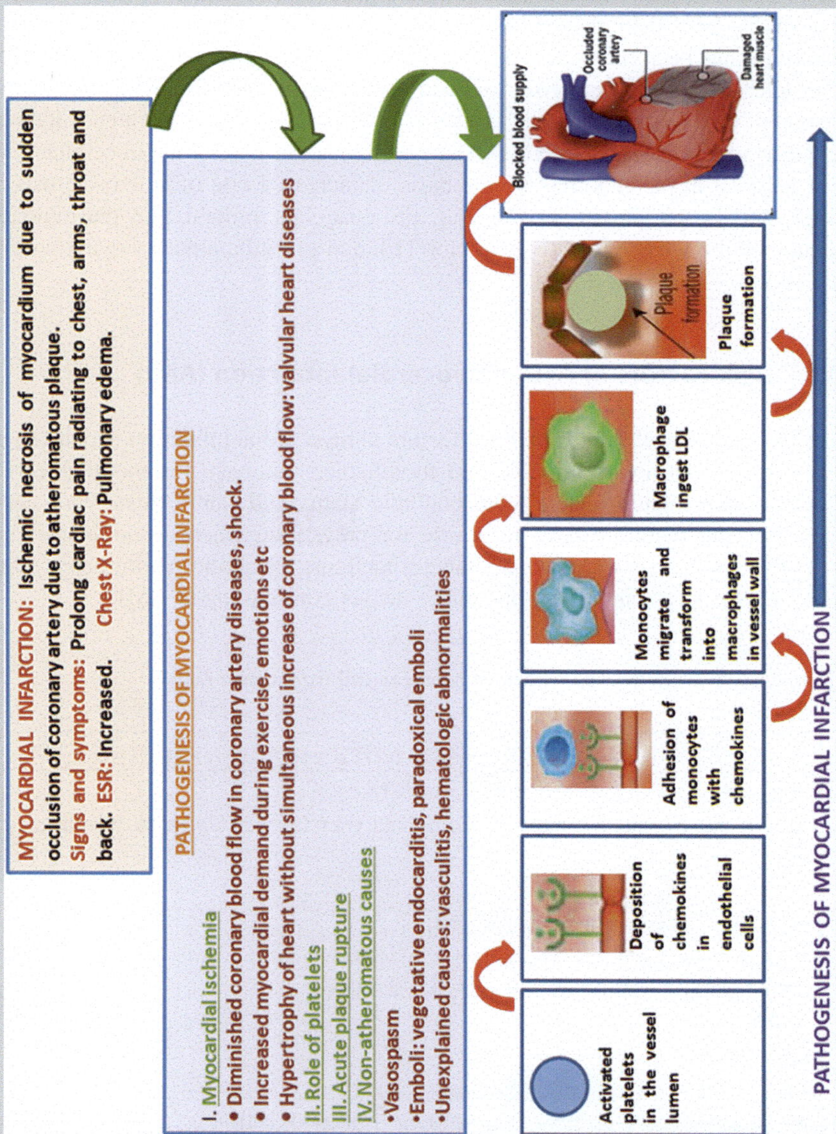

The symptoms, factors and steps of pathogenesis of myocardial infarction were noted.

4.4.2 Infective Theory of Acute Myocardial Infarction (AMI)

Chlamydia Pneumoniae is a common human pathogen causing a broad spectrum of infectious disease. It is also a widespread respiratory intracellular pathogen. This organism has been documented to cause endocarditis, myocarditis, and pericarditis. Recently, it has been implicated as a possible etiological agent in the development and progression of atherosclerosis and coronary artery disease (Gupta S. 1997). The transport of *C. pneumoniae* from the respiratory tract to the vascular wall are done by circulating monocytes, and it may persist within the monocytes without cellular lysis for at least 10 days. This may be the basis of increased rate of atherosclerosis in animal models, preferably by inducing pro-coagulant protein like plasminogen activator inhibitor-1(PAI-1), tissue factor(TF), and pro-inflammatory cytokine like interleukin-6 expression.

4.4.3 Risk Factors of Acute Myocardial Infarction (AMI)

There are several well-established risk factors of myocardial infarction which can be further divided as non-modifiable and modifiable. Amongst non-modifiable risk factors are age, male gender, socioeconomic status and family history of early onset AMI and therefore have little role for prevention whereas modifiable risk factors like smoking, hypertension, diabetes mellitus, metabolic syndrome, psychological factors, stress are among the major factors contributing to AMI.

Box 4.2: The Cardiac Markers of Myocardial Infarction (MI)

SERUM CARDIAC MARKERS OF MYOCARDIAL INFARCTION (MI)

➢**Markers:** should be specific and sensitive.

➢**Formed:** Intracellular macromolecules that leak out due to formation of infarction.

➢**Types:**

• Creatine Kinase (not very specific)

•CK- Different Isoenzymes

– **MM** in cardiacmuscles and skeletal muscles.

– **MB** exclusively in cardiac muscles (more specific for cardiac muscle than total CK).

– **BB** in brain, bowel and in bladder.

• Troponins (very sensitive and specific marker of MI).

• LDH (very nonspecific (in liver, lung, kidney, sk muscle, RBCs); High LDH1 isoenzyme more specific; Ratio of LDH1:LDH2 above 1 is helpful in making diagnosis of MI).

• Myoglobin (not widely used, but may be useful for early detection of MI).

The serum cardiac markers were mentioned.

4.4.4 Diagnostic Test

Myocardial infarction is diagnosed by the combination of the history of the presenting illness like chest pain or discomfort and physical examination with electrocardiogram findings and cardiac markers (Box 4.2). Commonly used cardiac markers include troponin tests, CK (Creatinine Kinase), CK-MB (Creatinine Kinase-MB fraction), SGOT (Serum glutamic oxaloacetic transaminase), and serum myoglobin tests.

Apart from several markers and tests used in the diagnosis and treatment of myocardial infarction, measurements of other parameters like CRP and markers of oxidative stress like serum malondialdehyde (MDA) and uric acid has a definite role in myocardial infarction.

4.4.5 Role of C-Reactive Protein (CRP) in Acute Myocardial Infarction

CRP, an inflammatory response indicator, is a budding risk marker for assessment of patients at primary cardiovascular risk and can also be used with limited scope, stable patients at secondary risk. Strong predictive value of CRP in angina pectoris (stable and unstable) independent of troponin can be demonstrated from several different clinical trials (Sabatine et al. 2002) and from the load of atherosclerosis (Zebrack et al. 2002a, b). The predictive nature of CRP is less well defined in the setting of AMI, in part due to its related acute phase reaction. After myocardial infarction, the level of CRP gradually rises within a few hours after infarction (Kushner et al. 1998; Baruah et al. 2012). Thus, if CRP levels are continuously monitored at random times after AMI, values are elevated above baseline at varying levels, and the strong predictive value for long-term prognosis is mislaid (Celik et al. 2001). If levels are measured in the early hours of AMI (Torzewski et al. 2000), at peak levels (Anzai T et al. 1997; Wang et al. 2004; Verma et al. 2003, 2004) or after several weeks (Bulkley 2002), CRP poses to provide a predictive value of weeks to several months (short-term prognosis) (Suleiman et al. 2003). The research by Suleiman et al. in their regard supports this association. In the setting of AMI, total circulating CRP levels likely represent an indicator of the composite of levels representing inflammatory response to acute myocardial injury as well as chronic vascular inflammation, which is dependent on immune response of host and size of infarct. In the process of myocardial infarction, complement and CRP are found to restrict only in areas of myocardial necrosis, escalating in concentration after 12 h (Lunec 1983). The production of nitric oxide which in turn inhibits angiogenesis is also reduced by CRP (Verma et al. 2002; McCord 1985). The level of CRP can also predict left ventricular thrombus after the event of myocardial infarction (Suwimiol Jetawattana 2005). Thus, there is increasing support that CRP is a risk marker for events in the early period after AMI and also might act as perpetrator toward it (Fig. 4.4).

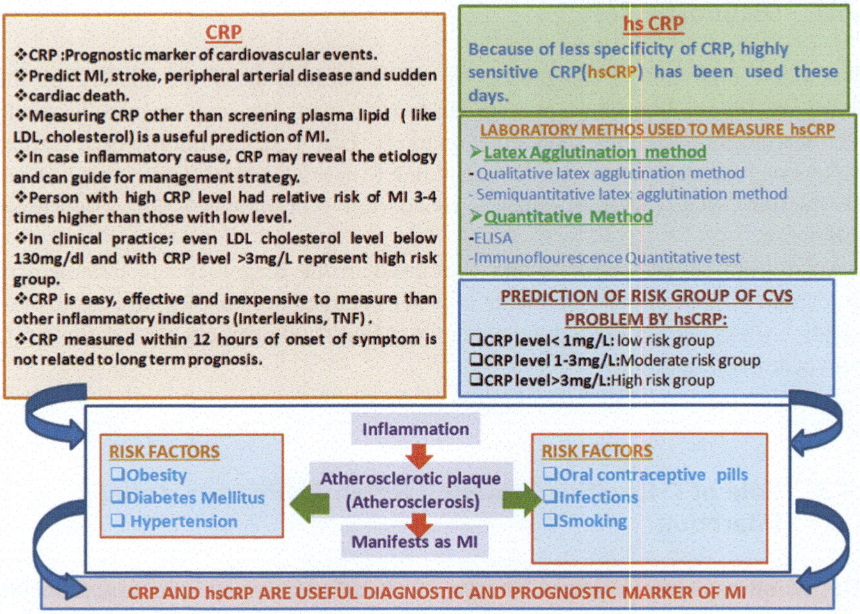

Fig. 4.4 The relation between CRP and myocardial infarction (MI). CRP and high sensitivity CRP (hsCRP) acts as diagnostic and prognostic marker in myocardial infarction. The methods used to measure hsCRP and the risk factors of MI were also noted. Low, moderate and high risk groups were ascertained based on the level of hsCRP

Although initially CRP was monitored exclusively a marker of the inflammatory response, recent clarification from several studies suggest that CRP poses a direct effect which support atherosclerotic processes such as vascular smooth muscle proliferation and endothelial inflammation (Johnson et al. 2003; Niskanen et al. 2004). CRP profoundly reduces nitric oxide (NO) synthesis while increasing the expression of vasoactive intermediaries with upregulation of chemokines and adhesion molecules, at levels that are known to predict diverse vascular insults (Niskanen et al. 2004). In addition, CRP facilitates endothelial cell apoptosis with activation of factor of transcription like NF-κB (nuclear factor kappa-light-chain-enhancer of activated B cells) (Fang and Alderman 2000) and attenuates endothelial progenitor cell survival (Bengtsson et al. 1988). Hence, CRP apart from being a biomarker of atherosclerotic events also seems to partake actively in plaque formation and cardiovascular morbidity.

4.4.6 Oxidative Stress in Acute Myocardial Infarction

Oxidative stress (OS) illustrates the steady-state level of oxidative damage in a cell, tissue, or organ, caused by the unpaired electron, i.e., free radicals mostly reactive oxygen species (ROS). In the last few decades, free radicals highly reactive and thereby destructive molecules have come to be accepted increasingly for their importance to disease and human health (Kroll et al. 1992; Nielsen et al. 1997). Many critical human diseases and emergency conditions, including cancer, atherosclerosis, ageing, and various injuries, have free radical reaction as a fundamental mechanism of injury. Over this period of time, our perception and understanding of such oxygen reactive species with living organisms has undergone a noteworthy evolution of how human knowledge of a scientific concept blooms and matures. Our thoughtful perception has gradually evolved to provide unprecedented opportunities for improving the quality and even time course of human life.

Free radicals are molecular species or uncharged atoms which do not have paired electrons in its outermost orbital but has independent existence. These unpaired electrons due to their highly reactive nature likely to take part in different chemical reactions. These unstable molecules (free radicals) in the body attack and cause damage and destruction to neighboring molecules, elaborate a chain reaction of free radical formation, and damage molecular structures. This chain reaction causes widespread damage of the cell including the DNA, cell membranes, and tissue proteins. Any free radical connecting with oxygen can be termed as ROS and these oxygen-centered free radicals contain two unpaired electrons in their outermost orbit. The inner mitochondrial membrane, the center of electron transport chain (ETC) in a cell, utilizes oxygen to liberate energy in the form of adenosine triphosphate (ATP). Here (in ETC), oxygen functions as a terminal electron acceptor and on reviewing various literature it suggests that of the total oxygen intake both during exercise and rest about 2–5% oxygen have the capability to form the highly damaging superoxide radical by escape of electrons from ETC, usually the escape is at the level of ubiquinone-cytochrome C.

Again various thrombolytic therapy targeted in myocardial infarction and the sequences of events that occur on reperfusion of ischemic myocardial tissue leading to further tissue damage. Oxygen-derived free radical generation is the basis of reperfusion injury (Kilgore KS 1993). These reactive oxygen species created within the initial moments of reperfusion are recognized to be cytotoxic to neighboring cells. Moreover, there is inflammatory system participation in mediating tissue damage upon reperfusion (Castelli et al. 1995).

During ischemia, depressed mitochondrial activity is accompanied by augmented dephosphorylation of ATP and AMP concentrations are greatly increased within cell. This AMP is in turn metabolized to adenosine and hypoxanthine. Cytosolic xanthine oxidase which exists as a dehydrogenase in healthy tissue. This enzymatic form is converted by ischemia into the oxidase form through calcium-dependent proteolytic enzyme which can act on the catabolic product of hypoxanthine, ATP, to liberate the $O_2^{\bullet -}$ anion. This takes place only once the tissue gets perfused since this enzyme system utilizes oxygen which converts hypoxanthine to uric acid. Iron decompartmentalization from iron stores may then give the required catalytic power to create the OH• radical (Kogure et al. 1999).

Other alternative mechanisms have also been suggested for the production of reperfusion injury. Reduction of oxygen takes place as a normal by-product during mitochondrial respiration. Under circumstances of failure or due to uncoupling of the mitochondrial electron transport chain during ischemia, this effect may be augmented. Various products derived from prostaglandins are generated during ischemia and also can generate free radical formation (Anker et al. 2003).

4.4.7 Malondialdehyde (MDA)

Malondialdehyde (MDA), a marker of oxidative stress is a three carbon dialdehyde, is a lipid peroxidation product is measured as an index of free radical mediated lipid peroxidation damage. MDA, which is a metabolically stable product of lipid peroxidation reactions in vivo, is now generally accepted as a biochemical indicator of lipid peroxidation injury through free radical. MDA is formed as a by-product of polyunsaturated fatty acid peroxidation and also produced in metabolism of arachidonic acid during the synthesis of prostaglandins.

MDA has the ability to bind to various functional groups of molecules including DNA, RNA, proteins, and lipoproteins. Thus, MDA monitoring in biological materials can be used as an important marker of in vitro and in vivo lipid peroxidation for different diseases (Suwimiol Jetawattana 2005).

4.4.8 Uric Acid

Various controversies loom regarding the role of uric acid as a risk factor for myocardial infarction. Recent evidence on uric acid suggests that they may be an important causal agent in cardiovascular disease by inducing renal disease thereby causing hypertension. With the dawn of ninteenth century, it was known that high uric acid levels are linked with hypertension. In spite of the dearth of experimental studies, raised uric acid levels were commonly considered a consequence rather than a cause of cardiovascular disease. However, various animal and human studies have recently recommended that high uric acid levels may have an impact on renal system via glomerular damage and pre-glomerular arteriolosclerosis, thus causing deranged kidney function and thus causing effects that may result in arterial hypertension (Nadkar and Jain 2008). Various cohort studies have shown uric acid to be an important independent risk factor for cardiovascular mortality (Berton et al. 2003a, b; Dimitrijevic and Blagica 2006). In fact, the position of uric acid in coronary heart disease is yet not well defined. Few studies have reported an independent association between coronary heart disease and uric acid (Castelli et al. 1995).

In cardiac tissue, adenosine synthesized locally by vascular smooth muscle gets rapidly degraded by the endothelium to uric acid, which undergo rapid efflux to the vascular lumen due to negative membrane potential and low intracellular pH (Kogure et al. 1999). Uric acid synthesis (Kogure et al. 1999) and xanthine oxidase activity (Castelli et al. 1995) are elevated in vivo under ischemic conditions, and therefore raised serum uric acid may act as an indicator of underlying tissue ischemia. Although the mechanisms of uric acid playing a pathogenetic role in cardiovascular disease is yet not clear, hyperuricemia is often linked with deleterious effects on endothelial dysfunction, platelet adhesiveness, oxidative metabolism, hemorheology, and aggregation. There is increasing evidence that elevated uric acid also act as a negative prognostic factor in mild to severe heart failure (Anker et al. 2003) although the development of hyperuricemia is mostly always correlated with worsening of renal failure in these patients (Ochiai et al. 2005). Therefore, while assessing prognosis of these patients it is not always easy to analyze the roles played by reduced renal function and high uric acid level. Few evidences propose that uric acid by stimulating inflammation may exert a negative effect on cardiovascular disease, and play an important role in the pathogenesis of cardiovascular disease (Ochiai et al. 2005).

4.5 Discussion

Dr. Baruah et al. (2012) studied a total of 72 cases admitted to Guwahati Medical College and Hospital, Guwahati, of which 42 patients were cases of AMI between 35 years and 81 years with a mean of 56.5 years and 30 were age and sex-matched healthy control between age 38 years and 77 years (25 males and 5 females) with a mean of 53.4 years. 6 AMI patients expired during the study (4 Males, 2 Females). Serum CRP, MDA, and uric acid were subjected to all participants of the study and their readings were taken on Day 1, 3, and 7 from the onset of symptoms. There were statistically significant elevation of serum CRP, MDA, and uric acid on all days (i.e., on days 1, 3, and 7) in study group (AMI and death subjects) compared to controls group as evident from Figs. 4.5, 4.6, and 4.7, respectively. Moreover, CRP, MDA,

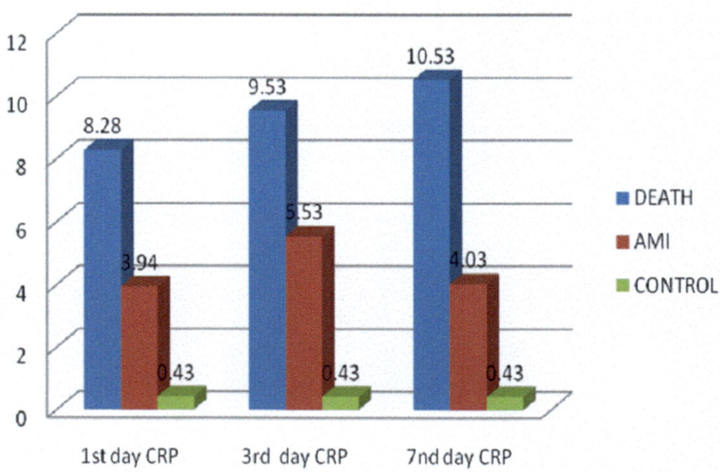

Fig. 4.5 Compare mean CRP in Day 1, 3 & 7 in Death, AMI and Control subjects. CRP is significantly elevated in Death patients

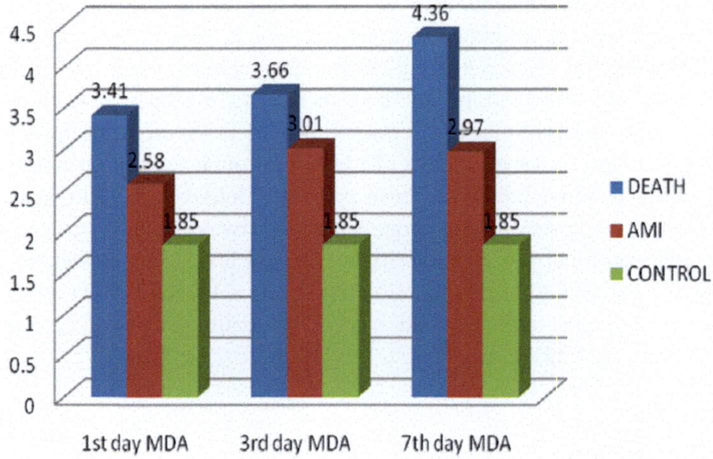

Fig. 4.6 Compare of Mean MDA in Day 1, 3 and 7 in Death, AMI and Control subjects. MDA is significantly elevated in Death patients

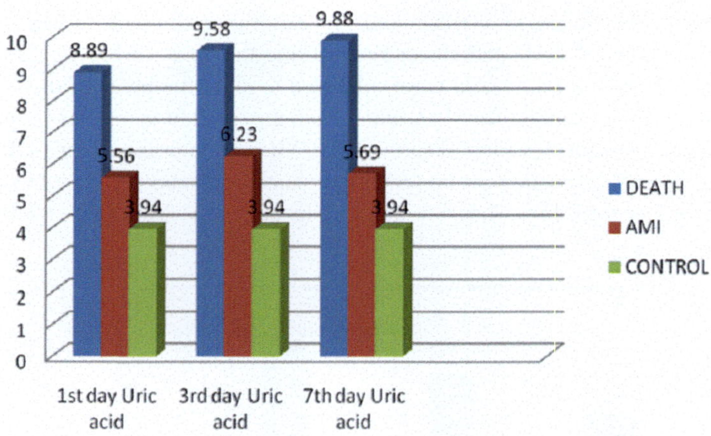

Fig. 4.7 Compare of mean uric in Day 1, 3 and 7 acid in death, AMI and control subjects. Uric acid is significantly elevated in death patients

Table 4.1 Distribution of sex in control and test group[a]

	Control		Myocardial infarction	
Sex	No.	Percentage	No.	Percentage
Female	5	16.60%	2	5.50%
Male	25	83.30%	34	94.40%
Total	30	100%	36	100%

[a]From the above table it can be seen that AMI is higher in males compared to females

and uric acid peaks around third day in patients of AMI with decline thereafter but Group I (death patients) does not show any decline but all the parameters show an elevated trend. The study also showed that the occurrence of myocardial infarction is more in males (94.4%) than in females (5.5%) among the admitted patients (Table 4.1) which corroborates with Nadkar et al. (2008), Giuseppe Berton et al. (2003a, b). This may be probably due to increased coronary heart disease risk factors, more associated with males and also due to the fact that the male hormone androgen decreases adiponectin levels which is anti-atherogenic and anti-inflammatory, thus the protective effect of adiponectin may be lost. Hence, Dr. Baruah et al. in their studied interpretated that an elevated level of CRP, uric acid, and MDA is statistically significant in AMI compared to control. The results showed increased levels of CRP, MDA, and uric acid in AMI on day 1 as compared to the control group. But the levels remain significantly elevated in death patient compared to the AMI and control subjects without subsequent decline as compared to other AMI subjects which match with the result obtained by Kushner et al. 1998. Hence, patients admitted with AMI should be subjected to serial measurement of CRP, uric acid, and MDA and their increasing rise should be taken cautiously in the management of AMI.

Nadkar et al. (2008) studied serum uric acid in AMI and a close association were observed between serum uric acid concentration and in AMI patients. 100 patients with AMI were compared with 50 controls subjects in their study. Serial measurements of serum uric acid on day 0, 3, and 7 of AMI were measured. The uric acid levels were comparable in AMI group on Day 0, 3, and 7; 5.23 ± 1.95, 5.20 ± 2.15, and 5.28 ± 2.52, respectively. Hence, the study observed statistically significant higher serum level of uric acid concentration in AMI as compared to controls.

Beer et al. in 1982 studied about “Measurement of CRP in Myocardial ischemia and Infarction” in 33 myocardial infarction patients. They measured CRP and CK-MB prospectively in patients with AMI. There were raised CRP levels in all individuals with infarction, and there were correlation between peak CRP and CK-MB values ($p < 0.001$), which is statistical significant. However, it was observed that around 50 h after onset of pain, CRP has a peak. Moreover, peak of CRP takes place at the time when CK-MB peak around 15 h. There were around 20 patients who recovered uneventfully and their CRP level declined, returning to baseline about 7d after infarction. But CRP continues to remain high in the rest 8 complicated cases including 4 who died within first 10 days.

Berton et al. in 2003a, b studied CRP in AMI association with heart failure in 269 patients admitted to the hospital with suspicion of AMI. Among these 220 patients were confirmed with diagnosis of AMI. Rest 49 subjects were then selected as controls. CRP was serially measured on the first, third, and seventh day after admission, and they found that CRP on admission to hospital can be taken for predicting the outcome heart failure (HF) in patients with AMI. The patients of AMI had significantly higher CRP than in the control individuals ($p = 0.001$) and it peaked on the third day. Moreover, within patients with AMI, CRP was significantly elevated in patients with HF than in patients without HF (adjusted $P = 0.008$,

$P = 0.02$ and $P = 0.03$ on first, third, and seventh day, respectively). Hence, peak CRP value is taken to be a strong independent predictor of global and mortality due to heart failure during the following year.

Dimitrijevic and Blagica in 2006 studied about—serial measurements of CRP after AMI in predicting 1-year outcome. In 31 patients with STEMI (ST-segment elevation myocardial infarction), CRP were serial measured, and the following values were determined from each patients: (i) values up to 12 h after onset of symptoms (at admission) (ii) values obtained 24–72 h after symptom onset (early acute values), and (iii) 96–120 h after symptom onset (late acute values). But only early and late acute CRP levels were found to be significantly elevated in patients with an adverse outcome compared to patients with a good outcome. Moreover, significantly increased rate of endpoint events was found in patients with elevated early CRP (Hazard ratio (HR) 5.54, 95%CI 2.05–25.40; $p = 0.007$) and late acute CRP (HR 9.01, 95% CI 1.66–19.56; $p = 0.005$). Moreover, CRP was found to be the independent predictor of an unfavorable outcome (Odds ratio, OR 8.00, 95%CI 1.15–55.60; $p = 0.04$), after adjustment for established risk factors. But CRP level measured after 24–72 h of symptom onset is an independent predictor of 1-year outcome.

Aznar et al. (1983) studied serum malondialdehyde-like material (MDA-LM), and serum cardiac enzymes (CK-MB, CK, LDH) in a group of 26 AMI patients, 7 patients with angina pectoris (AP), compared with to 94 normal control group. MDA-LM values showed a significant increase in AMI patients in the days following the acute event reaching a maximum within 6–8 d, when 90% of the patients had increased values than the upper normal limit (mean $\pm$ 2SD) of the control group. There were also significant correlations between them.

4.6 Conclusion

Inflammatory responses triggered by AMI play an important role in myocardial injury. Marker of inflammatory such as CRP and oxidative stress marker like MDA and uric acid reflects the degree of myocardial necrosis. It also correlates with cardiac outcome following AMI. This myocardial necrosis following AMI stimulates the generation of free radical and is responsible for the inflammatory cascade. Further escalation of the inflammatory reaction is done by reperfusion therapy with the recruitment of neutrophils into the reperfused myocardium (Frangogiannis et al. 2002).

Their role in various disease conditions like diabetes mellitus, stroke and vascular disease, ischemia, and reperfusion injury in human myocardium have been well established. Strong evidences exist implicating the involvement of CRP, MDA, and uric acid in the pathophysiology of AMI.

CRP, MDA, and uric acid have been demonstrated to have a significant role in AMI. Measurement of CRP, uric acid, and MDA at presentation may be considered alarming and should be used as an important marker in predicting outcome including death of the patients. Laboratory diagnosis and treatment should go on together to increase the life expectancy of the individuals. Moreover, regular monitoring of these parameters in AMI may improve patient outcome and is highly recommended.

References

Aguiar FJ, Ferreira-Júnior M, Sales MM, Cruz-Neto LM, Fonseca LA, Sumita NM, Duarte NJ, Lichtenstein A, Duarte AJ (2013) C-reactive protein: clinical applications and proposals for a rational use. Rev Assoc Med Bras (1992) 59(1):85–92

Anker SD, Doehner W, Rauchhaus M, Sharma R, Francis D, Knosalla C, Davos CH, Cicoira M, Osterziel KJ, Leyva F, Hetzer R, Ponikowski P, Coats AJ (2003) Uric acid and survival in chronic heart failure: validation and application in metabolic, functional, and hemodynamic staging. Circulation 107:1991–1997

Anzai T, Yoshikawa T, Shiraki H et al (1997) C-reactive protein as a predictor of infarct expansion and cardiac rupture after a first Q-wave acute myocardial infarction. Circulation 96:778–784

Aznar J, Santos MT, Valles J, Sala J (1983) Serum malondialdehyde-like material (mda-lm) in acute myocardial infarction. J Clin Pathol 36:712–715

Baruah M, Nath C, Goyal S et al (2012) Study of c-reactive protein, malondialdehyde and uric acid levels in predicting outcome in acute myocardial infarction. Int J Cur Res Rev 04(19):128–132

Bengtsson C, Lapidus L, Stendahl C, Waldenstrom J (1988) Hyperuricaemia and risk of cardiovascular disease and overall death. A 12-year follow-up of participants in The Population Study of Women in Gothenburg, Sweden. Acta Med Scand 224:549–555

Berton G, Cordiano R, Palmieri R et al (2003a) C-reactive protein in acute myocardial infarction: association with heart failure. Am Heart J 145:1094–1101

Berton G, Cordiano R, Palmieri R, Pianca S, Pagliara V, Palatini P (2003b) C-reactive protein in acute myocardial infarction: association with heart failure. Am Heart J 145(6):1094–1101

Bhatia ML (1995) Prevalence of CHD in India: a contemporary view. Indian Heart J 47:339–342

Boersma E, Mercado N, Poldermans D (2003) Acute myocardial infarction. Lancet 361 (9360):847–858

Bruetsch WL (1959) The earliest record of sudden death possibly due to atherosclerotic coronary occlusion. Circulation 20(3):438–441.

Bulkley GB (2002) Free radicals and reactive oxygen species. COSMOS J

Castelli P, Condemi AM, Brambillasca C et al (1995) Improvement of cardiac function by allopurinol in patients undergoing cardiac surgery. J Cardiovasc Pharmacol 25:119–125

Celik S, Baykan M, Erdol C et al (2001) C-reactive protein as a risk factor for left ventricular thrombus in patients with acute myocardial infarction. Clin Cardiol 24:615–619

CRP C-Reactive Protein (Homo sapiens) (2020) Entrez Gene. National Center for Biotechnology Information (NCBI), U.S. National Library of Medicine

de Beer FC, Hind CR, Fox KM et al (1982) Measurement of serum C-reactive protein concentration in myocardial ischaemia and infarction. Br Heart J 47:239–243

Dimitrijevic O, Blagica BS (2006) Serial measurements of C-reactive protein after acute myocardial infarction in predicting one-year outcome. Int Heart J 47(6):833–840

Du Clos TW (2000) Function of C-reactive protein. Ann Med 32(4):274–278

Du Clos TW, Mold C (2004) C-reactive protein: an activator of innate immunity and a modulator of adaptive immunity. Immunol Res 30(3):261–277

Fallon JT (1996) Pathology of myocardial infarction and reperfusion. In: Fuster V, Ross R, Topol EJ (eds) Atherosclerosis and coronary artery disease. Lippincott-Raven, Philadelphia, PA, pp 791–796

Fang J, Alderman MH (2000) Serum uric acid and cardiovascular mortality the NHANES I epidemiologic follow-up study, 1971–1992. National Health and Nutrition Examination Survey. JAMA 283:2404–2410

Falk E (1999) Plaque rupture with severe preexisting stenosis precipitating coronary thombosis: characteristics of coronary atherosclerotic plaques underlying fatal occlusive thrombi. Br Heart J 50:127–134
Frangogiannis NG, Smith CW, Entman ML (2002) The inflammatory response in myocardial infarction. Cardiovasc Res 53:31–47
Fusman G, Mardi T, Justo D et al (2002) Red blood cell adhesiveness/aggregation, C-reactive protein, fibrinogen, and erythrocyte sedimentation rate in healthy adults and in those with atherosclerotic risk factors. Am J Cardiol 90:561–563
Gupta R, Prakash H, Majumdar S et al (1995) Prevalence of CHD & coronary risk factors in urban population of Rajasthan. India Indian Heart J 47:331–338
Gupta S (1997) India. Indian Heart J Circ 96:404–407
Harrison's 17th edition (2008) Principal Internal Medicine (vol 1) 239:1486–9
Johnson RJ, Kang DH, Feig D, Kivlighn S, Kanellis J, Watanabe S, Tuttle KR, Rodriguez-Iturbe B, Herrera-Acosta J, Mazzali M (2003) Is there a pathogenetic role for uric acid in hypertension and cardiovascular and renal disease? Hypertension 41:1183–1190
Kilgore KS, Lucchesi BR (1993) Reperfusion injury after myocardial infarction: the role of free radicals and the inflammatory response. Clin Biochem 26(5):359–370
Kogure K, Ishizaki M, Nemoto M et al (1999) Evaluation of serum uric acid changes in different forms of hepatic vascular inflow occlusion in human liver surgeries. Life Sci 64:305–313
Kroll K, Bukowski TR, Schwartz LM, Knoepfler D, Bassingthwaighte JB (1992) Capillary endothelial transport of uric acid in Guinea pig heart. Am J Physiol 262:H420–H431
Kushner I, Kaplan MH (1961) Studies of acute phase pro-tein. I. An imrnunohistochemical method for the localization of Cx-reactive protein in rabbits. Association with necrosis in local inflammatory lesions. J Exp Med 114:961
Kushner I, Broder ML, Karp D (1998) Control of the acute phase response. Serum C-reactive protein kinetics after acute myocardial infarction. J Clin Invest 61:235–242
Lau DC, Dhillon B, Yan H, Szmitko PE, Verma S (2005) Adipokines: molecular links between obesity and atheroslcerosis. Am J Physiol Heart Circ Physiol 288(5):H2031–H2041. https://doi.org/10.1152/ajpheart.01058.2004
Li YP, Mold C, Du Clos TW (1994) Sublytic complement attack exposes CRP binding sites on cell membranes. J Immunol 152:2995–3005
Lloyd-Jones DM, Liu K, Tian L, Greenland P (2003) Narrative review: assessment of C-reactive protein in risk prediction for cardiovascular disease. Ann Intern Med 145 (1):35–42. PMID 16818927
Lunec J (1983) Free radical mediated aggregation of IgG. Agents Actions 15:37–38
Marnell L, Mold C, Du Clos TW (2005) C-reactive protein: ligands, receptors and role in inflammation. Clin Immunol 117(2):104–111
McCord JM (1985) Oxygen derived free radicals in post ischemic tissue injury. N Engl J Med 312:159–163
Mortensen RF, Osmand AP, Gewurz H (1975) Effects of CRP on the lymphoid system. I. Binding to thymus-dependent lymphocytes and alteration of their functions. J Exp Med 141:286
Nadkar MY, Jain VI (2008) Serum uric acid in acute myocardial Infarction exist. J Assoc Physicians India 56:759–761
Nielsen F, Mikkelsen BB, Nielsen JB, Andersen HR, Grandjean P (1997) Plasma malondialdehyde as biomarker for oxidative stress: reference interval and effects of life-style factors. Clin Chem 43(7):1209–1214
Nijmeijer R, Lagrand WK, Lubbers YT et al (2003) C-reactive protein activates complement in infarcted human myocardium. Am J Pathol 163:269–275
Niskanen LK, Laaksonen DE, Nyyssonen K, Alfthan G, Lakka HM, Lakka TA, Salonen JT (2004) Uric acid level as a risk factor for cardiovascular and all-cause mortality in middle-aged men: a prospective cohort study. Arch Intern Med 164:1546–1551
Ochiai ME, Barretto AC, Oliveira MT et al (2005) Uric acid renal excretion and renal in sufficiency in decompensated severe heart failure. Eur J Heart Fail 7:468–474

Pepys MB, Hirschfield GM (2003) C-reactive protein: a critical update. J Clin Invest 111 (12):1805–1812. https://doi.org/10.1172/JCI18921

Pepys MB (1981) C-reactive protein fifty years on. Lancet 21:653

Rentrop P, Blanke H, Kansch KR, Kaiser H, Kostering H, Leitz K (1998) Selective intracoronary thrombolysis in acute myocardial infarction and unstable angina pectoris. Circulation 63:307–317

Sabatine MS, Morrow DA, de Lemos JA et al (2002) Multimarker approach to risk stratification in non-ST elevation acute coronary syndromes: simultaneous assessment of troponin I, C-reactive protein, and B-type natriuretic peptide. Circulation 105:1760–1763

Suleiman M, Aronson D, Reisner SA et al (2003) Admission C-reactive protein levels and 30-day mortality in patients with acute myocardial infarction. Am J Med 115:695–701

Suwimiol Jetawattana (2005) Free radical & radiation biology; Malondialdehyde (MDA) a lipid oxidation product. For 77:222

Tillett WS, Francis T Jr (1930) Serological reactions in pneumonia with a nonprotein somatic fraction of pneumococcus. J Exp Med 52:561–585. https://doi.org/10.1084/jem.52.4.561

Torzewski M, Rist C, Mortensen RF, Zwaka TP, Bienek M, Waltenberger J, Koenig W, Schmitz G, Hombach V, Torzewski J (2000) C-reactive protein in the arterial intima: role of C-reactive protein receptor-dependent monocyte recruitment in atherogenesis. Arterioscler Thromb Vasc Biol 20:2094–2099

Tsujimoto M, Inoue K, Nojima S (1980) CRP induced agglutination of lipid suspensions prepared in the presence and absence of phosphatidylcholine. J Biochem 87:1531

Verma S, Wang CH, Li SH, Dumont AS, Fedak PW, Badiwala MV, Dhillon B, Weisel RD, Li RK, Mickle DA, Stewart DJ (2002) A self-fulfilling prophecy: C-reactive protein attenuates nitric oxide production and inhibits angiogenesis. Circulation 106:913–919

Verma S, Badiwala MV, Weisel RD, Li SH, Wang CH, Fedak PW, Li RK, Mickle DA (2003) C-reactive protein activates the nuclear factor-kappaB signal transduction pathway in saphenous vein endothelial cells: implications for atherosclerosis and restenosis. J Thorac Cardiovasc Surg 126:1886–1891

Verma S, Kuliszewski MA, Li SH, Szmitko PE, Zucco L, Wang CH, Badiwala MV, Mickle DA, Weisel RD, Fedak PW, Stewart DJ, Kutryk MJ (2004) C-reactive protein attenuates endothelial progenitor cell survival, differentiation, and function: further evidence of a mechanistic link between C-reactive protein and cardiovascular disease. Circulation 109:2058–2067

Volanakis JE, Kaplan MH (1971) Specificity of CRP for choline phosphatase residues of pneumococcal C-polysaccharide. Proc Soc Exp Biol Med 136:612

Volanakis JE, Narkates AJ (1981) Interaction of CRP with artificial phosphatidylcholine bilayers and complement. J Immunol 126:1820–1825

Volanakis JE, Wirtz KWA (1979) Interaction of CRP with artificial phosphatidylcholine mayers. Nature (London) 281:155

Wang CH, Li SH, Weisel RD, Fedak PW, Dumont AS, Szmitko P, Li RK, Mickle DA, Verma S (2004) C-reactive protein upregulates angiotensin type 1 receptors in vascular smooth muscle. Circulation 107:1783–1790

Wu Y, Potempa LA, El Kebir D, Filep JG (2015) C-reactive protein and inflammation: conformational changes affect function. Biol Chem 396(11):1181–1197. https://doi.org/10.1515/hsz-2015-0149

Wood HF (1951) Effect of CRP on normal human leukocytes. Proc Soc Exp Biol Med 76:843

Zebrack JS, Muhlestein JB, Horne BD et al (2002a) C-reactive protein and angiographic coronary artery disease: independent and additive predictors of risk in subjects with angina. J Am Coll Cardiol 39:632–637

Zebrack JS, Anderson JL, Maycock CA et al (2002b) Usefulness of high-sensitivity C-reactive protein in predicting long-term risk of death or acute myocardial infarction in patients with unstable or stable angina pectoris or acute myocardial infarction. Am J Cardiol 89:145–149

Zhang YX, Cliff WJ, Schoefl G, Higgins G (1999) Coronary CRP distribution: its relations to development of atherosclerosis. Atherosclerosis 145:375–379

Role of CRP in Monitoring of Acute Pancreatitis

5

Jawaid Ahmed Khan

Contents

5.1 Introduction ... 122
5.2 Disease Burden ... 122
5.3 Etiology ... 123
5.3.1 Gallstones ... 124
5.3.2 Alcoholism ... 125
5.3.3 Hypertriglyceridemia ... 126
5.3.4 Hypercalcemia ... 127
5.3.5 Post-Endoscopic Retrograde Cholangiopancreatography (ERCP) ... 127
5.3.6 Infections and infestations ... 128
5.3.7 Drug-Induced Pancreatitis ... 128
5.3.8 Genetic Causes of Pancreatitis ... 130
5.3.9 Autoimmune Pancreatitis ... 131
5.3.10 Traumatic Pancreatitis ... 132
5.3.11 Anatomic or Congenital Causes ... 132
5.4 Pathogenesis and Pathophysiology ... 133
5.5 Pathogenesis of Local Complications ... 136
5.6 Role of Gastrointestinal Tract (GIT) in AP ... 137
5.7 Obesity and Visceral Adiposity ... 138
5.8 Clinical Features of AP ... 139
5.9 Laboratory Diagnosis ... 140
5.9.1 Serum Pancreatic Enzyme Estimation ... 140
5.9.2 Serum Amylase ... 140
5.9.3 Serum Lipase ... 141
5.9.4 Other Hematological Tests ... 141
5.9.5 Imaging Studies ... 142
5.9.6 Plain X-Rays ... 142
5.9.7 Ultrasonography (US) ... 142
5.9.8 CT Scan ... 143
5.9.9 MRI Scan ... 143

J. A. Khan (✉)
G.D. Hospital and Diabetes Institute, Kolkata, West Bengal, India

W. Ansar, S. Ghosh (eds.), *Clinical Significance of C-reactive Protein*,
https://doi.org/10.1007/978-981-15-6787-2_5

5.10 Classification of AP 144
5.10.1 Definition of Types of AP 144
5.10.2 Definitions of the Severity of AP 145
5.10.3 Definitions of Local Complications 146
5.11 Local Determinants 149
5.12 Systemic Determinants 149
5.13 Disease Severity Prediction 150
5.13.1 Ranson Score 151
5.13.2 APACHE II 152
5.13.3 BISAP Score 152
5.14 Laboratory Markers 154
5.14.1 C-Reactive Protein 155
5.14.2 Blood Urea Nitrogen (BUN) 159
5.14.3 Hematocrit 160
5.14.4 Procalcitonin 161
5.14.5 Other Biomarkers 161
5.15 Radiological Scoring System 161
5.16 Management of AP 163
5.16.1 Fluid Resuscitation 164
5.16.2 Pain Control 165
5.16.3 Antibiotics 165
5.16.4 Nutrition 166
5.17 Conclusion 167
References 167

Abstract

Acute pancreatitis (AP) is an inflammatory disorder of the pancreas secondary to activation of its own enzymes, which can lead to local and systemic complications. The exact pathology of initiation of AP is still obscure but the inflammation can be severe enough to cause multiple organ failure. Gallstone diseases and alcohol abuse remain the two common cause of AP globally. The incidence of AP is increasing in last few decades without any change in its mortality rate. The course of disease is variable and is ranging from mild form, which generally needs only supportive care, to a very severe form, which progresses frequently into multiple organ failure and results in a very high mortality. It is of paramount importance to identify those patients who are likely to develop the severe form early in the course of the disease. There are many scoring systems and biomarkers which help in identifying high risk patients but most of them do not perform adequately enough very early in the course of the disease. C-reactive protein (CRP) estimation is a cheap and widely available method to identify those patients who are likely to develop severe AP (SAP). Further refinement in the timing of CRP estimation early in the course of AP has the potential to become the simplest and the least cumbersome method of identifying high risk patients in the first 24 h of the onset of the disease.

Keywords

Acute pancreatitis · C-reactive protein · Gallstone · Alcohol abuse · Sphincter of Oddi dysfunction · Multiple organ dysfunction · SIRS · Multifactorial scoring system

Abbreviations

5-ASA	5-aminosalicylic acid
ACE	Angiotensin-converting enzyme
ACG	American College of Gastroenterology
AIDS	Acquired immunodeficiency syndrome
ALT	Alanine transaminase
ANC	Acute necrotic collection
AP	Acute pancreatitis
APA	American Pancreas Association
APACHE II	Acute Physiology and Chronic Health Evaluation II
APFC	Acute peripancreatic fluid collection
AST	Aspartate aminotransferase
ATP	Adenosine triphosphate
AUC	Area under the curve
BISAP	Bedside Index for Severity in Acute Pancreatitis
BMI	Body mass index
BP	Blood pressure
BT	Bacterial translocation
BUN	Blood urea nitrogen
CASR	Calcium sensing receptor gene
CBD	Common bile duct
CECT	Contrast-enhanced computed tomography
CFTR	Cystic fibrosis transmembrane regulator
CP	Chronic pancreatitis
CRP	C-reactive protein
CTSI	CT severity index
CT	Computerized tomography
CTRC	Chymotrypsin-C
DAMPs	Damage-associated molecular patterns
DBC	Determinant-based classification
DIP	Drug-induced pancreatitis
DOR	Diagnostic odds ratio
dsDNA	Double-stranded DNA
EMBASE	Excerpta Medica dataBASE

ERCP	Endoscopic Retrograde cholangiopancreatography
ESR	Erythrocyte sedimentation rate
EUS	Endoscopic ultrasound
FIO_2	Fraction of inspired oxygen
GALT	Gut-associated lymphoid tissue
GCS	Glasgow Coma Score
GGT	Gamma-glutamyl transferase
GI	Gastrointestinal
GIT	Gastrointestinal tract
gp50b	Glycoprotein 50 b
gp60a	Glycoprotein 60 a
HAPS	Harmless AP Score
HDU	High dependency unit
HES	Hydroxyl ethyl starch
HIV	Human immunodeficiency virus
HMGB1	High mobility group protein B1
HP	Hereditary pancreatitis
HR	Heart rate
IAP	International Association of Pancreatology
ICU	Intensive care unit
ICUA	ICU admission
IDCP	Idiopathic duct-centric pancreatitis
IgA	Immunoglobulin A
IL-1	Interleukin 1
IL-1β	Interleukin 1β
IL-6	Interleukin 6
IL-8	Interleukin 8
IPN	Infected pancreatic necrosis
IPN	Isolated peripancreatic necrosis
JSS	Japanese Severity Score
LDH	Lactate dehydrogenase
LOHS	Length of hospital stay
LPSP	Lymphoplasmacytic sclerosing pancreatitis
M/F ratio	Male/Female ratio
MAP	Mean arterial pressure
mCRP	Monomeric C-reactive protein
MOD	Multiple organ dysfunction
MOP	Mesenteric oedema and peritoneal fluid
MRCP	Magnetic resonance cholangiopancreatography
MRI	Magnetic resonance imaging
nCRP	Native C-reactive protein

NF-κB	Nuclear factor kappa B
NO	Nitric oxide
NSAID	Non-steroidal anti-inflammatory drugs
NSF	Nephrogenic systemic fibrosis
OF	Organ failure
PAF	Platelet-activating factor
PAI-1	Plasminogen activator inhibitor-1
PaO_2	Partial Pressure of Oxygen in the arterial blood
PCA	Patient-controlled analgesia
PKC	Protein kinase C
PNec	Pancreatic necrosis
PO_2	Partial pressure of oxygen
POF	Persistent organ failure
POP	Pancreatitis Outcome Prediction
NPV	Negative predicative values
PPV	Positive predicative values
PSTI	Pancreatic secretory trypsin inhibitor
RP	Regional pancreatectomy
RR	Respiratory rate
SAP	Severe acute pancreatitis
SIRS	Systemic inflammatory response syndrome
SLE	Systemic lupus erythematosus
SO	Sphincter of Oddi
SOD	Sphincter of Oddi dysfunction
SOFA	Sequential Organ Failure Assessment
SPINK1	Serine protease inhibitor Kazal type 1
TAP	Trypsinogen activation peptide
TJ	Tight junctions
TLC	Total leucocyte count
TLRs	Toll-like receptors
TNF	Tumor necrosis factor
TNF-α	Tumor necrosis factor alpha
TPN	Total parenteral nutrition
US	Ultrasonography
VLDL	Very low density lipoprotein
WBC	White blood cell count
WHO	World Health Organization
WON	Walled-off necrosis

5.1 Introduction

Acute pancreatitis (AP) is acute inflammation of pancreas and is one of the commonest causes of abdominal pain, and frequently requires hopitalization. The clinical course of the AP is variable and ranges from mild disease, which resolves without much intervention, and the severe form which carries a very high morbidity and mortality. In majority of cases, it settles over few days but in about 20% of the cases, it may progress to pancreatic necrosis and persistent organ failure (POF), needs nursing in critical care unit and prolonged hospitalization. Owing to its variable course, it is of paramount importance to find out a way of early identification of patients, who have the highest risk of developing the severe disease. There are several reasons for stratification of AP early in the disease. The most important is to triage patients because of the implications for management, prognostication, and the allocation of healthcare resources. More specifically, it allows the identification of patients who require early intravenous fluid in an intensive care unit, or an expeditious transfer to a center of expertise (Windsor 2008). In fact, in last 3–4 decades most of the researchers have focused on this problem and a galaxy of scoring methods, serum markers, and imaging methods have been devised to identify these high risk patients early. But unfortunately, most of these methods achieve accuracy after 48 h of the onset, and a very early marker of disease severity remains elusive till date.

It is most commonly caused by gallstone diseases and alcoholism. In most patients, the disease takes a mild route, where moderate fluid resuscitation, management of nausea and pain, and timely oral feeding result in rapid clinical progress. The incidence of the disease is increasing gradually in last few decades. Many cases of recurrent AP progress to chronic pancreatitis, which is a devastating condition and most often happens due to pancreatitis secondary to alcoholism. It is often difficult to assess the course of the disease in the initial stage, and many scoring systems have been developed and imaging studies and serological markers have been tried to gauge the disease progress and early identification of those progressing into severe disease. Treatment is essentially supportive and few patients who developed severe AP may require interventional procedure to tide over complications. The mechanism by which AP progresses into pancreatic necrosis and organ failure is still obscure. Due to relative lack in clear understanding of its initial pathogenesis, there have been practically no medications, which can reverse the initiation to severe AP.

5.2 Disease Burden

Worldwide, the incidence of AP is between 4.9 and 73.4 cases per 100,000 populations (Fagenholz et al. 2007). The impact of AP at 20–80 per 100,000 per annum is substantial, with varying incidence rates reported from different countries, all increasing over the last 40 years (Yadav and Lowenfels 2013). There are scattered data relating to the incidence of AP in India. Overall, the prevalence of the disease is around 7.9 per 100,000. Prevalence rate for men and women 8.6 and 8.0 per 100,000, respectively.

The incidence in the United Kingdom is approximately 56 cases per 100,000 people per year and the overall mortality rate around 5%. About 50% of the cases of AP are caused by gallstones, 25% by alcohol and 25% by other factors (NICE Guidelines on Pancreatitis 2018). In some cases, AP may progress to chronic pancreatitis, particularly after recurrent attacks. Chronic pancreatitis is an inflammatory process of the pancreas that results in fibrosis, cyst formation, and stricture of the pancreatic duct.

The annual incidence in Western Europe is about 5 new cases per 100,000 people although this is probably an underestimate. In the Netherlands, overall incidence rate of AP increased during the 2000–2005 period from 13.2 in 2000–14.7 per 100,000 (Spanier et al. 2013).

In Japan, the estimated number of patients with AP showed a 1.8-fold increase in the last decade. The research committee of intractable pancreatic diseases, supported by the Japanese Ministry of Health, conducted a nationwide survey of AP patients in 2011. Based on the survey, the prevalence in Japan was 49.4 per 100,000 populations. Alcohol was a major cause of AP in male patients, whereas gallstone AP was dominant in female patients. Sepsis, cardiovascular failure, and respiratory failure were seen in 22.8%, 21.1%, and 12.3% of the total death, respectively. Renal failure and disseminated intravascular coagulation accounted for 7.0% of the deaths. There has also been a recent increase in overall mortality in patients with severe AP (8.0–10.1%) (Hamada et al. 2014).

AP is the most common diagnosis for hospitalization among the gastrointestinal conditions in the United States, accounting for as many as 230,000 hospitalizations per year. The incidence is on an increasing trend during the past decades and has ranged from approximately 5–35 per 100,000 populations per year. AP with its associated complications is a major cause of morbidity and mortality worldwide; mortality ranges from approximately 1–20% in mild to severe cases, respectively. As a result, AP poses a huge financial healthcare burden as well.

The reasons for the changing incidence and severity of AP are not well understood. Increased obesity with its complications of gallstones, immoderate alcohol consumption, and metabolic disorders may play a role (Yadav and Lowenfels 2013). Another possible contributor is the increased use of measurements of serum pancreatic enzymes in emergency departments which may pick up milder cases of pancreatitis.

5.3 Etiology

About 60–80% cases of AP are due to gall bladder stones (or biliary diseases) and alcohol abuse (Sakorafas and Tsiotou 2000). Both these etiologies are dependent on the population evaluated. Biliary pancreatitis is more common in women whereas alcoholic pancreatitis seems to affect middle aged men more commonly. This pattern is more or less same in both Eastern world and the West. About 10% of the cases are caused by diverse conditions like hypertriglyceridemia, trauma, hypercalcemia, malignancy, viral infections, drugs and iatrogenic, etc. However, in nearly 10–30% of the cases, the cause of AP remains unknown (Boxes 5.1 and 5.2).

BOX 5.1 CAUSES OF ACUTE PANCREATITIS

•ESTABLISHED CAUSES
•Biliary or Gallstone diseases
•Alcohol abuse
•Hypertriglyceridemia
•Post ERCP
•Duct Obstruction due to Choledochocele or Duodenal Diverticula
•Trauma : Accidental
•Post-Surgical operations
•Infections and Parasitic infestations
•Hyperparathyroidism and Hypercalcemia
•Drug Induced
•Autoimmune
•Genetic
•Pancreatic duct narrowing due to Pancreas Divisum
•Scorpion Bites
•Posterior penetrating duodenal ulcer
Idiopathic

•DOUBTFUL CAUSES
•Sphincter of Oddi Dysfunction (SOD)
•Gallbladder Sludge or Microlithiasis
•Panaceas Divisum without any duct narrowing

5.3.1 Gallstones

Gallstone disease is the cause of AP in almost 40% of the cases (Cappell 2008). The AP secondary to gallstone is usually mild; however, it recurs frequently if the cause is not removed. Although gallstones are the most common cause of AP, only 3–7% of the gallstone patients develop the disease. Gallstone AP is more common in women and tends to occur in patients who harbor small stones (<5 mm). The pancreatic duct and bile duct open through a common channel at the sphincter of Oddi (SO) in the second part of the duodenum. The pathogenesis is believed to be related to increase in pancreatic ductal pressure, possibly with biliopancreatic reflux that occurs when the gallstone passes or gets impacted at the ampulla. Further attacks of AP are prevented if the gall bladder is removed surgically, confirming the cause-and-effect relationship (Cucher et al. 2008).

BOX 5.2 ETIOLOGY OF ACUTE PANCREATITIS

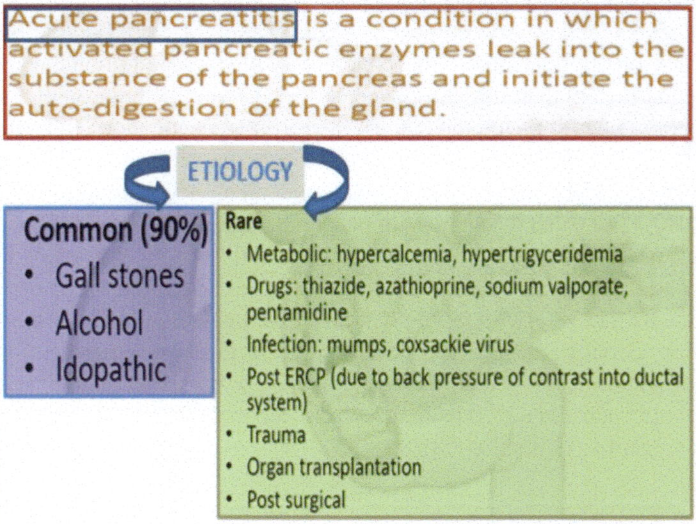

Multiple small stones (<5 mm), a dilated cystic duct, a good postprandial gall bladder emptying and the presence of sludge with or without stones are contributing factors for acute biliary pancreatitis (Scott and William 2010).

Sludge is a liquid suspension which contains very small stones (<3 mm), cholesterol monohydrate crystal, or calcium bilirubinate granules. On ultrasonography sludge is identified as a low amplitude layer in the most dependent part of the gall bladder which does not produce acoustic shadow and shifts on changing the position.

Obstructive causes of AP in addition to gallstones include pancreas divisum, SO disease or stenosis, pancreatic cancer, parasites, and blood clots. All these would be discussed later in this chapter. Pancreatic cancer can cause AP because of temporary duct obstruction by clots within the pancreatic duct and occasionally can mimic chronic pancreatitis because of chronic malignant obstruction of the pancreatic duct. A very thick and viscid secretion in the pancreatic duct can also cause AP due to obstruction and ductal hypertension. The examples of this latter type are cystic fibrosis and intraductal papillary mucinous neoplasm of the pancreas (Cappell 2008).

5.3.2 Alcoholism

Alcohol causes about 30–35% of AP, and it is more common in men (M/F ratio 2.5:1). The difference is probably due to tendency of men to drink more rather than any gender-based difference in susceptibility. Alcohol-induced AP is more common in the western world, and the mean age range is 40 years. Alcoholic pancreatitis generally requires more than 100–150 g of alcohol every day for at least 5 years. Curiously, only about 10% of these heavy drinkers develop AP. These observations suggest that alcohol alone may not cause pancreatitis unless accompanied by additional genetic and/or environmental factors (Sakorafas and Tsiotou 2000); it is probable that alcohol sensitizes the pancreas, with these additional genetic and environmental factors then initiating pancreatitis. Among the environmental factors studied, smoking has attracted the highest interest. A recent cohort study shows that smoking was a dose-dependent risk factor for alcoholic pancreatitis after controlling for age, gender, body mass index (BMI), and alcohol consumption (Scott and William 2010). The exact mechanism of pancreatic damage by ethyl alcohol remains unclear and most of the researchers suggest possibility of alcohol-related injuries, including perturbations in exocrine function, changes in cellular lipid metabolism, induction of oxidative stress, and activation of stellate cells. Most of the patients, who develop alcohol-induced AP, slowly progress to chronic pancreatitis. About 70% of the patients developed chronic pancreatitis 10 years after an initial episode of acute alcoholic pancreatitis. Abstinence from alcohol retards the progression to chronic pancreatitis (Cappell 2008).

5.3.3 Hypertriglyceridemia

Hypertriglyceridemia is the third most common identifiable cause of AP, and its incident ranges from 2% to 10%. Hypertriglyceridemia has also been implicated in nearly half of the gestational AP. A causal association of AP with hypertriglyceridemia was first reported by Speck in 1865 (Munigoti and Rees 2011). A serum triglyceride level more than 1000 mg/dL (11 mmol/L) is suggestive and above 2000 mg/dL is diagnostic. The triglyceride level should be measured early after the disease because this level tends to decline rapidly during the hospitalization due to fasting, restoration of fluid and electrolyte balance, and insulin therapy. The serum in these patients may be opalescent due to high level of very low density lipoprotein (VLDL) or milky due to high level of serum chylomicrons. AP from hypertriglyceridemia in childhood is usually due to congenital types I, II, and V hyperlipidemia. Adults can develop an acquired hypertriglyceridemia because of alcoholism, obesity, poorly controlled diabetes mellitus, and hypothyroidism in association with a mild form of genetically inherited type I or V hyperlipidemia. Chylomicrons are triglyceride-rich lipoprotein particles believed to be responsible for pancreatic inflammation. They usually present in the circulation when serum triglyceride levels exceed 10 mmol/L (>900 mg/dL). These largest of lipoproteins might impair circulatory flow in capillary beds; if this occurs in the pancreas, the resulting ischemia might disturb the acinar structure and exposes these triglyceride-rich particles to pancreatic lipase. The proinflammatory non-esterified free fatty acids generated from the enzymatic degradation of chylomicron-triglycerides may lead to further damage of pancreatic acinar cells and microvasculature. Subsequent amplification by the release of inflammatory mediators and free radicals may ultimately lead to necrosis, edema, and inflammation (Gan et al. 2006).

5.3.4 Hypercalcemia

Hypercalcemia is one of the rarer causes (1–4%) of AP. It tends to occur more frequently with primary hyperparathyroidism, in widespread bony secondaries of a malignant disease and in sudden or acute increase in the serum calcium level. It is a rare cause of AP in patients who have chronic hypercalcemia. The pathophysiology of hypercalcemic AP is not clear. The deposition of precipitated calcium in pancreatic duct and premature activation of pancreatic digestive enzymes are the most popular theories. Acute hypercalcemia also increases the permeability of the pancreatic ductal membrane, allowing enzymes to leak and injure the pancreatic parenchyma. Due to its common association with primary hyperparathyroidism suggests some role of paratharmone also. Acute hypercalcemia results from calcium infusion during total parenteral nutrition and seen in patients with myeloma, leukemia, vitamin D poisoning, disseminated cancer, severe hyperthyroidism, or infusion of calcium in high doses perioperatively during cardiopulmonary bypass, all of which have been associated with AP. AP secondary to hypercalcemia is usually mild, and rapid correction of the serum calcium level leads to resolution.

5.3.5 Post-Endoscopic Retrograde Cholangiopancreatography (ERCP)

Post-endoscopic retrograde cholangiopancreatography (ERCP) is the cause of AP in nearly 2% of the cases. AP happens to be the most common complication of ERCP and occurs in about 5% of the cases. It occurs far more frequently if pancreatic duct cannulation is attempted or SO studies are carried out. It is diagnosed by the presence of typical abdominal pain after the procedure and at least three-fold increase in the serum levels of amylase and lipase. Mild to moderate elevations in the serum levels of these enzymes is usual after the procedure, but the absence of typical pain generally rules out the disease. High levels of serum amylase and lipase 2–3 h after the procedure have almost 100% positive predictive value for post-ERCP pancreatitis. There are few high risk predictors of post-ERCP pancreatitis and include young age, female gender, suspected sphincter of Oddi dysfunction (SOD), normal bilirubin level, pancreatic duct cannulation, dye injection or balloon dilatation, and inexperience operator. The pancreatic duct hypertension due to spasm of the sphincter or dye injection may be the principal cause; however, post-ERCP pancreatitis is believed to be multifactorial, involving a combination of chemical, hydrostatic, enzymatic, mechanical, and thermal factors. Many medications have been tried to attenuate or prevent post-ERCP pancreatitis but only two non-steroidal anti-inflammatory drugs (NSAID), namely indomethacin and diclofenac sodium, have been found worthy of widespread use. Both drugs have been used in rectal suppository forms and should be used in high risk cases (Dumonceau et al. 2014).

5.3.6 Infections and infestations

Infections are rare cause of AP and the exact incidence is unknown. There are many types of infections implicated in AP. Viruses have shown most consistent association. Bacterial, fungal, and parasitic infestations have also been implicated rarely.

Viruses are the largest infective agents associated with AP. The diagnosis is confirmed by the identification of viral antibodies coupled with clinical diagnosis of AP. Mumps (single-stranded DNA paramyxovirus) was the first to be identified as a confirmed cause of AP in year 1905. Due to widespread childhood vaccination against this disease, it is rarely reported nowadays. Hepatitis causing viruses have also been reported to cause AP; of these hepatitis B has been most consistently implicated with AP. HIV may be a direct cause of AP in AIDS patients, but there are many confounding factors like antiretroviral drugs, alcohol, opportunistic viral infections, and various malignancies blur the clear-cut association.

Other viruses which may cause AP are coxsackie virus type B, cytomegalovirus, varicella-zoster, and herpes simplex.

Among bacterial infections *Mycoplasma pneumoniae* has been most extensively studied with AP. There association is established by the presence of high antibody titer. The exact pathologic mechanism is unknown. Other pathogenic bacteria, such as *Leptospira interrogans, Campylobacter jejuni, Salmonella typhi, Brucella, Yersinia enterocolitica, Y. pseudotuberculosis, Legionella, Nocardia, Mycobacterium tuberculosis,* and *M. avium,* have been reported as causes of sporadic cases of AP.

Fungal etiology of AP is seen in immunocompromised patients and among those who are on immunosuppressants and have been treated by broad spectrum antibiotics. *Aspergillus* happens to be the most commonly associated with AP although the occurrence is rare.

Among parasitic infestations, *Ascaris lumbricoides* (roundworms) is the most commonly reported organism, particularly, from developing countries. AP is likely triggered by the obstruction of the pancreatic duct. They most commonly cause AP in pediatric population due to their narrower pancreatic duct. Other parasites implicated in causing AP by similar mechanisms include *Clonorchis sinensis, Opisthorchis spp., Dicrocoelium dendriticum, Fasciola hepatica, Schistosoma haematobium and S. mansoni, Toxoplasma, Cryptosporidium,* and *Paragonimus westermani.*

5.3.7 Drug-Induced Pancreatitis

Drug-induced pancreatitis (DIP) is reported around in 1–4% of the cases but due to lack of clear diagnostic tests may be <1%. Many drugs have been described as cause of AP; however, clear evidence is lacking. In fact, a WHO database has listed 525 different drugs which can cause AP. Mallory and Kern have described four criteria for DIP in 1980 and they are: (1) Pancreatitis develops during the treatment with a suspected drug. (2) There is no evidence of any other etiologic factors. (3) Symptoms disappear after withdrawal of the drug. (4) Recurrence of pancreatitis on reintroduction of the pharmacologic agent.

Many of the published case reports suffer from a combination of inadequate criteria for the diagnosis of DIP. In the absence of well-designed clinical trials, clinicians largely rely on these published case reports for the determination that a drug may have caused AP.

DIP tend to occur 4–6 weeks after continuous use of the offending drug and is commonly seen in elderly and pediatric age groups, females, and patients with inflammatory bowel disease or HIV.

Very little is known about the precise causes of DIP, and most theories center on a few mechanisms. Accumulation of a toxic metabolites and hypersensitivity reactions cause immune-mediated injuries and pancreatic duct constriction, localized angioedema effect in the pancreas, and arteriolar thrombosis. Adverse effects of drugs causing hypercalcemia or hypertriglyceridemia are also mechanisms for DIP. The family of drugs often reported to cause AP are the angiotensin-converting enzyme (ACE) inhibitors, antidiabetic agents, statins, 5-aminosalicylic acid (5-ASA) and derivatives, antibiotics, and valproic acid. Generally, the DIP is mild and resolves on simple measures and discontinuation of the drug which is critical (Table 5.1).

Table 5.1 Drugs commonly implicated as causes of pancreatitis

A. *Drug-induced hypersensitivity reaction*
• 5-Aminosalicylic acid/sulfasalazine
• Azathioprine
• 6-Mercaptopurine
• Metronidazole
• Tetracycline
B. *Toxic metabolite*
• Pentamidine
• Valproic acid
• Didanosine
C. *Drug-induced hypertriglyceridemia*
• Thiazides
• Isoretinoin
• Tamoxifen
D. *Overdose reaction*
• Acetaminophen
• Erythromycin

5.3.8 Genetic Causes of Pancreatitis

Few genetic mutations have been found to be closely linked with cases of recurrent AP and chronic pancreatitis (Table 5.2). Pancreatitis having genetic basis have defined by two types:

1. *Hereditary Pancreatitis* (HP) where two or more individuals with pancreatitis in two or more generations of the family or disease-causing germ line mutation in the cationic trypsinogen gene *PRSS1* (*Serine protease-1*).
2. *Familial Pancreatitis* is defined as when more than one person in the family has recurrent AP or chronic pancreatitis—regardless of etiology—since the incidence is above the expected rate in the population by chance alone.

HP has been conclusively linked with gain-of-function mutation of *PRSS1*. This mutation leads to an increase in autocatalytic conversion of trypsinogen into active trypsin, which triggers autodigestion of pancreatic parenchyma and subsequently to widespread inflammation; the details would be discussed further in the relevant section.

Another gene which is the hallmark of cystic fibrosis, an autosomal recessive disorder, is cystic fibrosis transmembrane regulator (*CFTR*) gene. The *CFTR* protein forms a regulated anion channel that facilitates transport of chloride and bicarbonate across the apical membrane of epithelial cells during active secretion. The secretion of bicarbonate-rich fluid flushes out pancreatic enzyme out of ducts since longer stay of zymogen granules can lead to their activation. Mutation in the gene encoding for this protein can cause recurrent AP. There are more than 1200 mutations that have been described for the CFTR gene. Some of these are considered severe and others mild. Homozygous severe mutations produce a thick, concentrated, and acidic pancreatic juice leading to ductal obstruction and pancreatic insufficiency in infanthood. Heterozygotes of minor or major mutations may lead to acute recurrent or chronic pancreatitis by altering acinar or ductal cell function. Loss or diminished *CFTR* causes recurrent AP.

Serine protease inhibitor Kazal type 1 (*SPINK1*) is a trypsin inhibitor protein and an acute phase reactant, secreted by pancreatic acinar cells. It is encoded by *SPINK1*

Table 5.2 Genetics of pancreatitis

Types	Related genes or proteins
Hereditary	*PRSS1* (Autosomal Dominant) Unknown
Familial	*SPINK1/SPINK1* (Autosomal Recessive)
	CFTR/CFTR (Recessive)
Atypical cystic fibrosis	*CFTR/SPINK1* (Complex)
Tropical pancreatitis	*SPINK1/unknown* (Complex)
Sporadic	*CTRC*
	CASR
Hyperlipidemias	Lipoprotein Lipase
	Apolipoprotein C-II

CASR calcium-sensing receptor, *CFTR* cystic fibrosis transmembrane conductance regulator, *CTRC* chymotrypsin-C gene, *SPINK1* serine protease inhibitor Kazal type 1 which is located on the short arm of chromosome 5. Its mutation causes loss-of-function and blunted response in inhibiting trypsin (Box 5.3).

Chymotrypsin C (*CTRC*) gene which codes for chymotrypsin C enzyme destroys trypsin and it's mutation causes impaired clearance of trypsin. Calcium sensing receptor gene (*CASR*) plays a major and important role in maintaining the calcium homeostasis through its effect on renal tubules and parathyroid gland. A high calcium level in the pancreatic cells activates the trypsinogen and a low level maintains it in an inactive state. Mutation in *CASR* genes results in hyperstimulation or dysregulation in the calcium homeostasis.

BOX 5.3 ABOUT TRYPSIN

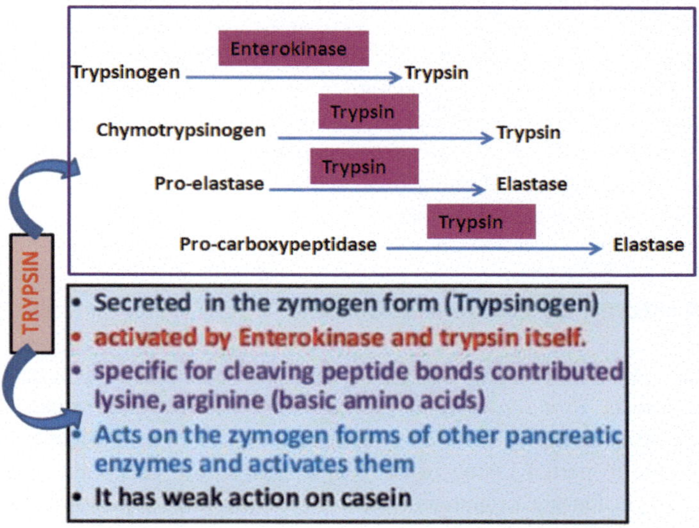

5.3.9 Autoimmune Pancreatitis

Autoimmune pancreatitis is a recently recognized pancreatic inflammatory disease that is further classified into two subtypes. Type 1, called *lymphoplasmacytic sclerosing pancreatitis* (*LPSP*), is a systemic disease affecting the pancreas, bile ducts (sclerosing cholangitis), nephritis, salivary glands (sclerosing sialadenitis), retroperitoneum (retroperitoneal fibrosis), and other organs. It is also associated with infiltration of these organs by IgG4-bearing plasma cells, with elevations in serum levels of IgG4. The IgG4 is not felt to be pathogenic but is useful for diagnosis. AIP type 1 is most commonly seen in middle-aged men presenting with painless or mild pain, obstructive jaundice and weight loss, a similar picture of much more common, pancreatic cancer. Type 2 affects only the pancreas, is called *idiopathic duct-centric pancreatitis (IDCP)*, is not associated with elevated IgG4, and is commonly seen in younger patients presenting with AP. Furthermore, type 2 has no biomarker and is strongly associated with *inflammatory bowel disease*. Both subtypes are corticosteroid responsive; however, type I AIP is poorly responsive and relapses commonly, whereas, type 2 AIP responds dramatically to corticosteroids and relapses rarely. Thus, maintenance therapy with either an immunomodulator or rituximab is often necessary for patients with type 1 AIP. Nevertheless, prior to initiating a therapeutic trial, it is essential to rule out pancreatic malignancy (Krasinskas et al. 2007).

5.3.10 Traumatic Pancreatitis

AP can occur following abdominal trauma. Due to its retroperitoneal location, it is seen in about 2% of the blunt trauma cases; however, it occurs in about 12% cases of penetrating injury of the abdomen. It is recognized by the triad of abdominal pain, high leukocyte count, and elevated serum amylase. However, this triad of signs suffers from the lack of specificity and high index of suspicion is essential for correct and early diagnosis. The damage can range from minor contusion to crush injury to complete transaction. The latter can occur where the pancreas crosses over the spine. Pancreatic trauma with ductal injury heals with scarring and stricture of the pancreatic duct and can cause segmental pancreatitis of the distal part.

5.3.11 Anatomic or Congenital Causes

Many cases of AP and recurrent pancreatitis can be caused by anatomical or congenital defects. *Pancreas divisum* is the commonest developmental defects of the pancreas. It is variously reported to be present in 7–12% of the general population and is due to partial fusion of the dorsal and ventral pancreatic buds during embryonic stage. During organogenesis, the proximal dorsal pancreatic duct normally regresses while the rest of the dorsal duct fuses with the ventral pancreatic duct to produce a single, continuous, relatively wide and long pancreatic duct that opens through the major papilla. In pancreas divisum, the normal fusion fails to occur: the dorsal duct drains most of the pancreatic secretions through the minor papilla, and the ventral duct drains only the head of the pancreas through the major papilla. Ductal hypertension during stimulated secretion can cause AP. However, a vast majority, nearly 95%, of people with pancreas divisum never develop pancreatitis. It is thought that there may be an associated duct narrowing which makes few of them prone to develop AP. More recently, CFTR mutations associated with pancreas divisum have suggested a synergistic effect in the pathogenesis of AP. It may be that those persons who also harbor dysfunction in the CFTR transporter are at risk of developing pancreatitis when both are present in the same host (Lee et al. 2016).

Annular pancreas is another rare condition resulting in the encirclement of both the CBD and duodenum by the annular growth of the pancreas. Approximately one third of patients with annular pancreas have pancreas divisum, but it is not clear whether pancreatitis depends on the annular variant or the pancreas divisum (Lee et al. 2016).

SOD refers to the abnormality of SO contractility that can manifest clinically as pain, pancreatitis, or deranged liver function tests. It is also a controversial cause of pancreatitis and is diagnosed on manometry when the basal pressure exceeds 40 mm of Hg and SO fails to relax on smooth muscle relaxants administration.

Beside these other anatomic causes which can be responsible for AP, include choledochal cyst, particularly type 3 which is also known as choledochocele, and duodenal duplication cysts.

5.4 Pathogenesis and Pathophysiology

The molecular mechanism of AP is triggered by intra-acinar activation of the key enzyme trypsinogen and some key nuclear factors. Many elegant experiments on mice have led to understanding of the basic mechanism of AP in last decade; however, few issues are still hotly debated and subject of active research. The basic model has been depicted in Fig. 5.1 (Box 5.4) (Dawra et al. 2011) and steps of current understanding are summarized below.

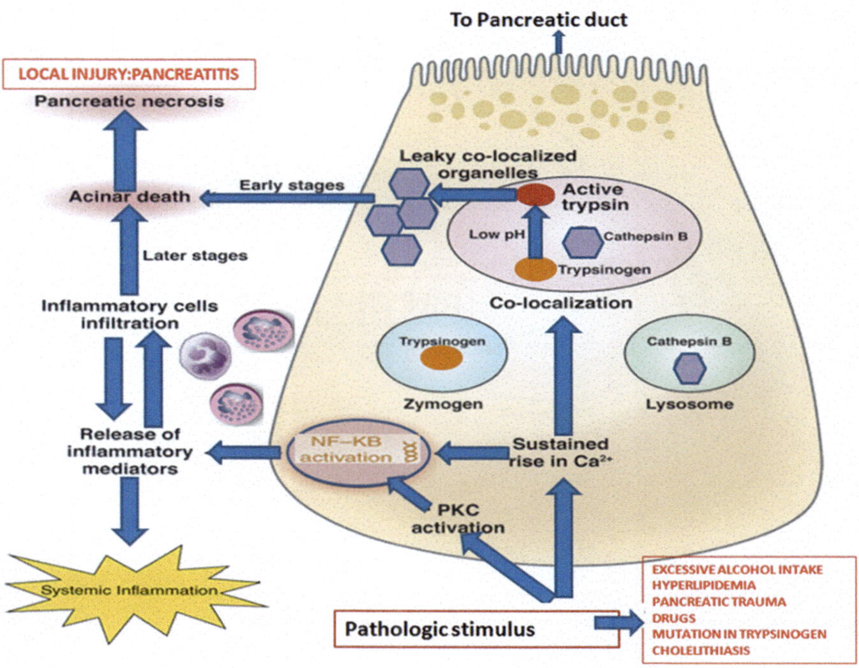

Fig. 5.1 Schematic description of pathogenesis of AP. Activation of trypsinogen in the organelles formed from co localization of zymogen and lysosomal compartments leads to leakage of cathepsin B in the cytosol resulting in acinar death in the early stages of pancreatitis. Activation of NFkβ occurs early independent of trypsinogen activation and leads to release of inflammatory mediators and recruitment of inflammatory cells, which causes acinar cell death in later stages of pancreatitis and drives the systemic inflammatory response seen in pancreatitis. *PKC* Protein Kinase C. [Saluja A, Dudeja V, Dawra R and Sah RP: *Early Intra- Acinar Events in Pathogenesis of AP*: Gastroenterology 2019;156:1979–1993]

The first step is the injury of cell membrane of acinar cells which leads to rapid infusion of calcium intracellularly. A low pH may be contributory factor and depletion of intracellular ATP causes blunting of energy-driven calcium extrusion process from the cell (Gukovskaya et al. 2016).

The digestive enzymes synthesized in the acinar cell are stored in an inactive form or zymogen. These zymogens are stored in zymogen granules, which is important to keep them safe. The enzyme which has been consistently shown to be prematurely activated in AP is *trypsinogen*. The membrane of zymogen granules insulates trypsinogen from various intracellular-activating factors. Normally, small amounts of trypsinogen are spontaneously activated within the pancreas, but intrapancreatic mechanisms quickly remove the trypsin. Pancreatic secretory trypsin inhibitor (PSTI, now called *SPINK1*) binds and inactivates about 20% of the trypsin activity. Other mechanisms for removing trypsin involve mesotrypsin, enzyme Y, and trypsin itself, which splits and inactivates other trypsin molecules (Scott and William 2010).

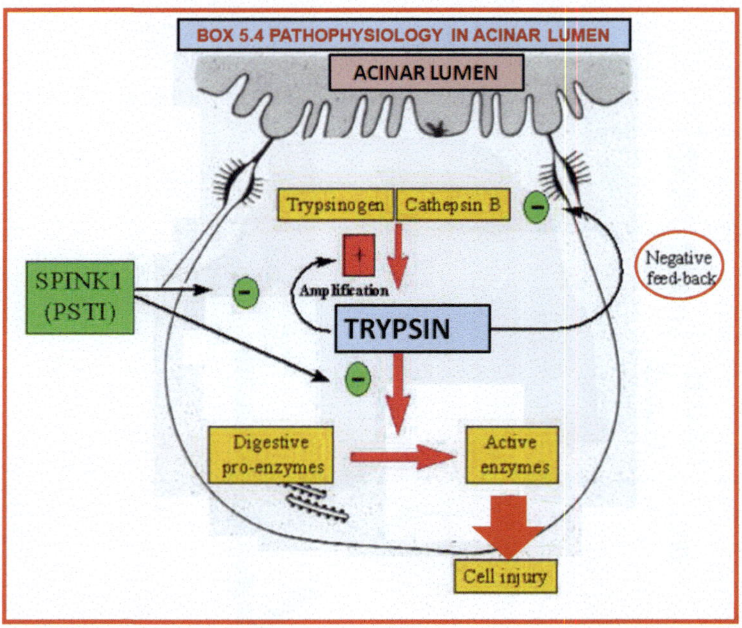

BOX 5.4 PATHOPHYSIOLOGY IN ACINAR LUMEN

The rise in intracellular calcium causes zymogen granules to merge or co-localize with cytoplasmic vacuoles that contain lysosomal enzymes *cathepsin B*. Cathepsin B then activates trypsinogen to trypsin. The trypsin also dissolves the membrane of vacuoles which contains it and releases cathepsin B in the cytosol. Once in the cytosol the cathepsin B triggers the process of apoptotic cell death (Halangk et al. 2000; Van Acker et al. 2002).

One significant event which was identified in parallel with initiation of AP is activation of nuclear factor ƙB (NF-ƙβ) in the acinar cell. Rise in cytosolic calcium level causes activation of protein kinase C (PKC) which in turn activates NF-ƙB. Activation of NF-ƙB also initiates cell death and mainly responsible for the release of cytokines that attract inflammatory response cells which mediate local and systemic inflammation cascades in AP (Dawra et al. 2011; Fisher William et al. 2015).

This paradigm demonstrates pathogenic events in AP. Activation of trypsinogen in the organelles formed from co-localization of lysosomal and zymogen compartments leads to leakage of cathepsin B in the cytosol resulting in acinar death in the early stages of pancreatitis. Activation of NF-kB takes place early independent of trypsinogen activation and leads to the release of inflammatory mediators and recruitment of inflammatory cells, which causes acinar cell death in later stages of pancreatitis and drives the systemic inflammatory response seen in pancreatitis.

The damaged and dying acinar cells expose many *self* molecules called damage-associated molecular patterns (DAMPs), which are normally not exposed to immune systems. Various molecules that act as DAMPs are high mobility group protein B1 (HMGB1), S100 proteins, extracellular double-stranded DNA (dsDNA), ATP, and many others. HMGB1 belongs to a protein family of non-histone nuclear proteins that maintain chromosomal structure and regulate DNA-associated activities. Remarkably, in apoptotic cells HMGB1 binds tightly to chromatin remnants, but in necrotic cells HMGB1 is passively released and triggers inflammatory response. These DAMPs molecules serve as toll-like receptor (TLR) ligands. TLRs are present on the immune cells, particularly macrophages, and initiates the release of pro-inflammatory cytokines and chemokines. Active granulocytes and macrophages release pro-inflammatory cytokines in response to transcription factors such as nuclear factor ƙB (NF-ƙB). Pro-inflammatory cytokines include TNF, IL-1, IL-6, and IL-8, and platelet-activating factor (PAF). Pro-inflammatory cytokines frequently are followed by anti-inflammatory cytokines (IL-2, IL-10, IL-11) that attempt to downregulate inflammation. Other mediators of inflammation include arachidonic acid metabolites (prostaglandins, PAF, and leukotrienes), nitric oxide, proteolytic and lipolytic enzymes, and reactive oxygen species that overwhelm scavenging by endogenous antioxidant systems. The release of these substances leads to vascular injury causing capillary leak and edema with thrombosis of small vessels which subsequently result in local microcirculatory failure and amplification of the pancreatic injury (Saluja et al. 2019).

In most of the cases of AP, this initiation of inflammation resolves on its own. However, in about 20% of the cases it continues inexorably to involve many end organs (cardiovascular, respiratory, gastrointestinal, and renal systems) and subsequently to persistent multiple organ dysfunction and failure. It is helpful to describe AP in three phases; (1) Initiation, when the acinar damage is ensued by activation of self-digestion and initial inflammation, (2) Perpetuation, due to persistent inflammation with more and more of inflammatory cells recruitment and in turn mediators release, and (3) Secondary escalation, when settling inflammation escalates again due to infection of the pancreatic necrosis.

5.5 Pathogenesis of Local Complications

AP begins with local inflammation of variable intensity and generally results in local and systemic complications. The common type of edematous AP is due to capillary leak which is due to their increased permeability. It is seen on CT scan as swollen pancreas with homogenous enhancement and peripancreatic fat stranding. Pancreatic necrosis is due to failure of microcirculation mainly due to stasis or microthrombi obstructing the capillaries (Windsor 2008; Escott et al. 2017). The local vasoconstriction is caused by *endothelin* release and results in progressive exclusion of capillaries from the pancreatic circulation causing discrete areas of pancreatic hypoperfusion and necrosis. Furthermore, leucocytes form plaques and obstruct venules. The activation of coagulation cascade, as part of inflammation, leads to intravascular microthrombi and clogs small vessels. And lastly rise in hematocrit due to hemoconcentration secondary to third space fluid loss causes increasingly viscous blood in the microcirculation. Peripancreatic necrosis can occur with or without pancreatic necrosis. It is generally thought to be due to the effects of active lipolytic enzymes of pancreas (Isenmann et al. 2001).

The local complications have been classified according to the duration, contents, and the presence or absence of infection. Acute peripancreatic fluid collection is formed due to aggregation of local edema fluid and does not have a well-defined capsule around it. By definition, it does not have any solid component (necrotic material) in it and homogenous on imaging studies. It resolves on its own and very rarely can get infected. If it persists for several weeks and acquires a well-defined capsule, then it is referred to as pancreatic pseudocyst. The pseudocyst formation and its failure to spontaneous resolution are due to its connection with a disrupted pancreatic duct, which can happen during the course of pancreatic inflammation. This is likely to be the reason of elevated pancreatic enzymes in pseudocyst fluid (Windsor 2008; Escott et al. 2017).

Pancreatic necrosis, develops in 5–10% of the cases, can resolve, persist, liquefy, or can get infected or remain sterile. Initially, they are termed acute necrotic collection (ANC) and if persistent they can acquire a well-define capsule and are known as walled-off necrosis (WON). Both ANC and WON can get infected (Windsor 2008; Escott et al. 2017).

There are few very dreaded vascular complications associated with AP, but happily they are very rare. The vascular injuries cause rapid blood loss in a compromised patient and generally unsalvageable. Most commonly, these injuries lead to pseudoaneurysm formation, where blood collects in a contained peritoneal space or organ. Less commonly, there occurs arterial rupture into a pancreatic pseudocyst (converting it into a pseudoaneurysm) and rarely, arterial rupture into the gastrointestinal (GI) tract. The patients with severe AP and those with extensive necrosis are at increased risk. Previous pancreatic necrosectomy, long-term anticoagulation therapy, and underlying vasculitis also increase the probability of vascular complications (Barge and Lopera 2012). It is speculated that the cause of this phenomenon is the local spread of pancreatic proteolytic and lipolytic enzyme-rich fluids. These fluids cause weakening and elastolytic erosions of the vessel wall, which may result in the formation of a pseudoaneurysm (if there is continued bleeding into a contained space or organ), contained hematoma (if the pseudoaneurysm becomes thrombosed or active leak stops), or frank intraperitoneal hemorrhage (if the pseudoaneurysm ruptures) (Flati et al. 2003).

The pancreas is intimately associated with the splenic vein, inferior mesenteric vein, superior mesenteric vein, and portal vein. Splenic vein thrombosis, caused by local inflammation or compression from an adjacent peripancreatic fluid collection, may result in isolated left-sided portal hypertension, gastroesophageal variceal formation, and eventual GI bleeding. Thus, splenic vein thrombosis is more likely to occur in the presence of pseudocysts and severe necrotizing pancreatitis. Complications related to portal vein involvement are also most commonly due to thrombosis (Flati et al. 2003).

5.6 Role of Gastrointestinal Tract (GIT) in AP

It has been increasingly recognized over last three decades that the intestine is not just a passive bystander in the course of AP, in fact in all critically ill patients, but it is an end organ in the disease process and a significant contributor to the resultant sepsis and MODS. Although the intestinal failure is an extremely critical event and leads to very high morbidity and mortality, it has not been considered in any of the organ failure scoring systems (e.g., SOFA, Marshall, and APACHE II) (Windsor 2008; Escott et al. 2017).

The gastrointestinal tract is the largest organ of the human body and forms the largest interface between the internal milieu and the external environment. The complexity of its function could be understood by considering that the intestine must serve two opposite functions at the same time; the selective permeability for the required nutrients from the intestinal lumen into the circulation and, on the other hand, the prevention of the penetration of harmful entities including microorganisms, luminal antigens, and luminal pro-inflammatory factors. The latter function is known as barrier function (Assimakopoulos et al. 2018).

The gut barrier function can collectively be described by three major lines of defense: (1) The biological barrier, which is made up of normal intestinal flora (gut microbiota) responsible for colonization resistance; (2) The immune barrier, which is composed of gut-associated lymphoid tissue (GALT), effector and regulatory T cells, IgA producing B (plasma) cells, group 3 innate lymphoid cells, and, resident macrophages and dendritic cells in the lamina propria; and (3) The mechanical barrier, consisting of the closed-lining intestinal epithelial cells and by the capillary endothelial cells. The epithelial and endothelial cells come into the closest possible contact in the most apical part of the lateral cell membranes by specific structures named "tight junctions" (TJs), which interconnect the cells and restrict the passage of ions, molecules, and cells through the paracellular space (Assimakopoulos et al. 2011; Assimakopoulos et al. 2007).

The term bacterial translocation (BT) was first described by Berg and Garlington in 1979, as the phenomenon of passage of *viable bacteria* from the gastrointestinal tract through the epithelial mucosa into the lamina propria and then to the mesenteric lymph nodes and possibly other normally sterile organs. There was a strong evidence of bacterial translocation in post-laparotomy patients where the lymph nodes were biopsied to confirm the presence of gut-origin bacteria. However, in patients with severe sepsis and MODS the failure to consistently find gut-derived bacteria or bacterial products, such as endotoxin, in the blood led many to speculate that there must be another mechanism of this phenomenon. Subsequently, these observations led to the *gut-lymph hypothesis* of MODS, where gut-derived factors present in intestinal (mesenteric) lymph serve as the triggers that initiate the systemic inflammatory and tissue injurious responses observed after major trauma or episodes of shock or sepsis (Deitch et al. 2006). This septic response appears to occur through the release of *non-microbial factors* into the intestinal lymphatics which, upon reaching the systemic circulation, are sufficient to cause both distant organ injury and a septic state. Edwin Deitch stated that the process of bacterial translocation should be dissociated from the pathophysiology of gut-derived sepsis and MODS although these two conditions may occur together, gut-origin sepsis and MODS do not require bacterial translocation and bacterial translocation by itself may not lead to MODS (Deitch 2012).

The development of intestinal dysfunction is the consequence of ischemia-reperfusion injury. Hypotension at the onset of AP causes reflex vasoconstriction of the splanchnic post-capillary venules. This happens to maintain perfusion of the vital organs. The resultant mucosal ischemia sets off damage to the intestinal barriers. Since intestine is the last organ to receive perfusion after resuscitation, the delay causes significant breach in the barrier system. Increased enterocytes' apoptosis, decreased proliferative response, and loss of tight junctions' integrity are some of the most important cellular alterations associated with intestinal barrier dysfunction in this setting. The reperfusion injuries are mainly due to free oxygen radicals and activated leucocytes entering the systemic circulation to cause injury to remote organs. Early fluid resuscitation, therefore, has a pivotal role in prevention of this deleterious effect originating from the intestine (Windsor 2008; Escott et al. 2017).

5.7 Obesity and Visceral Adiposity

Fatty or adipose tissues also have a role in promoting systemic inflammation. In many instances, it has been consistently observed that obese people have worse outcome in AP. Researchers have proposed that obesity is associated with a low-grade inflammatory status in normal circumstances. The chronic low-grade inflammatory state is characterized by high levels of leptin, TNF-a, IL-6, IL-10, IL-1β, plasminogen activator inhibitor-1 (PAI-1), and CRP (Lee and Pratley 2005).

Table 5.3 Diagnostic criteria for *Systemic Inflammatory Response Syndrome*[a]

Body temperature	>38 °C (100.4 °F) OR <36 °C (96.8 °F)
Heart rate	>90 beats/min
Respiratory rate	>20 breaths/minute OR $PaCO_2$ <32 mm of Hg
Total leukocyte count	>12,000/cmm or <4000/cmm or >10% immature forms

[a]SIRS defined by persistence of two or more of the above criteria

When obese patients are afflicted by AP, obesity augments the inflammatory response to pancreatic injury by upregulating these pro-inflammatory mediators. The cytokines produced by the adipose tissue are called adipokines and include visfatin, leptin, resistin, and adipsin, which play prominent roles in the pathogenesis of multisystem organ failure. Leptin is generally considered to be a pro-inflammatory adipokine, whereas adiponectin functions as a potent anti-inflammatory molecule. High levels of leptin contribute to the inflammatory state of adipose tissue associated with obesity (Tilg and Moschen 2006). Sharma et al. found that serum adiponectin level in AP patients was inversely correlated with BMI and organ dysfunction (Sharma et al. 2009). The dysregulation between inflammatory and anti-inflammatory mediators causes a greater systemic inflammatory response (SIR) and results in worse or even fatal outcome in obese AP patients (Table 5.3).

5.8 Clinical Features of AP

AP generally presents with constellation of symptoms and signs, but none of them is specific or pathognomonic. The diagnosis is established by the presence of abdominal pain and one of the following features: (1) an elevated (>3 times of the upper limit of normal level) serum amylase or lipase, and (2) the presence of typical features of AP on imaging studies.

The cardinal feature of AP is abdominal pain, which varies in intensity and is parallel to the severity of the disease. The pain generally starts after heavy meal or binge drinking, occurs mainly in the upper abdomen and poorly localized with radiation to the back in a band-like fashion in upper abdomen and lower thorax. The pain generally makes patients restless, and they tend to sit in forward leaning position or curl up while lying on the bed to find relief. This is in contrast to intra-abdominal events with bowel perforation where patients prefer to lie still. In most of the cases, the pain is followed by nausea and vomiting.

Patients generally have tachycardia and low pulse pressure due to dehydration. Their breathing may be rapid and shallow due to limited excursion of the diaphragm.

The signs of peritonitis are absent and muscle guarding and rebound tenderness are not seen. Many patients have low-grade pyrexia due to inflammation causing the release of many pyrogenic cytokines in the circulation. The presence of jaundice generally points toward bile duct stone in the acute setting.

In cases where there is intense inflammation, patients may experience abdominal distension due to paralytic ileus. If there is significant hemorrhage in the retroperitoneum, it can give rise to flank ecchymoses known as Gray-Turner's sign or Cullen's sign, which is periumbilical ecchymoses, if the bleeding is intraperitoneal.

These features are helpful in the diagnosis, but they fail to identify the small minority of the patients who progress to severe form of the disease. Therefore, it is of paramount importance to identify these patients and institute aggressive treatment early and referral to a tertiary care center. To address this issue, various classification and grading systems have been developed to stratify the disease and they would be discussed further down in this chapter.

5.9 Laboratory Diagnosis

5.9.1 Serum Pancreatic Enzyme Estimation

AP causes spillage of many of pancreatic enzymes into the serum. Estimation of amylase and lipase, however, have stood the test of time and used widely in the diagnosis of AP. Estimation of phospholipase A_2, carboxypeptidase A, elastase, and ribonuclease were found to be much less useful than amylase or lipase and have been abandoned. Tests that are more specific for AP but less widely available, evaluate levels of trypsinogen activation peptide and trypsinogen 2 (Whitcomb 2017).

5.9.2 Serum Amylase

The serum amylase is most frequently ordered to diagnose AP. It is cheap and widely available. It rises within 6 h of the onset of the disease and has a half-life of about 10 h. Amylase is cleared mainly by renal excretion but may remain elevated in uncomplicated disease for 3–5 days. In spite of its usefulness, it lacks specificity and rises in hosts of unrelated conditions and diseases. The pancreatic contribution of serum amylase is less than half; rest is contributed by salivary glands. The general consensus is that the level of amylase should rise by more than three times of the upper limit for diagnosis of AP (Whitcomb 2006). In many cases, urinary clearance of pancreatic enzymes from the circulation increases during pancreatitis; therefore, urinary levels may be more sensitive than serum levels. For these reasons, it is recommended that amylase concentrations also be measured in the urine. Urinary amylase levels usually remain elevated for several days after serum levels have returned to normal (Windsor 2008; Fisher William et al. 2015). Serum amylase may be elevated in diseases of salivary glands (mumps, acute parotitis), bowel perforation or strangulation, ruptured ectopic pregnancy, ovarian cysts, macroamylasemia, and renal failure. On the other hand, in few conditions it fails to show elevation in spite of AP. The serum amylase may be normal or minimally elevated in fatal pancreatitis (where massive necrosis of pancreatic parenchyma fails to produce any amylase), during a mild attack or an attack superimposed on chronic pancreatitis (because the pancreas has little remaining acinar tissue), or during recovery from AP. Another limitation to the sensitivity of serum amylase is that the level may return to normal quickly, in just a few days. Hypertriglyceridemia competitively interferes with amylase assay, so a false low level of serum amylase can be found in patients having hypertriglyceridemia. (Matull et al. 2006)

5.9.3 Serum Lipase

Lipase is another pancreatic enzyme whose serum estimation is frequently undertaken to establish the diagnosis. The serum concentration of lipase increases within 3–6 h of onset of disease and peaks within 24 h. The increased serum level stays for around 7–14 days before it comes down to the normal level. Its sensitivity is almost same as amylase but it is more specific. However, lipase also rises in bowel perforation or strangulation and patients with renal failure have elevated levels due to reduce clearance. Diabetic patients appear to have higher median lipase levels and higher cut-offs may be required for the diagnosis of AP in this population (Steinman et al. 2014). Pancreatic lipase is four times more active than amylase, and it is less affected by exocrine pancreatic deficiency occurring in patients of chronic pancreatitis. Hypertriglyceridemia does not influence the serum lipase assay as happens in the case of serum amylase. Patients taking frusemide can show increased lipase activity (Meher et al. 2015).

The combination of both serum amylase and lipase does not appear to increase the accuracy of the diagnosis. The degree of elevation of both serum amylase and lipase does not correlate with severity of AP, nor does the daily estimation of serum pancreatic enzymes help to determine clinical deterioration or resolution (Conwell and Banks 2016).

Urinary trypsinogen-2 dipstick (positive if >50 ng/mL) is a quick bedside diagnostic test which has been shown to be as accurate as serum amylase or lipase in diagnosis of AP. Although this test is not currently in the widespread use, it may be alternative to serum pancreatic enzyme estimation given its convenience (Kemppainen et al. 1997; Mayumi et al. 2012).

5.9.4 Other Hematological Tests

The total leucocyte count, which is an acute phase reactant, is frequently elevated with neutrophil predominance in AP. It generally parallels the extent of inflammation and does not denote infection in the early phases. Blood glucose is frequently elevated due to an increase in glucagon secretion and suppression of insulin. The hepatic enzymes are frequently abnormal in biliary AP and less frequently in alcoholic pancreatitis. A mild elevation of aspartate transaminase (AST) and alanine transaminase (ALT) with a significant increase in gamma-glutamyl transferase (GGT) points toward alcoholic etiology. An isolated rise in ALT to almost three times of the normal level has a positive predictive value of 95% in diagnosing acute biliary pancreatitis (Tenner et al. 1994).

5.9.5 Imaging Studies

Abdominal imaging studies like ultrasonography (US), both transabdominal and endoscopic, computerized tomography (CT) and magnetic resonance imaging (MRI) are very important in diagnosis. Moreover, they are useful in diagnosing the cause and complications of AP and help identify other intra-abdominal conditions which may cause similar features and biochemical abnormalities. CECT has a very useful role in predicting the severity and this would be discussed later.

5.9.6 Plain X-Rays

Plain abdominal films, in the emergency department, help to diagnose bowel perforation. Plain films can also show some subtle signs of pancreatitis, like distended bowel due to ileus, multiple air fluid levels in severe ileus, the overlying mid portion of transverse colon devoid of any gas due to spasm (colonic cut-off sign) and widening of duodenal C-loop due to edema of the head of pancreas.

The chest X-rays may show pleural effusion which is more common on the left side in AP and those patients who show pleural effusion at the time admission are likely to have a severe disease. There may be raised hemidiaphragm on the left side due to retroperitoneal fluid collection. Chest X-ray may also show basal atelectasis and pulmonary infiltrates (Raghu et al. 2007).

5.9.7 Ultrasonography (US)

Transabdominal US is usually the first imaging investigation in evaluation of abdominal pain. It has the advantage of easy availability, low cost, nil radiation, and can be carried out at the bedside in a critically ill patient. Moreover, it helps in diagnosis of calculus cholecystitis with a sensitivity of 95%; however, it can diagnose bile duct stones in 50% of the cases only. The pancreas appears hypoechoic and enlarged on US, in cases of AP and it may also show acute peripancreatic fluid collection. US is operator dependent and overlying gas shadow, in a distended abdomen, obscures the view, and pancreatic features may not be delineated clearly. In patients with truncal obesity, the fat attenuates the penetration of sound waves causing poor view of the pancreas in overweight patients (Bennett and Hann 2001).

Endoscopic ultrasonography (EUS) is a relatively new technique in which the ultrasonic transducer is mounted on the endoscope. The procedure is minimally invasive but has all the advantages of US. The close proximity of the endoscopic ultrasound probe to the pancreas results in superior spatial resolution compared to CT scan and MRI. This technique is rapidly evolving into a significant method for evaluation of idiopathic AP (Vila 2010). EUS has high 94% sensitivity and 95% specificity for the detection of common bile duct stones, and is superior to transabdominal ultrasound, CT, and magnetic resonance cholangiopancreatography (MRCP), with MRCP being inferior for small stones in particular. In fact, EUS has a greater than 95% negative predictive value for choledocholithiasis [Teshima CW, Sandha GS]. EUS is also useful in diagnosis of pancreas divisum and small tumors of biliary tract and pancreas which are generally missed by other methods of imaging.

5.9.8 CT Scan

CT scan of the abdomen with contrast enhancement is the gold standard for the diagnosis of AP and its local complications. However, it is usually not required for initial diagnosis since a combination of typical abdominal pain, estimation of serum enzymes, and US serve this purpose. Abdominal CT is highly useful to determine the severity and complications of pancreatitis. Pancreatic inflammation is recognized by pancreatomegaly, a smooth pancreatic margin, parenchymal inhomogeneity, peripancreatic fluid, or peripancreatic inflammation visualized as peripancreatic streakiness (Cappell 2008). CT scan is generally advised 3–4 days after the onset of symptoms because early CT may not show pancreatic necrosis. Contrast CT is not advised in patients with renal insufficiency and in those who have hypersensitivity to the intravenous contrast medium.

The CT has acquired a significant role in the prediction of severity. E J Balthazar and colleagues have devised a scoring system, called CT severity index (CSI), which is based on both unenhanced and contrast CT findings (Balthazar et al. 1990; Balthazar 2002). This would be discussed later in this chapter.

5.9.9 MRI Scan

MRI is as good as CT in detecting the acute pancreatic changes and associated local complications. MR cholangiopancreatography (MRCP) is far more sensitive than CT scan and abdominal US in detecting common duct stones. However, they are less sensitive than EUS in detecting stones in common duct. MRI is not frequently used because they are more expensive, less available, and more cumbersome. MRI is not done if the patient has a metal implant. Gadolinium contrast which is used in MRI scan is contraindicated in patients with renal insufficiency since they have been implicated in nephrogenic systemic fibrosis (NSF). This condition is a serious disabling illness with frequent fatal outcome (Thomsen 2006). MRI is valuable in patients who are hypersensitive to CT contrast media. MRI, though, useful in evaluation of the patients with recurrent idiopathic AP its usefulness is getting eroded rapidly by endoscopic US (Vila 2010; Somani et al. 2017).

5.10 Classification of AP

It is most important for the treating physician to identify at-risk patients who can develop severe AP early in the course of the disease. Many classification systems with definitions of the various stages of the disease developed in last few decades to help identify these patients. The first classification system was developed by Marseille in 1965, but in later years the Atlanta classification in year 1992, became the standard tool for this purpose. In this system, AP was divided into *mild, interstitial,* and *severe pancreatitis* based on clinical, pathologic, and radiological criteria. This system, though useful, failed to recognize the *number* of organs failing and *duration* of organ failure; the two most important prognosticators of AP. The system also does not distinguish between the predicted (based on APACHE II and Ranson Criteria) and actual severity (based on the markers of organ failure) of severe AP.

Owing to these shortcomings, the Atlanta system was revised in 2012. In this revision, AP is classified into *mild AP, moderately severe AP,* and *severe AP* (Fig. 5.2). Moreover, definitions of pancreatic and peripancreatic collections and the distinction between collections composed of fluid versus those arising from necrosis, containing solid components were also made. The Revised Atlanta 2012 also recognized that AP is an evolving disease, and the severity may change during the course of the disease. Following definitions were proposed under various headings (Banks et al. 2013).

5.10.1 Definition of Types of AP

Two types were described

1. *Interstitial edematous pancreatitis* have diffuse, or less often localized, enlargement of the pancreas due to inflammatory edema. In contrast enhanced CT scan (CECT), there is homogenous enhancement and haziness in the peripancreatic fat. There may be some amount of peripancreatic fluid collection.

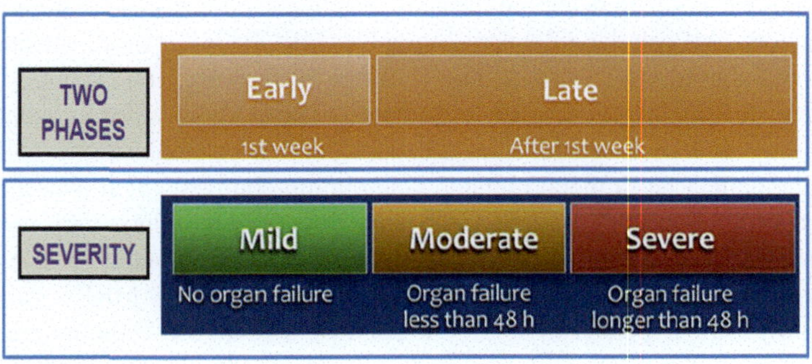

Fig. 5.2 Types of acute pancreatitis

2. *Necrotizing pancreatitis* when pancreatic or peripancreatic tissues show necrosis. The necrosis can involve both the pancreas and peripancreatic tissues and less commonly as necrosis of only the peripancreatic tissue, and rarely of the pancreatic parenchyma alone. In CECT, it manifests itself is as patchy areas of non-enhancement. CECT, early in the course of the disease, does not show up these changes; therefore, to pick necrosis accurately CT scan should be done after the first week of the onset of the disease. Pancreatic and peripancreatic necrosis can remain sterile or become infected; most of the evidence suggests no absolute correlation between the extent of necrosis and the risk of infection and duration of symptoms. Infected necrosis is rare during the first week. Pus accumulates after liquefaction of the infected necrosis and rare in occurrence. Accumulation of pus is treated in the same manner as infected necrosis; therefore, the term "*pancreatic abscess*" was dropped from this classification.

5.10.2 Definitions of the Severity of AP

In Atlanta 2012, severity has been divided into three types based on the presence or persistence of organ failure, local complications, and systemic complications. Organ failure is defined as a score of two or more of the three organ systems assessed using the modified Marshall scoring system (Tables 5.4 and 5.5). The modified Marshall scoring system has the merit of simplicity, universal applicability, and the ability to stratify disease severity easily and objectively. The modified Marshall scoring system is preferred to the SOFA (Sequential Organ Failure Assessment) scoring system, which is for patients managed in a critical care unit and which takes into account the use of inotropic and respiratory support. Both scoring methods have the

Table 5.4 Modified Marshall's scoring system for grading organ failure

Organ systems	Grade 0	Grade 1	Grade 2	Grade 3	Grade 4
Respiratory (PaO_2/FIO_2)	> 400	301–400	201–300	101–200	≤101
Renal (creatinine in mg/dL	<1.5	1.5–1.9	1.9–3.5	3.5–4.9	>5
Cardiovascular (systolic BP)	>90	≤90 Fluid responsive	<90 Not fluid responsive	<90, pH < 7.3	<90, pH < 7.2

Table 5.5 For modified Marshall's scoring system

Supplements oxygen (L/min)	FIO_2 (%)
Room air	21
2	25
4	30
6–8	40
9–10	50

advantage of being able to be used on presentation and repeated daily. *Transient* organ failure is organ failure that is present for <48 h. POF is defined as organ failure that persists for >48 h. Local complications include *peripancreatic fluid collections* and *acute necrotic collections*, while systemic complications can be related to exacerbations of underlying comorbidities related to the AP. The three definitions based on severity are as follows:

1. *Mild AP*: Mild AP is characterized by the absence of organ failure and the absence of local or systemic complications. Patients with mild AP will usually be discharged during the early phase. Patients with mild AP usually do not require pancreatic imaging, and mortality is very rare.
2. *Moderately Severe AP*: Moderately severe AP is characterized by the presence of transient organ failure or local or systemic complications in the absence of POF. Example of local complication is acute fluid collection which can cause prolonged leucocytosis, fever abdominal pain, and distension. Systemic complications mean exacerbation of a comorbid disease-like ischemic heart disease or any respiratory pathology.
3. *Severe AP*: AP with persistent organ failure (POF). Organ failure can develop anytime during the course of the disease and caused by continuous and overwhelming systemic inflammation which is characterized by SIRS Score. When SIRS is present and persistent, there is an increased risk that the pancreatitis will be complicated by POF, and the patient should be treated as if they have severe AP.

5.10.3 Definitions of Local Complications

Local complications are the fall out of the pancreatic and peripancreatic inflammation and represented by either pure fluid collection or fluid with varying amount of solid components, which are generally secondary to necrosis. Early fluid collections and necrosis have been divided on the basis of the presence of definite wall or capsule around them. Since it takes nearly 4–6 weeks for a capsule to develop around them, pancreatic pseudocyst (Fig. 5.3) and walled-off necrosis are late features of local complication.

1. *Acute Peripancreatic Fluid Collection (APFC)*: Fluid collections usually develop in the early phase of pancreatitis. In CECT, APFCs do not have a well-defined wall, are homogeneous, are confined by normal fascial planes in the retroperitoneum, and may be multiple (Fig. 5.4). Most acute fluid collections remain sterile and usually resolve spontaneously without intervention.
2. *Pancreatic Pseudocyst*: A pancreatic pseudocyst is surrounded by a well-defined wall and contains essentially no solid material. APFC if persists for more than 4 weeks it tends to develop into a pseudocyst. Occasionally, the pseudocyst is

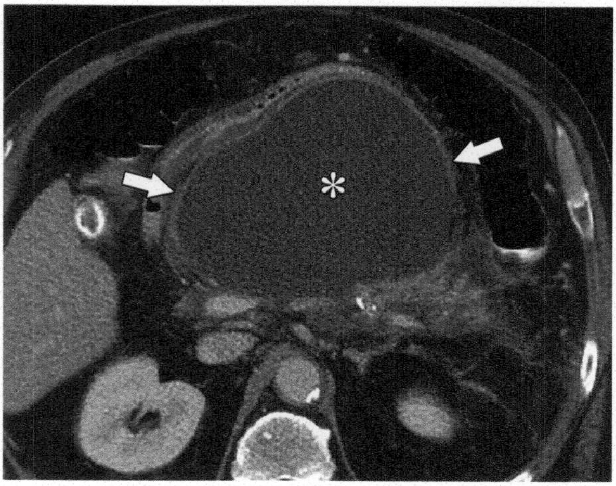

Fig. 5.3 A Pseudocyst 6 week after attack of AP (Foster et al. 2016)

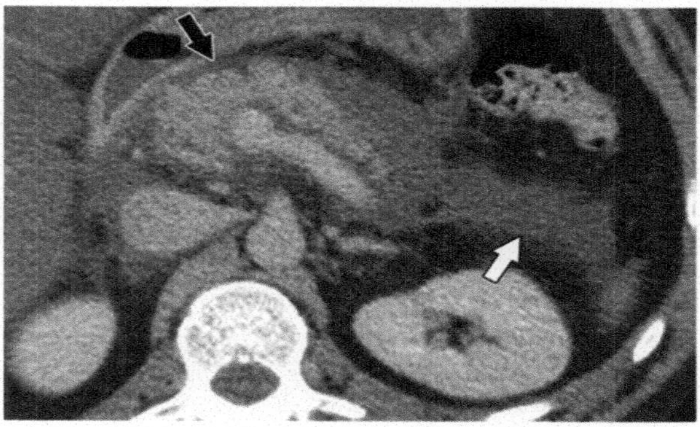

Fig. 5.4 Acute Interstitial pancreatitis with peripancreatic oedema (black arrow) and APFC (white arrow) (Foster et al. 2016)

wholly intrapancreatic. It is thought to develop due to disruption of pancreatic duct or one of its branches and contains high level of amylase which is diagnostic. CECT shows it clearly but to rule out any solid contents MRI or ultrasonography is better modality (Fig. 5.4).

3. *Acute Necrotic Collection (ANC)*: Collection of fluid with varying amount of solid materials which forms within the first 4 weeks and have CECT evidence of necrosis of pancreatic or peripancreatic tissues. They can be single or multiple or may be multiloculated. ANC must be differentiated from APFC and which can only happen after a week of the onset of disease and requires an MRI or US to rule out the presence of solid components.
4. *Walled-off Necrosis (WON)*: It is a mature, encapsulated collection of pancreatic and/or peripancreatic necrosis (Fig. 5.5) and has a well-defined inflammatory wall; usually, this maturation occurs ≥4 weeks after onset of necrotizing pancreatitis. WON is derived from necrotic pancreatic parenchyma and/or necrotic peripancreatic tissues and may be infected, may be multiple, and may be present at sites distant from the pancreas (Fig. 5.6). A US or MRI is necessary for the diagnosis because CECT may not readily distinguish solid from liquid content.

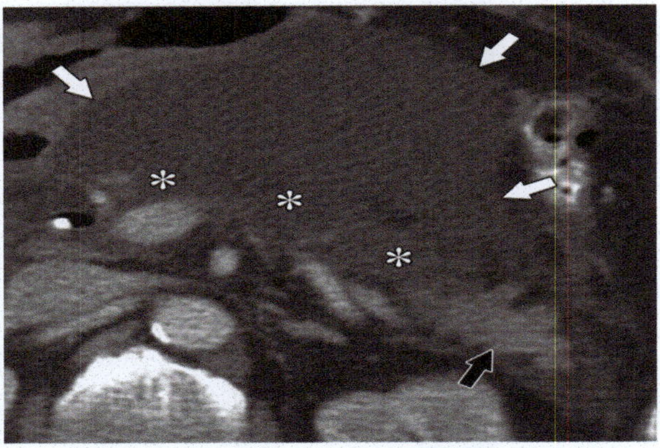

Fig. 5.5 Image of pancreatic and peripancreatic necrosis. The large Acute Necrotic Collection (white arrows) and normal enhancement of the tail (black arrow) were noted (Foster et al. 2016)

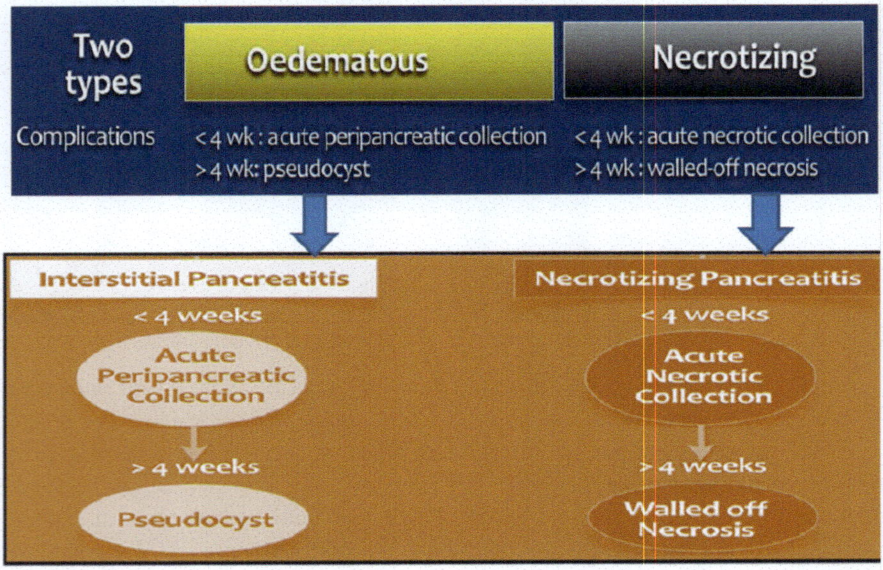

Fig. 5.6 Fluid collection in acute pancreatitis*
*Classifications of acute pancreatitis based on fluid collections

Infected Pancreatic Necrosis: It has been defined as clinical or radiological evidence of infection of ANC or WON. So infection can appear early or late in the necrotic tissues. It is manifested in CECT with the presence of gas bubbles within necrotic area or clear-cut air fluid level (Sarr et al. 1991). It may be difficult to identify them on CECT early in the course of disease due to lack of liquefaction. Hence, a CT-guided fine-needle aspiration may clinch the diagnosis (Banks 2005).

The enhancement of capsule on CECT does not indicate infection since this can happen after the maturity of capsule in WON (Foster et al. 2016).

Almost in the same time a new system, called determinant-based classification (DBC) of classification was developed in year 2013 based on a large international survey (Dellinger et al. 2012). This system was developed to address two limitations of the original Atlanta System (1992). First, it is based on actual factors of severity, such as necrosis and organ failure, rather than predictive factors of severity, such as the APACHE II and Ranson Criteria. Second, the severity stratification in this system is based on key factors that are *causally* associated with severity (pancreatic or peripancreatic necrosis with or without infection), rather than those events which may correlate with severity but are not causally associated (prolonged hospitalization, exacerbation of comorbid disease, need for endoscopic intervention).

The group classified principal determinants into two, local and systemic and defined following terms (Dellinger et al. 2012).

5.11 Local Determinants

The local determinant of severity is necrosis of the pancreas and/or peripancreatic tissue. This is covered by the term *(peri)pancreatic necrosis*.

Definitions

- *(Peri)pancreatic necrosis* is nonviable tissue located in the pancreas alone, or in the pancreas and peripancreatic tissues, or in peripancreatic tissues alone. It can be solid or semisolid (partially liquefied) and is without a radiologically defined wall.
- *Sterile (peri)pancreatic necrosis* is the absence of proven infection in necrosis.
- *Infected (peri)pancreatic necrosis* is defined when at least one of the following features is present:
  - Gas bubbles within (peri)pancreatic necrosis on computed tomography
  - A positive culture of (peri)pancreatic necrosis obtained by image-guided fine-needle aspiration
  - A positive culture of (peri)pancreatic necrosis obtained during the first drainage and/or necrosectomy.

5.12 Systemic Determinants

The systemic determinant of severity is a certain degree of distant organs dysfunction due to AP. This is covered by the term *organ failure*.

Definitions

- *Organ failure* is defined for three organ systems (cardiovascular, renal, and respiratory) on the basis of the *worst* measurement over a 24-h period. In patients without preexisting organ dysfunction, organ failure is defined as either a score of

Table 5.6 Classification of AP based on severity[a]

	Mild AP	Moderate AP	Severe AP	Critical AP
(Peri) pancreatic Necrosis	No	Sterile	Infected	Infected
	And	And/Or	Or	And
Organ failure	No	Transient	Persistent	Persistent

[a]Determinant of the classification of AP severity. AP = AP (Dellinger et al. 2012)

two or more in the assessed organ system using the SOFA (Sepsis-related Organ Failure Assessment) Score 48 or when the relevant threshold is breached, as shown below:

- *Cardiovascular:* need for inotropic agent
- *Renal:* creatinine $\geq$171 μmol/L ($\geq$2.0 mg/dL)
- *Respiratory:* $PaO_2/FiO_2 \leq 300$ mmHg ($\leq$40 kPa)

- *POF* is the evidence of organ failure in the same organ system for 48 h or more.
- *Transient organ failure* is the evidence of organ failure in the same organ system for <48 h (Table 5.6).

Both the above systems are being subjected to intensive investigations the results of which are coming in at present. However, most of the results show that DBC and Atlanta 2012 are better than Atlanta 1992 in predicting outcomes (Acevedo-Piedra et al. 2014). Few studies have shown that the DBC performed better than the Atlanta 2012 in predicting *need for interventions*, whereas the Atlanta 2012 performed better than the DBC in predicting *length of hospital stay* (Nawaz et al. 2013). The infected pancreatic necrosis (IPN) was significantly associated with hospital mortality; therefore, IPN should be a causal association with severity which the Atlanta 2012 lacks. The DBC defined only two "determinants," pancreatic necrosis and organ failure. They did not mention other potential "determinants" of severity, such as obesity, diabetes mellitus, and hypertriglyceridemia (Chen et al. 2015). Further evaluation would make it clear but for now both systems are extremely useful in stratifying patients with AP.

5.13 Disease Severity Prediction

The inflammation in AP is highly variable and has a broad range of outcomes. The wide range covers mild self-limited disease to severe form with very high morbidity and mortality. It is imperative to predict the severity of pancreatitis early in the course of disease to identify those patients who need to be treated in a specialized unit. To address this issue, many scoring systems have been devised and many biochemical and hematological markers have been studied. However, none of these scoring systems are accurate enough to predict the disease severity in the early stages. It is recommended that several clinical parameters must be recorded at the time of admission and those include age, BMI, history of duration and quantity of alcohol intake, raised hematocrit, elevated serum creatinine or blood urea nitrogen (BUN), and the presence of pleural effusion or pulmonary infiltrates at the time of admission. People older than 55 years are generally more prone to severe disease.

Obesity (BMI above 30) is an important risk factor for the disease severity and has been confirmed by many researchers (Hong et al. 2011; Martinez et al. 2006; Moran et al. 2018).

5.13.1 Ranson Score

John Ranson developed the first scoring system in 1974 which was based on 11 criteria. It is one of the most widely used scoring systems in Europe and the United States. The original paper was based on alcoholic AP, but it was modified later on to include gallstone pancreatitis as well. A low Ranson of <2 carries a low mortality of less than 2.5% and score above 3 significantly increase the mortality to more 60%. In fact, higher scores predict the necrosis and POF pretty accurately. It has few drawbacks of which the first and foremost is that it requires a second assessment after 48 h of admission for completion and has not been validated beyond 48 h. Second, it is cumbersome and need two lists on suspected etiologies. Third, the estimation of serum lactate dehydrogenase (LDH) is not available widely. Fourth, the sensitivity (40–88%) and specificity (43–90%) of predicting severe disease are low and the positive predictive value is only 50% and the negative predictive value is 90%. Therefore, it is good in predicting mild disease (Papachristou et al. 2010). The Imrie or Glasgow Score is a slightly simplified list (8 criteria) that is used commonly in the United Kingdom (Tables 5.7 and 5.8) (Ranson et al. 1974; Ranson 1982). It has similar drawbacks to the Ranson Score.

Table 5.7 Ranson's criteria for AP[a]

At admission	During the initial 48 h
Age >55 y	Heamatocrit fall >10 points
WBC >16,000/mm^3	BUN elevation >5 mg/dL
Blood glucose >200 mg/dL	Serum calcium <8 mg/dL
Serum LDH >350 IU/L	Arterial PO_2 <60 mm Hg
Serum AST >250 U/dL	Base deficit >4 mEq/L
	Estimated fluid sequestration >6

[a]AP without gall stone

Table 5.8 For acute billiary pancreatitis

At admission	During the initial 48 h
Age >70 y	Heamatocrit fall >10 points
WBC >18,000/mm	BUN elevation >2 mg/dL
Blood glucose>220 mg/dL	Serum calcium <8 mg/dL
Serum LDH >400 IU/L	Base deficit >5 mEq/L
Serum AST >250 U/dL	Estimated fluid sequestration >4 L

AST aspartate transaminase, *BUN* blood urea nitrogen, *LDH* Lactate dehydrogenase, PO_2 partial pressure of oxygen, *WBC* white blood cell count

Less than 3 positive criteria predict mild, uncomplicated disease whereas more than 6 positive criteria predict severe disease with a mortality risk of 50%

5.13.2 APACHE II

The scoring system, other than Ranson Score, which has the widest use in severity assessment of AP, is APACHE II. It stands for Acute Physiology and Chronic Health Evaluation and was originally developed for patients in ICU. The original APACHE system estimated 34 parameters for classification. It was simplified to APACHE II where 12 variables, age, and 5 organ-based chronic health points are taken into account for stratification (Knaus et al. 1985; Larvin and McMahon 1989). The advantage of APACHE II system is that it can be estimated daily. Because obesity has been shown to be an important prognostic factor of mortality from pancreatitis, the APACHE-O scale has been proposed as an improvement to APACHE II and has been shown to improve its prognostic value (Papachristou et al. 2006). A rise in value over 48 h of admission denotes progress to a severe disease. An APACHE II Score of 9 or less during the first 48 h denotes a mild disease. Patients with APACHE II scores of 13 or more have a high likelihood of severe disease and high mortality. At admission, sensitivity is 34–70%, and specificity is 76–98% in predicting a severe disease. At 48 h, sensitivity remains less than 50%, but specificity is close to 90–100%. Its drawbacks are complexity, low sensitivity on admission, and the fact that at 48 h the score is no better than other scoring systems. Like the Ranson Criteria, the APACHE II Score has its highest value in predicting mild disease (Scott and William 2010).

5.13.3 BISAP Score

The Bedside Index for Severity in AP (BISAP), a relatively new scoring system developed at Brigham & Women's Hospital in 2008 by Wu and colleagues (2008, 2009). They retrospectively developed this scoring which was based on five variables (The first letter of each parameter can also be abbreviated as BISAP): *b*lood urea nitrogen (BUN) level >25 mg/dL, *i*mpaired mental status, development of *s*ystemic inflammatory response syndrome (SIRS), *a*ge >60 years, and the presence of *p*leural effusion. The presence or absence of each parameter is recorded as 1 or 0 point. Impaired mental status was defined as any Glasgow Coma Score (GCS) of <15 and Table 5.3 as the presence of any two variables (Table 5.9). The cut-off point is assigned at three or more. Using this score, Singh et al. have found high specificity and negative predictive value for mortality. The development of organ failure, POF, and pancreatic necrosis were also predicted pretty accurately in their study (Singh et al. 2009). BISAP Score carries several important advantages over other prognostic scoring systems in AP. The first is that the score is simple to calculate, requiring only those vital signs, laboratories, and imaging that are commonly obtained at the time of presentation or within 24 h of presentation. The second is that the score was initially derived and tested using a large number of patients of AP across more than 300 hospitals of all levels; reflecting the full spectrum of

Table 5.9 Glasgow Coma Scale

Glasgow Coma Scale		
Response	**Scale**	**Score**
Eye opening response	Eyes open spontaneously	4 Points
	Eyes open to verbal command, speech, or shout	3 Points
	Eyes open to pain (not applied to face)	2 Points
	No eye opening	1 Point
Verbal response	Oriented	5 Points
	Confused conversation, but able to answer questions	4 Points
	Inappropriate responses, words discernible	3 Points
	Incomprehensible sounds or speech	2 Points
	No verbal response	1 Point
Motor response	Obeys commands for movement	6 Points
	Purposeful movement to painful stimulus	5 Points
	Withdraws from pain	4 Points
	Abnormal (spastic) flexion, decorticate posture	3 Points
	Extensor (rigid) response, decerebrate posture	2 Points
	No motor response	1 Point
Minor brain injury = 13-15 points; **Moderate brain injury** = 9-12 points; **Severe brain injury** = 3-8 points		

healthcare delivery. The third is that the score predicts in-hospital mortality (Singh et al. 2009). Accurate, yet much easier to use, this new simple scoring system appears to be useful in the early identification of patients who are at risk of developing complications and mortality.

Estimation of Table 5.3 Score can also be used as one of the scoring systems. Measurement of SIRS values, if persistent, increases the risk of organ failure and a transient or, more seriously, POF increase the risk of morbidity and mortality quite significantly (Johnson and Abu-Hillal 2004). There are several other types of scoring systems like POP (Pancreatic Outcome Prediction Score) Panc 3, HAPS (Harmless AP Score), and JSS (Japanese Severity Score of AP) were also developed (Table 5.10). There are still not a single scoring systems which can be used reliably in the assessment of AP. Few authors have speculated that the scoring systems have reached their maximum efficacy (Mounzer et al. 2012). In fact, all major guidelines, including International Association of Pancreatology (IAP)/American

Pancreas Association (APA) guidelines (Working Group IAP/APA 2013) and the American College of Gastroenterology (ACG) guidelines (Tenner et al. 2013) presently do not recommend the use of scoring systems. Only the recently published Japanese guidelines have recommended the JSS scores in their 2015 edition (Yokoe et al. 2015).

Table 5.10 Various scoring systems (in alphabetical order) with associated parameters and year of initial reports[a] (Mounzer et al. 2012)

Scoring systems	Year	Parameters
APACHE II	1989	At admission and at 48 h Temperature, MAP, heart rate, PaO_2, Arterial pH, HCO_3, Na, K, Creatinine, hematocrit, WBC, Glasgow Coma Scale (GCS), age, chronic health points
BISAP	2008	At admission and at 48 h BUN (>25 mg/dL), Impaired mental status (GCS <15), SIRS (≥2), age (>60 yrs), Pleural effusion
GLASGOW	1984	At admission and at 48 h Age (55 yrs), WBC (>15,000/mL), glucose (>180 mg dL), BUN (>45 mg/dL), PaO_2 (<60 mm Hg), Ca (<8 mg/dL), albumin (<3.2 g/dL), LDH (>600 IU/L)
HAP	2009	At admission and at 48 h Abdominal tenderness, hematocrit (>43 mg/dL for men and >39.6 mg dL for women), creatinine (>2 mg/dL)
JSS	2009	At admission and at 48 h Base excess (≤3 mEq/L), PaO_2 (≤60 mmHg or respiratory failure), BUN (≥40 mg/dL) or creatinine (≤2 mg/dL), LDH (≥2 × upper limit of normal), platelet (≤100,000/mm^3), calcium (≤7.5 mg/dL), CRP (≥15 mg/dL), SIRS (≥3), age (≥70 yrs)
Panc 3	2007	At admission and at 48 h Hematocrit (>44 mg/dL), BMI (>30 kg/m^2), pleural effusion
POP	2007	At admission and at 48 h Age, MAP, PaO_2/FiO_2, arterial pH, BUN, calcium
Ranson	1974	*At admission*: age (>55 yr), WBC (>16,000/mL), glucose (>200 mg/dL), LDH (>350 IU/mL), AST (>250 IU/mL) *At 48 h*: hematocrit (decrease >10%), BUN (increase >5 mg/dL), calcium (<8 mg/dL), PaO_2 (>60 mm Hg), base deficit (>4 mEq/L), fluid sequestration (>6 L)
SIRS	2006	At admission and at 48 h Temperature (<36 °C or >38 °C), heart rate (>90/min), respiratory rate (>20/min or $PaCO_2$ <32 mmHg), WBC (<4000/mm^3, >12,000/mm^3 or >10% bands)

[a]*AST* aspartate aminotransferase, *BMI* body mass index, *CRP* C-reactive protein, *FiO_2* fraction of inspired oxygen, *LDH* lactate dehydrogenase, *MAP* mean arterial pressure, *POP* pancreatitis outcome prediction, *JSS* Japanese Severity Score, *HAPS* harmless AP score, *WBC* white blood cell count

5.14 Laboratory Markers

Estimation, either serial or onetime, of serum amylase and lipase does not correlate with severity of the disease. Many tests have been evaluated to predict the severity. The important ones are discussed here.

5.14.1 C-Reactive Protein

C-reactive protein is a homopentameric acute phase inflammatory protein, a highly conserved plasma protein that was initially discovered in 1930 by Tillet and Francis while investigating the sera of patients suffering from the acute stage of pneumococcal infection and was named for its reaction with the capsular C-polysaccharide of *Diplococcus pneumoniae.* CRP is a nonspecific acute phase reactant that is a member of the pentraxin proteins, which are pattern recognition proteins and an integral part of the innate immune system. CRP is synthesized primarily in liver hepatocytes but also by smooth muscle cells, macrophages, endothelial cells, lymphocytes, and adipocytes. CRP is produced as a homopentameric protein, termed native CRP (nCRP), which can irreversibly dissociate at sites of inflammation, tissue damage, and infection into five separate monomers, termed monomeric CRP (mCRP). Evidence indicates that nCRP often tends to exhibit more anti-inflammatory activities compared to mCRP. The nCRP isoform activates the classical complement pathway, induces phagocytosis, promotes apoptosis, and inhibits nitric oxide (NO) production, while mCRP stimulates chemotaxis and recruitment of circulating leukocytes to areas of inflammation, induces NO production, and can delay apoptosis (Sproston and Ashworth 2018).

CRP is synthesized initially as monomers and then assembled into the pentamer in the endoplasmic reticulum of the source cell. In hepatocytes, this protein is stored in the endoplasmic reticulum by binding to two carboxylesterases, gp60a and gp50b (Macintyre et al. 1994). In a resting state, CRP is released slowly from the endoplasmic reticulum, but following an increase in inflammatory cytokine levels, the binding CRP to the carboxylesterases decreases and CRP is secreted rapidly (Du Clos and Mold 2004). The stimulation of CRP synthesis and secretion mainly occurs in response to pro-inflammatory cytokines, most notably IL-6 and to a lesser extent by IL-1 and tumor necrosis alpha (TNF-α) (Sproston and Ashworth 2018).

CRP gene in human can be located at 1q23.2 on the long arm of chromosome 1, and there have been no allelic variations or genetic deficiencies discovered for this gene till now although some polymorphisms have been identified. For example, up to 50% of the baseline variance in CRP is associated with the number of dinucleotide repeats found in an intronic region of the gene (Hage and Szalai 2007).

Quantitative estimation of CRP level in serum is in widespread clinical use as a sensitive marker of inflammation. CRP is generally measured by *immunomephelometric* assay on a Behring Nephelometer II analyzer. The detection limit for CRP was 0.17 mg/L, and the measuring range was 0.175–1100 mg/L, according to the manufacturer. Another method which is used in detection of a very low level of CRP, the so-called hs-CRP, is *immunoturbidimetric* assay, performed on a Hitachi 717 automated analyzer. The manufacturer claimed the detection limit to be 0.1 mg/L. It should be noted that the hs-CRP is not a different analyte from the conventional CRP. It is just the CRP level measured with a very sensitive assay method to detect a very low level (Pepys and Gideon 2003).

The sole determinant of CRP level in the serum is its rate of synthesis which is directly proportional to the intensity of infection. In normal person without any disease the level of CRP is less than 2 mg/dL but in few conditions with low level of inflammation and tissue injury the level may be up to 10 mg/dL. This is seen in smokers, people with cardiac ischemia, uremia, and obesity. The higher than normal CRP in obese subjects may be due to secretion of IL6 by adipocytes. The baseline IL6 production has a small contribution from adipocytes normally, and in obese person this contribution becomes substantial due substantial adiposity. The level of CRP in people with higher BMI is also associated with insulin resistance. The CRP level returns to normal after weight reduction (Pepys and Gideon 2003). CRP level start rising within 12 h of start of inflammation and reach a peak in 48–72 h. In condition with mild inflammations like, skin infections, bronchitis, periodontitis, and cystitis, the level rises moderately and stays between 50 and 100 mg/dL. In severe infections with tissue necrosis, the level may rise up to 1100 mg/dL (Markanday 2015).

CRP has a role in the clearance of bacteria and of dying and altered cells and might also have more complex immunomodulatory functions. The plasma half-life is about 18 h and complexed-CRP is catabolized by hepatocytes in vivo and rapidly cleared from the circulation (Ansar and Ghosh 2013). CRP assays have a distinct advantage over other inflammatory markers as it remains unaffected by drugs and does not show variation with eating or diurnal rhythm. However, it may fail to rise in severe hepatic diseases. It is mainly used in clinical medicine as a screening tool for inflammatory diseases, monitoring of the response to the treatments, detection of intercurrent infection in immunocompromised individuals, and in few specific diseases with absent or feeble acute phase response (e.g., systemic lupus erythematosus (SLE) and Crohn's disease) (Pepys and Gideon 2003).

The main role of CRP in inflammation is the activation of the C1q molecule in the complement pathway leading to the opsonization of pathogens. It can also initiate cell-mediated pathways by activating complement as well as to binding to Fc receptors of IgG. CRP binds to Fc receptors with the resulting interaction leading to the release of pro-inflammatory cytokines. CRP also has the ability to recognize self and foreign molecules based on the pattern recognition, something that other activators of complement such as IgG cannot achieve because these molecules only recognize distinct antigenic epitopes (Sproston and Ashworth 2018).

It is the most frequently used single biomarker for assessment of severity in AP today and regarded as gold standard. This is because it is inexpensive, widely available, and easy to measure. The rise in level of CRP after initiation of AP is proportional to the degree of inflammation and hence relates directly to the severity of the disease. The main stimuli for CRP synthesis and secretion in AP are cytokines especially IL6; however, rise in serum level of CRP lags behind these biomarkers. The CRP level starts rising soon after the onset of AP and reaches its peak in 48–72 h. Due to its lack of specificity CRP is not used as diagnostic test for AP or infected pancreatic necrosis. However, it is used extensively for prediction of severity and has been regarded as gold standard by many authors in this respect.

In one of the earliest studies, in 1984, comprising of 55 patients with AP, CRP estimation was compared with α_1 protease inhibitor and antichymotrypsin, the other two acute phase reactants. They found that CRP is much more accurate in identifying the patients who are at the higher risk of developing the severe form of the disease. They also noticed that the estimation of CRP is only useful if done late in the first week after admission (Mayer et al. 1984).

After this there were many studies citing the usefulness of CRP in assessment of severity of AP. Almost all studies in this period showed that the efficacy of CRP estimation in predicting the course is maximum if measured between 48 and 72 h of the onset of disease. The cut-off values cited by different authors measured from 125 mg/L to 192 mg/L. The *Santorini Consensus (1999)*, comprising of a group of experts in the subject from all over the globe, recommended 150 mg as the cut-off value to identify the patients who are most likely to develop severe AP. They also recommended that this estimation must be made 48 h from the *onset of symptoms* (Beger and Isenmann 1999). This value was further endorsed by the *Japanese Severity Score* (JSS) published in year 2008. JSS has been shown to be most accurate in predicting severe AP (Mounzer et al. 2012).

In the year 2000, CRP was compared with serum amyloid A, an acute phase reactant in a study consisting of 66 patients. The authors concluded that Amyloid A is useful in discriminating acute necrotizing pancreatitis from the acute edematous variety; however, CRP provided an earlier differentiation between both of these varieties and better overall accuracy (Rau et al. 2000).

A comparative study of various scoring systems and disease markers were carried out at IMS, BHU, India, in year 2012. The authors studied a cohort of 72 patients with AP, and the results were analyzed after comparing different prognostic markers in prediction of severe acute pancreatitis (SAP), organ failure (OF), pancreatic necrosis (PNEC), length of hospital stay (LOHS), requirement of ICU admission (ICUA), and mortality in acute pancreatitis. Their conclusions include that a cut-off level of 150 mg/L within the first 48 h of symptom onset has sensitivity and specificity of 80–86% and 61–84%, respectively, for SAP an accuracy >80% for necrotizing pancreatitis. The incidence of SAP, OF, PNEC, ICUA, MORT, and prolonged LOHS was found to be 100%, 76.0%, 68.0%, 24.0%, and 24.0% with an average length of hospital stay of 13.8 days, respectively. CRP had the highest sensitivity (100%), NPV (100%), and specificity (81.4%) for pancreatic necrosis, followed by sensitivity of 86.2% and specificity and PPV of 100% for prediction of SAP in that study. The conclusion drawn by authors stated that as a whole CRP is a good marker for prediction of complications and mortality in acute pancreatitis. The AUC for prediction for PNEC was higher for CRP 0.90 (0.82–0.77) (Khanna et al. 2013).

In papers published before year 2000, the cut-off mark for CRP was not agreed upon, neither there was a consensus on the most appropriate time for CRP estimation. Keeping these difficulties in mind C J Young et al. carried out a systematic review of the published papers from 1950 to January 2013. They made a search of EMBASE, Cochrane Database, and Pubmed. In their search, the end points of the disease were POF and infected pancreatic and/or peripancreatic necrosis (IPN); both were subsets of severe AP. On comparing the highest values of CRP, they concluded that among the predictors of IPN in all patients with acute pancreatitis, the accuracies of procalcitonin and C-reactive protein (CRP) levels were most frequently evaluated (in 5 and 3 cohorts, respectively) and no significant difference between the two predictors was observed in terms of sensitivity, specificity, positive likelihood ratio, negative likelihood ratio, and diagnostic odds ratio (DOR). But for the prediction of POF and IPN together, they found CRP with a cut-off of 430 mg/L had a sensitivity of 0.50, a specificity of 0.99, a positive likelihood ratio of 50, and a negative likelihood ratio of 0.51. Whereas, procalcitonin with a cut-off of 3.5 ng/mL had a sensitivity and specificity of 0.90 and 0.89, respectively, a positive likelihood ratio of 8.18 and negative likelihood ratio of 0.11 (Young et al. 2014).

In a single center study, Cardoso et al. noted that the best time for CRP estimation in AP is at 48 h. This retrospective cohort of 379 patients also showed that a value of 190 mg/dL at 48 h is the best predictor of severe AP and pancreatic necrosis, whereas a value of 170 mg/dL predicts the in-hospital mortality (Cardoso et al. 2015).

There is substantial evidence of CRP estimation as a helpful marker in AP across the literature. However, they also differ substantially in terms of cut-off absolute value and the best time for estimation after the onset of the disease. The Atlanta classification was revised in 2012, as a result of improved understanding of the pathophysiology of AP. Consequently, the inclusion of a moderate AP subtype has been adapted to include those patients that have transient organ dysfunction. Since most of the studies were done prior to Atlanta 2012, they are less applicable in predicting severity in AP. Very recently a retrospective study, carried out in a tertiary care hospital In Australia, tried to answer this question in light of most acceptable definitions. The authors noted that *interval change* or rise of CRP value by >90 mg/dL from the time of admission to 48 h (*Positive Likelihood Ratio* 2.15, *Negative Likelihood Ratio* 0.26) or an absolute value of >190 mg/dL at 48 h (*Positive Likelihood Ratio* 2.72, *Negative Likelihood Ratio* 0.24) predict severe disease most accurately. The authors argued that the previous consensus of 150 mg/dL at 48 h was decided when the presence of local complications, and transient (<48 h) organ failure, would have constituted severe AP. Currently, the revised Atlanta classification would consider this moderately severe pancreatitis; a prognostically distinct subgroup (Stirling et al. 2017).

Several serum inflammatory markers have been studied individually as predictors of severe disease, including interleukin-1, 6, 8, procalcitonin, polymorphonuclear elastase, trypsinogen activation peptide, and others. Among all, CRP is the most promising one, with many studies showing a correlation of its high levels with pancreatic necrosis development and severe AP course (Koutroumpakis et al. 2015). However, there are at least three major drawbacks of CRP; first it takes 72 h to reach its peak after the onset of AP; second, it is nonspecific marker of inflammation and all possible infections must be excluded before interpreting the result; third, it may remain low or normal in hepatic parenchymal diseases like cirrhosis irrespective of the degree of inflammation. The prognostic role of CRP in AP is noted in Table 5.11.

Table 5.11 Prognostic role of CRP

Prognostic role of CRP in AP	References
The first study on the function of CRP in the prediction of the outcome of AP. CRP can be used as a tool in assessment and management of AP They concluded that increased levels of CRP may predict the severity of inflammation in AP. CRP level are better predictor than WBC count, ESR, temperature, and the concentrations of antiproteases	Mayer et al. (1984)
CRP is a better laboratory marker than ESR in predicting severity of AP	Pongprasobchai et al. (2010)
High CRP level and increased computed tomography (CT) severity index assessed on 150 patients presented on admission into the ICU ward. CRP and CT severity indexes are good predictors of the mortality risk in patients with AP. High CRP level, respiratory and renal failure, and male gender are correlated with longer ICU stay	Makela et al. (2007)
The early risk assessment of AP by comparing biochemical markers and various scores were evaluated. Both APACHE II and BISAP are comparable in predicting outcome in AP. They also calculated a predictive value of CRP, with an AUC of 0.8218 for severe AP	Vasudevan et al. (2018)
CRP and IL-6 have been assessed in early detection of severity of AP and pancreatic necrosis with better results. Ranson Score and APACHE II was used in predicting AP-associated mortality	Khanna et al. (2013)
CRP and IL-6 were evaluated clinically to demonstrate and differentiate between mild and non-mild AP early in the course of AP. Harmless Acute Pancreatitis Score (HAPS) is an inferior predictor than CRP and IL-6	Sternby et al. (2017)
CRP and BISAP have been assessed in early risk evaluation of patients with AP. CRP at 24 h revealed good prognostic precision for in-hospital mortality in AP. CRP was selected in prediction of lowest risk patients	Cardoso et al. (2015)
CRP was the best discriminating marker among different serum proteins. CRP has better detection rates for pancreatic necrosis in patients suffering from AP	Büchler et al. (1986)
CRP assessed clinically as the predictors of severe outcome of AP in day 3. The serum CRP, alpha 2 macroglobulin (A2M), alpha 1 antiprotease (A1AP), amylase, and WBC count were measured on days 1, 3, and 7. When associating the relation between severe clinical outcomes and mild outcomes in AP, serum CRP levels were higher on all 3 days	Leese et al. (1988)
The CRP/albumin ratio is a unique promising marker which is easy-to-measure, repeatable, noninvasive inflammation-based prognostic score in AP. Patients with CRP/albumin ratio above >16.28 had a 19.3 times greater chance of death	Kaplan et al. (2017)

5.14.2 Blood Urea Nitrogen (BUN)

Blood urea nitrogen (BUN) actually measure the renal function along with creatinine level. Urea is synthesized in the liver to detoxify the amino acid (i.e., protein) metabolism products and excreted by kidneys. They rise in renal insufficiency and in dehydrated patients. In cases of AP, the rise in BUN means inadequate rehydration and regarded as one of the earliest severity markers. Early institution of rigorous fluid resuscitation is the single most important step in restoring the pancreatic microcirculation and prevention of pancreatic necrosis. Few patients may remain refractory to fluid therapy and need advanced setup to monitor their cardiovascular system so that further escalation in fluid infusion does not lead to cardiac or respiratory failure. BUN and hematocrit both measure the intravascular volume and routinely done in all patients. Therefore, either test could be useful in monitoring early response to initial fluid resuscitation.

Many studies have employed serial BUN estimation to stratify patients, and it has figured as one of the parameters in almost all of the major scoring systems (exception: APACHE II). In a landmark study, in year 2009, Wu and colleagues performed an observational cohort study on hospital-based adult patients with the diagnosis of AP. The study included patients from 69 of US hospitals. Their salient findings were: (1) they demonstrated that mean BUN levels were persistently elevated among non-survivors versus survivors of AP during the first 48 h of hospitalization. (2) An elevated admission BUN and a rise in BUN within the first 24 h of hospitalization were both independently associated with increased mortality after controlling for the effect of age, gender, and hemoglobin. And, (3) serial BUN measurement was the most accurate single prognostic marker for in-hospital mortality compared with serial measurement of alternative routinely collected laboratory tests at admission and 24 and 48 h. In this study, an important finding was the strong association between the extent of BUN increase at 24 h and the risk of mortality, irrespective of admission BUN. For every 5 mg/dL increase in BUN during the first 24 h, the age- and gender-adjusted odds ratio for mortality increased by 2.2 (Wu et al. 2008, 2009). In year 2015, another study was published by similar group. In this study, clinical data of 1612 patients from three independent cohorts (two hospitals from the United States and rest from Dutch Pancreatitis Study Group) were taken and analyzed. This study also confirms their earlier findings and confirms serial BUN estimation along with hematocrit as the most accurate predictor of POF and pancreatic necrosis (Koutroumpakis et al. 2015).

5.14.3 Hematocrit

Hematocrit estimation also measures the status of fluid in the intravascular compartment. It is the percentage of volume of total red cells in the vascular compartment. A rapid loss of fluid in acute inflammation due to capillary leak causes the elevation in the value of hematocrit. A hematocrit above 44% at the time of admission is regarded by many authors as significant and its accuracy rises many folds if combined with the value of BUN. In a large study, which has been cited above, the authors have concluded that half of the patients who have an admission hematocrit of >44% and rise in BUN level in first 24 h develop pancreatic necrosis organ failure. On the other end, approximately a tenth of patients with admission hematocrit <44% and no rise in BUN developed POF and pancreatic necrosis (Koutroumpakis et al. 2015). In 2007, Brown et al. combined an admission value of hematocrit as >44%, with BMI > 30 kg/m^2 and pleural effusion to develop Panc 3 Score (Brown et al. 2007).

5.14.4 Procalcitonin

Procalcitonin is the prohormone of calcitonin, which is normally produced by type C cells of the thyroid gland. It presents a trace amount (<0.05 ng/mL) in a normal person's serum. However, in a bacterial infection and several inflammations it rises exponentially to 100–150 folds. In these situations, some inflammatory chemokines and cytokines like TNF-α, IL 6, and IL 1b can induce many extra thyroidal tissues including liver, kidney pancreas, intestine, and leucocytes to produce calcitonin. Interestingly, its secretion is suppressed in viral infections probably due to action of interferons. Therefore, procalcitonin is a very good marker of bacterial infection.

The rise in procalcitonin parallels with bacterial infection, and it quite accurately predicts infected necrosis in patients with severe pancreatitis. The best cut-off levels for predicting infected necrosis or persisting pancreatic sepsis were 1.8 ng/mL for procalcitonin. If these cut-off levels were reached on at least two days, sensitivity, specificity, and accuracy for the prediction of infected necrosis were 94%, 91%, and 92% (Rau et al. 1997). Monitoring of procalcitonin level also helps in assessment of effectiveness of antibiotic and other therapy for infected pancreatic necrosis. Since level of procalcitonin rises in any kind of bacterial infection, one needs to exclude other infections.

5.14.5 Other Biomarkers

Many studies have cited the usefulness of many other inflammatory markers in an effort to predict the course of AP; of these IL6 and IL8 have been studied most extensively. Among enzymes trypsinogen and urinary trypsinogen activation peptide (TAP) have been the focus of many researches. Their usefulness has not been established yet and moreover, their estimation is not widely available except for few selected centers.

5.15 Radiological Scoring System

Contrast-enhanced dynamic CT scan has two significant applications in the diagnosis and staging of AP. First, it helps in diagnosis of doubtful cases, and the second is identification of pancreatic or peripancreatic fluid collection and the presence and extent of necrosis. The CT features of AP take few days to evolve, hence in early phase, interstitial edema and fluid collections are recognized. Subsequently, with progression of disease, pseudocysts, acute necrotic collections, and walled-off pancreatic necrosis may develop. The CT defined features of AP have been incorporated in the 2012 Atlanta Classification also.

Besides many clinical and biochemical scoring as briefed earlier, there were few radiological marking systems were also developed over last 3 decades. Among them, the CT severity index (CTSI), designed by Emil Balthazar et al. (1990), is the most widely adopted system for clinical purposes (Balthazar et al. 1990). The CTSI is a numeric scoring system that combines a quantification of pancreatic and extrapancreatic inflammation with the extent of pancreatic necrosis. The CTSI was initially based on the unenhanced CT features of AP. In 1990, the same group included the features of pancreatic necrosis by using contrast-enhanced films. The severity of AP is classified into five grades (0-4) on unenhanced CT, whereas the degree of necrosis is measured and given a score of 0–6. They found an excellent correlation between necrosis, length of hospitalization, development of complications, and death: patients with a CTSI > 3 showed a mortality rate of 3% and a morbidity rate of 8%. In patients with the highest classification (CTSI > 7), however, the mortality rate was 92% with a 17% morbidity rate. In 2004, a modified scoring system was proposed by few investigators Mortele (2004). The modified CTSI or MCTSI added two extra points for extrapancreatic complications (one or more of the following: pleural effusion, ascites, vascular complications, parenchymal complications, or gastrointestinal tract involvement) and change the maximum score of pancreatic necrosis to 4 (i.e., any necrosis > 30%) instead of 6 as in original CTSI. However, further researches did not find this modified system superior to the original CTSI (Bollen et al. 2011).

The subtle inflammatory changes in extrapancreatic tissues were often ignored in these scoring systems. To address this issue, a new system was designed in 2007 and was named as extrapancreatic inflammation on CT Score (EPIC) (Tables 5.12 and 5.13).

Besides not requiring the contrast enhancement, this CT-based scoring has the added advantage that it can be done within first 24 h of the onset of the disease. It was shown to be 100% sensitive and 71% specific in predicting the severe AP and mortality if the score is 4 or more (De Waele et al. 2007).

Besides the above-mentioned three CT-based scoring systems, few other systems were also reported. Pancreatic Size Index (PSI) Score, mesenteric oedema, peritoneal fluid (MOP) scores, and extrapancreatic scores; all these three are based on unenhanced CT evaluation.

Table 5.12 CT Severity index (Balthazar et al. 1990)

CT severity index	Points
1. Grade of AP based on *non contrast* CT findings	
A. Normal pancreas	0
B. Focal or diffuse enlargement, including contour irregularities and inhomogeneous parenchymal attenuation	1
	2
C. Grade B plus peripancreatic inflammation	3
D. Grade C plus a single fluid collection	4
E. Grade C plus two or more fluid collections or retroperitoneal gas	
2. Degree of pancreatic necrosis based on *contrast* CT findings	
A. No pancreatic necrosis	0
B. Necrosis of up to one third of pancreas	2
C. Necrosis of up to one half of pancreas	4
D. Necrosis of more than one half of the pancreas	6

Table 5.13 EPIC score (non contrast CT scan) (Delrue et al. 2010)

Prognostic Indicators	Points
Pleural effusion	
None	0
Unilateral	1
Bilateral	2
Ascites in either one of these locations: perisplenic, perihepatic, interloop, or pelvis	
None	0
One location	1
More than one location	2
Retroperitoneal inflammation	
None	0
Unilateral	1
Bilateral	2
Mesenteric Inflammation	
Absent	0
Present	1

In a series of 307 patients, TL Bollen et al. (2012) carried out an evaluation all these 6 CT-based systems with APACHE II and BISAP. They concluded that all these scoring systems had been shown to correlate with morbidity and mortality; it remained difficult to accurately identify individual patients who would develop clinically severe disease on admission or early in the course of their hospitalization. They further opined that CT should not be used in initial evaluation of AP; instead a clinical scoring system (BISAP, SIRS, and APACHE II) should be used serially. Presently, CT is recommended in patients with equivocal diagnosis, for those with POF, for those who have the persistent SIRS or sepsis, for those who do not improve within first week of the onset, and for those with probable infected pancreatic necrosis.

5.16 Management of AP

Most of the cases of AP are mild and resolve without much medical support; however, about 20% of the cases may progress into a severe course which is hard to identify in the early phases after the onset. It is helpful to identify three phases in the evolution of this disease: I. *Initiation*, which starts in pancreatic acinar cells leading to widespread inflammation within pancreas; II. *Perpetuation*; is the inflammation spilling over to other organs especially to intestine and adipose tissues causing persistent organ dysfunctions and III. *Secondary escalation*; which is due to infection of the pancreatic necrosis (Garg 2017).

Owing to its divergent and unpredictable course, it is extremely important to identify those patients who are likely progress into a severe form. Those who persistently score higher on simple scoring system, like SIRS and BISAPS, are likely to develop organ dysfunction or failure and should be managed in the intensive care unit (ICU) or high dependency unit (HDU), depending upon the level of monitoring and support they require. The need for critical care management may occur early in the clinical course, usually as a result of the acute inflammatory response progressing to organ failure, or late, often as a consequence of superimposed sepsis either pancreatic or extrapancreatic. Pathophysiology of the disease, rather than anatomical details of the pathology, is the key determinant of the need for a higher level of support.

5.16.1 Fluid Resuscitation

Vigorous fluid resuscitation is the cornerstone of therapy during the early management of AP. It is important to start the fluid as early as possible. Intravenous fluid resuscitation is necessary to help support the pancreatic microcirculation, as alteration in the flow to this complex network of perfusing arterioles and capillaries plays an important role in the pathogenesis of AP. Several causes of pancreatic microcirculation disruption have been identified and include hypovolemia, increased capillary permeability, hypercoagulability causing microthrombi and damage of vascular endothelium from oxidative stress and free radicals. The volume and rate of fluid infusion is still debated, but many studies have shown that an intensive early infusion is beneficial in terms of outcome.

It is known that aggressive fluid therapy is most effective in the first 24–48 h. Almost all guidelines on AP have recommended isotonic crystalloid infusion in this period. The initial recommended volume varies between 250 mL and 500 mL/h (Tenner et al. 2013) or 5–10 mL/kg of body weight/h (Working Group IAP/APA 2013). This aggressive infusion of fluid must be done after checking the cardiovascular, renal, and pulmonary status of the patient since fluid overload in the compromised subjects can cause serious hemodynamic problems. To overcome this, *goal directed therapy* is recommended which means titration of intravenous fluids to specific clinical and biochemical targets of perfusion (e.g., heart rate, mean arterial pressure, central venous pressure, urine output, BUN, and hematocrit). It is important to check adequacy of fluid therapy very early during the treatment and if the patient seems to be fluid refractory after 6 h of initiation, the need for advance invasive or noninvasive cardiovascular support may be needed (Jin et al. 2018). These should be undertaken in well-equipped ICU setup.

Colloids use has been consistently discouraged and many guidelines and authors have cautioned against the use of hydroxyl ethyl starch (HES). Although use of HES was found to have detrimental effects in patients with severe sepsis, there are sufficient concerns of its use in AP also. Among isotonic crystalloid Ringer's lactate or lactated Ringer is favored over isotonic or normal saline (Wu et al. 2011). If large volume of normal saline is used in resuscitation, it may cause hyperchloremic acidosis which cause further increase in pathologic zymogen activation in the acinar cells and further deterioration of the disease (Wu et al. 2011).

5.16.2 Pain Control

It is extremely important to control the pain in any form of a disease, but unfortunately it is often neglected and junior most staffs are generally given the responsibility of managing it. A good pain control has a direct bearing on the outcome of the disease. Besides its humanitarian value, it has several other physiological functions. First, an uncontrolled pain increases the secretion of many stress-related hormones like catecholamines and glucagon which could add to the metabolic derangement. Second, adequate pain control makes lung expansion easier and reduces the chance of pulmonary collapse and infection. Third, an increase in heart rate secondary to anxiety adds to cardiac stress and increases the myocardial oxygen demand. And lastly, good pain control helps in early mobilization of the patients.

Opioids have been the most useful drug in pain control. There was a concern about the use of opioids, particularly morphine, in AP due to its spasmodic effects on sphincters. However, A Cochrane Review in 2013 examined five randomized trials comparing different analgesics in AP and found no evidence of increased complications related to opioid use (Basurto Ona et al. 2013). However, in a recent retrospective cohort study of 4307 patients over a period of 18 years, Bechien Wu and colleagues found that increase baseline use of opioids increases the hospital stay of the patients. It is speculated that it may be the severity of the disease itself which causes both the longer hospital stay and increase the requirement of opioids. Another reason is that increased use of opioids leads to prolonged paresis of the gut causing delay in the recovery. In fact, animal studies have shown that opioids exacerbate the inflammation and inhibit the tissue recovery in the setting of AP (Wu et al. 2019).

Supplementing opioids with paracetamol or non-steroidal analgesics reduces the required dosage. Patient-controlled analgesia (PCA) is the most appropriate method of administering the drugs. Thoracic and upper abdominal epidural analgesia is also very attractive in these patients, but its invasiveness and need for a close respiratory monitoring have dampen the initial enthusiasm.

5.16.3 Antibiotics

Prophylactic antibiotics use in AP is highly controversial topic, and the debate is still going on about its advisability. The only consensus all guidelines and most of the studies have reached is that antibiotics have no role in mild AP and does not expedite the course of this self-limiting form. However, there are conflicting reports about usefulness of antibiotic prophylaxis in severe AP.

The severe AP has two phases: the first two weeks is characterized by POF and persistent SIRS. Most of these patients harbor pancreatic necrosis, and about 40% of them develop infection of the necrosis; this in turn causes a second escalation of the inflammation and carries a significant mortality. It is not very clear how pathogens reach sterile necrosis. Hematogenous route is suspected but the paradox is that necrotic tissues lack blood circulation. Other routes suspected are through the biliary system, ascending from the duodenum via the main pancreatic duct, or through transmural colonic migration via translocation of the colonic bacteria to the lymphatics (Mourad et al. 2017). Many authors have recommended prophylactic antibiotics in patients with severe AP. The use of antibiotics should be instituted as soon as there is a secondary elevation in inflammatory markers (Mourad et al. 2017). In summary, antibiotic therapy should only be considered as treatment in patients with necrotizing pancreatitis and a strong clinical suspicion of infected necrosis or proven sepsis.

The antibiotics are important once an infected pancreatic necrosis is established; however, it is only one facet of management, and these patients require some form of intervention to debride and drain the infected necrosis.

5.16.4 Nutrition

Early oral or enteral nutrition is recommended by all major guidelines on AP. In mild AP, oral feedings can be started immediately if there is no nausea and vomiting, and the abdominal pain has resolved. Initiation of feeding with a low-fat solid diet appears as safe as a clear liquid diet in mild AP.

The routine use of nasogastric tube suction is not indicated unless there is paralytic ileus with abdominal distension and persistent vomiting.

Severe AP is associated with a significant catabolic state, and nutritional support is of paramount importance early in the course of the disease. The enteral route has several advantages over parenteral nutrition in critically ill patients for several reasons:

1. Enteral nutrition is much cheaper than parenteral both in terms of formulation and maintenance.
2. Enteral nutrition is thought to contribute to gut barrier function and can reduce the incidence of infected pancreatic necrosis and organ failure by at least two folds (Petrov et al. 2008).
3. Gastric colonization by pathogenic bacteria, which may also increase the risk of septic complications, is reduced with enteral nutritional support.
4. Enteral nutrition can better regulate the acute phase response and maintain visceral protein metabolism, while also potentially inhibiting the toxic reaction of splenic cells (Hegazi and DeWitt 2014).
5. There are more complications associated with the parenteral route, including line sepsis and several specific metabolic disorders.

For enteral nutrition to exert its immune and other beneficial effects, the patient's tolerance to the fed formula is the key. Tube feeding-associated intolerance is common, occurring in approximately 50% of the patients. Due to the associated exocrine pancreatic insufficiency, patients with severe AP are at even higher risk of feeding intolerance. The nutrient composition of enteral nutrition formulas may help enhance the tolerance to the formula and increases the likelihood of adherence for patients to achieve their nutritional goal. Major clinical and scientific societies consistently recommend feeding patients with severe AP with peptide-based and high medium chain triglycerides formulas. Interestingly, medium chain triglycerides have been shown to exert anti-inflammatory effects in animal models of inflammatory bowel diseases (Hegazi and DeWitt 2014).

In a minority of the patients with severe AP, the enteral may not be tolerated even after manipulation of the formulae. In these patients, total parenteral nutrition (TPN) should be considered. Patients who have developed enteric fistulae also need TPN to manage their nutrition.

5.17 Conclusion

AP is a prime cause of acute abdominal pain. The clinical presentation of the acute pancreatitis is changeable. The most crucial aspects in terms of management and outcome are diagnosis, prediction and classification of the severity of AP. Most of these patients recover without any instead of complications. Severe AP occurring in approximately 20% of cases, carries a high morbidity and mortality and defined by the presence of pseudocyst, POF or pancreatic necrosis.

The roles of different prognostic markers have not shown much promise. Besides regular laboratory tests CRP estimation is a very useful marker in prediction of disease severity. The added attraction of CRP estimation is its easy availability and low cost. In this section, we have strived to ascertian the importance of CRP estimation for early prediction of severe AP. The clinical features, diagnosis, classifications, management, and outcome of AP are described in detail.

References

Acevedo-Piedra NG, Moya-Hoyo N, Rey-Riveiro M et al (2014) Validation of the determinant-based classification and revision of the Atlanta classification systems for acute pancreatitis. Clin Gastroenterol Hepatol 12:311–316. https://doi.org/10.1016/j.cgh.2013.07.042. PMID:23958561; Epub 2013 Aug 16

Ansar W, Ghosh S (2013) C-reactive protein and the biology of disease. Immunol Res 56:131–142. https://doi.org/10.1007/s12026-013-8384-0. PMID: 23371836

Assimakopoulos SF, Scopa CD, Vagianos CE (2007) Pathophysiology of increased intestinal permeability in obstructive jaundice. World J Gastroenterol 13:6458–6464. https://doi.org/10.3748/wjg.v13.i48.6458. PMID: 18161914

Assimakopoulos SF, Papageorgiou I, Charonis A (2011) Enterocytes' tight junctions: from molecules to diseases. World J Gastrointest Pathophysiol 2(6):123–137. https://doi.org/10.4291/wjgp.v2.i6. PMID: 22184542

Assimakopoulos SF, Triantos C, Thomopoulos K, Fligou F, Maroulis I, Marangos M, Gogos CA (2018) Gut-origin sepsis in the critically ill patient: pathophysiology and treatment. Infection 46 (6):751–760. https://doi.org/10.1007/s15010-018-1178-5. PMID: 30003491

Balthazar EJ (2002) Acute pancreatitis: assessment of severity with clinical and CT evaluation. Radiology 223:603–613. https://doi.org/10.1148/radiol.2233010680. PMID: 12034923

Balthazar EJ, Robinson DL, Megibow AJ et al (1990) Acute pancreatitis: value of CT in establishing prognosis. Radiology 174(2):331–336. https://doi.org/10.1148/radiology.174.2.2296641. PMID: 2296641

Banks PA (2005) Pro: computerized tomographic fine needle aspiration (CT-FNA) is valuable in the management of infected pancreatic necrosis. Am J Gastroenterol 100:2371–2372. https://doi.org/10.1111/j.1572-0241.2005.00328_1.x

Banks PA, Bollen TL, Dervenis C, Gooszen HG, Johnson CD, Sarr MG, Tsiotos GG, Vege SS (2013) Classification of Acute Pancreatitis—2012: Revision of the Atlanta Classification and Definitions by International Consensus. Gut 62:102–111. https://doi.org/10.1136/gutjnl-2012-302779. PMID: 23100216

Barge JU, Lopera JE (2012) Vascular complications of pancreatitis: role of interventional therapy. Korean J Radiol 13(S1):S45–S55. https://doi.org/10.3348/kjr.2012.13.S1.S45. PMID: 22563287

Basurto Ona X, Rigau Comas D, Urrutia G (2013) Opioids for acute pancreatitis pain. Cochrane Database Syst Rev (7):CD009179. https://doi.org/10.1002/14651858.CD009179.pub2. PMID: 23888429

Beger H, Isenmann R (1999) Diagnosis, objective assessment of severity, and management of acute pancreatitis: Santorini consensus conference. Int J Int Assoc Pancreatol 26:1–2. https://doi.org/10.1007/bf2925968. PMID: 10453421

Bennett GL, Hann LE (2001) Pancreatic ultrasonography. Surg Clin North Am 81(2):259–281. https://doi.org/10.1016/s0039-6109(05)70117-1. PMID: 11392416

Bollen TL, Singh VK, Maurer R, Repas K, van Es HW, Banks PA, Mortele KJ (2011) Comparative evaluation of the modified CT severity index and CT severity index in assessing severity of acute pancreatitis. AJR Am J Roentgenol 197(2):386–392. https://doi.org/10.2214/AJR.09.4025. PMID: 21785086

Bollen TL, Singh VK, Maurer R, Repas K, van Es HW, Banks PA, Mortele KJ (2012) A comparative evaluation of radiologic and clinical scoring systems in the early prediction of severity in acute pancreatitis. Am J Gastroenterol 107(4):612–619. https://doi.org/10.1038/ajg.2011.438. PMID: 22186977

Brown A, James-Stevenson T, Dyson T, Grunkenmeier D (2007) The Panc 3 score: a rapid and accurate test for predicting severity on presentation in acute pancreatitis. J Clin Gastroenterol 41 (9):855–858. https://doi.org/10.1097/01.mcg.0000248005.73075.e4. PMID: 17881932

Büchler M, Malfertheiner P, Schoetensack C et al (1986) Sensitivity of antiproteases, complement factors and C reactive protein in detecting pancreatic necrosis. Results of a prospective clinical study. Int J Pancreatol 1:227–235. https://doi.org/10.1007/BF02795248

Cappell MS (2008) Acute pancreatitis: etiology, clinical presentation, diagnosis, and therapy. Surg Clin N Am 92:889–923. https://doi.org/10.1016/j.mcna.2008.04.013. PMID: 18570947

Cardoso FS, Ricardo LB, Oliveira AM, Horta DV, Papoila AL et al (2015) C-Reactive protein at 24 h after hospital admission may have relevant prognostic accuracy in acute pancreatitis: a Retrospective Cohort Study. Portugese J Gastroenterol 22(5):198–203. https://doi.org/10.1016/j.jpge.2015.03.002. PMID: 28868408

Chen Y, Ke L, Tong Z, Li W, Li J (2015) Association between severity and the determinant-based classification Atlanta 2012 and Atlanta 1992 in acute pancreatitis—a clinical retrospective study. Medicine 94(13). https://doi.org/10.1097/MD.0000000000000638. PMID: 25837754

Conwell DL, Banks PA (2016) Acute pancreatitis. In: Greenberger N (ed) Current diagnosis and treatment gastroenterology, hepatology and endoscopy, 3rd edn. McGraw and Hill Education, pp 335–342

Cucher D, Kulvattunyou N, Green DJ, Jie T, Ong ES (2008) Gallstone pancreatitis: a review. Surg Clin N Am 92:257–280. https://doi.org/10.1016/j.suc.2014.01.006. PMID: 24679420; Epub 2014 Feb 20

Dawra R, Sah PR, Dudeja V, Rishi L, Talukdar R, Garg P, Saluja AK (2011) Intra-acinar trypsinogen activation mediates early stages of pancreatic injury but not inflammation in mice with acute pancreatitis. Gastroenterology 141:2210–2217. https://doi.org/10.1053/j.gastro.2011.08.033. Epub 2011 Aug 27

De Waele JJ, Delrue L, Hoste EA, De Vos M, Duyck P, Colardyn FA (2007) Extrapancreatic inflammation on abdominal computed tomography as an early predictor of disease severity in

acute pancreatitis—evaluation of a new scoring system. Pancreas 2:185–190. https://doi.org/10.1097/mpa.0b013e31802d4136. PMID: 17312456
Deitch EA (2012) Gut-Origin sepsis; evolution of a concept. Surgeon 10(6):350–356. https://doi.org/10.1016/j.surge.2012.03.003. Epub 2012 Apr 23
Deitch EA, Xu D, Kaise VL (2006) Role of the gut in the development of injury- and shock induced SIRS and MODS: the gut-lymph hypothesis, a review. Front Biosci 11:520–528. https://doi.org/10.2741/1816. PMID: 16146750
Dellinger EP, Forsmark CE, Layer P et al (2012) Determinant-based classification of acute pancreatitis severity: an international multidisciplinary consultation. Ann Surg 256:875–880. https://doi.org/10.1097/SLA.0b013e318256f778
Delrue LJ, De Waele JJ, Duyck PO (2010) Acute pancreatitis: Radiologic scores in predicting severity and outcome. Abdom Imaging 35:349–361. https://doi.org/10.1007/s00261-009-9522-y. Epub 2009 May 13
Du Clos TW, Mold C (2004) C-reactive protein: an activator of innate immunity and a modulator of adaptive immunity. Immunol Res 30(3):261–277. https://doi.org/10.1385/IR:30:3:261. PMID: 15531769
Dumonceau J, Andriulli A, Elmunzer JB, Mariani A, Meister T, Deviere J et al (2014) Prophylaxis of post-ERCP pancreatitis. European Society of Gastrointestinal Endoscopy (ESGE) Guideline—Updated June 2014. Endoscopy 46:799–815. https://doi.org/10.1055/s-0034-1377875. Epub 2014 Aug 22
Escott ABJ, Phillips AJ, Windsor JA (2017) Locoregional pathophysiology in acute pancreatitis: pancreas and intestine in acute pancreatitis. In: Adams DB, Cotton PB, Zyromski NJ, Windsor JA (eds) Pancreatitis: medical and surgical management, 1st edn. John Wiley and Sons, pp 19–28
Fagenholz PJ, Castillo CF, Harris NS et al (2007) Increasing united states hospital admissions for acute pancreatitis, 1988–2003. Ann Epidemiol 17:491–497. https://doi.org/10.1016/j.annepidem.2007.02.002. Epub 2007 April 19
Fisher William E, Anderson Dana K, Windsor John A, Saluja Ashok K, Brunicardi FC (2015) Pancreas (Ch 33, pp 1341–1422). In: Schwartz's principles of surgery, 10th edn. Mc Graw and Hill Education
Flati G, Andrén-Sandberg A, La Pinta M, Porowska B, Carboni M (2003) Potentially fatal bleeding in acute pancreatitis: pathophysiology, prevention, and treatment. Pancreas 26:8–14. https://doi.org/10.1097/00006676-200301000-00002. PMID: 12499910
Foster BR, Jensen KK, Bakis G, Shaaban AM, Coakley FV (2016) Revised Atlanta classification for acute pancreatitis—a pictorial essay. Radiographics 36:675–687. https://doi.org/10.1148/rg.2016150097. PMID: 27163588
Gan SI, Edwards AL, Symonds CJ, Beck PL (2006) Hypertriglyceridemia-induces pancreatitis: a case-based review. World J Gastroenterol 12(44):7197–7202. https://doi.org/10.3748/wjg.v12.i44.7197. PMID: 17131487
Garg PK (2017) Pathophysiology of systemic inflammatory response syndrome and multiorgan dysfunction syndrome in acute pancreatitis. In: Adams DB, Cotton PB, Zyromski NJ, Windsor J (eds) Pancreatitis: medical and surgical management, 1st edn. John Wiley and Sons, pp 29–37
Gukovskaya AS, Pandol SJ, Gukovsky I (2016) New insights into the pathways initiating and driving pancreatitis. Curr Opin Gastroenterol 32(5):429–435. https://doi.org/10.1097/MOG.0000000000000301. PMID: 27428704
Hage FG, Szalai AJ (2007) C-reactive protein gene polymorphisms, C-reactive protein blood levels and cardiovascular disease risk. J Am Coll Cardiol 50(12):1115–1122. https://doi.org/10.1016/j.jacc.2007.06.012. Epub 2007 Sep 4
Halangk W, Lerch MM, Brandt-Nedelev B et al (2000) Role of cathepsin B in intracellular trypsinogen activation and the onset of acute pancreatitis. J Clin Invest 106(6):773–781. https://doi.org/10.1172/JCI9411. PMID: 10995788

Hamada S, Masamune A, Kikuta K et al (2014) Nationwide epidemiological survey of acute pancreatitis in Japan. Pancreas 43:1244–1248. https://doi.org/10.1007/s00535-019-01658-7. Epub 2019 Dec 23; PMID: 31872350

Hegazi RA, DeWitt T (2014) Enteral nutrition and immune modulation of acute pancreatitis. World J Gastroenterol 20:16101–16105. https://doi.org/10.3748/wjg.v20.i43.16101. PMID: 25473161

Hong S, Qiwen B, Ying J et al (2011) Body Mass Index and the risk and prognosis of acute pancreatitis: a meta-analysis. Eur J Gastroenterol Hepatol 46:1136–1143. https://doi.org/10.1097/MEG.0b013e32834b0e0e. PMID: 21904207

Isenmann R, Rau B, Beger HG (2001) Early severe acute pancreatitis: characteristics of a new subgroup. Pancreas 22(3):274–278

Jin T, Jiang K, Deng LGJ, Wu Y, Wang Z, Shi N, Zhang X, Lin W, Asrani V, Jones P, Mittal A, Phillips A, Sutton R, Huang W, Yang X, Xia Q, Windsor JA (2018) Response and outcome from fluid resuscitation in acute pancreatitis—a prospective cohort study. HPB 20 (11):1082–1091. https://doi.org/10.1016/j.hpb.2018.05.018. 2018 Published by Elsevier Ltd on behalf of International Hepato-Pancreato-Biliary Association Inc

Johnson CD, Abu-Hillal M (2004) Persistent organ failure during the first week as a marker of fatal outcome in acute pancreatitis. Gut 53:1340–1344. PMID: 15306596. https://doi.org/10.1136/gut.2004.039883

Kaplan M, Ates I, Akpinar MY et al (2017) Predictive value of CRP/Albumin ratio in acute pancreatitis. J Hepato Pancreat Dis Intl 16(4):424–430. https://doi.org/10.1016/s1499-3872(17)60007-9

Kemppainen EA, Hedstrom JI, Puolakkainen PA et al (1997) Rapid measurement of Urinary trypsinogen-2 as a screening test for acute pancreatitis. N Engl J Med 336:1788–1793. https://doi.org/10.1056/NEJM199706193362504

Khanna A, Meher S, Prakash S, Tiwary SK, Singh U, Srivastava A, Dixit VK (2013) Comparison of Ranson, Glasgow, MOSS, SIRS, BISAP, APACHE-II, CTSI Scores, IL-6, CRP, and procalcitonin in predicting severity, organ failure, pancreatic necrosis, and mortality in acute pancreatitis. HPB Surg:367581. https://doi.org/10.1155/2013/367581

Knaus WA, Draper EA, Wagner DP, Zimmerman JE (1985) APACHE II: a severity of disease classification system. Crit Care Med 13(10):818–829

Koutroumpakis E, Wu BU, Bakker OF, Dudekula A, Singh VK, Besselink MG, Yadav D, van Santvoort HC, Whitcomb DC, Gooszen HG, Banks PA, Papachristou GI (2015) Admission hematocrit and rise in blood urea nitrogen at 24 h outperform other laboratory markers in predicting persistent organ failure and pancreatic necrosis in acute pancreatitis: a post hoc analysis of three large prospective databases. Am J Gastroenterol 110:1707–1716. https://doi.org/10.1038/ajg.2015.370. PMID: 26553208

Krasinskas AM, Raina A, Khalid A, Tublin M, Yadav D (2007) Autoimmune pancreatitis. Gastroenterol Clin North Am 36:239–257. https://doi.org/10.1016/j.gtc.2007.03.015

Larvin M, McMahon MJ (1989) Apache II score for assessment and monitoring of acute pancreatitis. Lancet. https://doi.org/10.1016/s0140-6736(89)90381-4. PMID: 2568529

Lee YH, Pratley RE (2005) The evolving role of inflammation in obesity and the metabolic syndrome. Curr Diab Rep 5:70–75. https://doi.org/10.1007/s11892-005-0071-7

Lee SY, Goh BKP, Chan CY (2016) Etiology, pathogenesis and diagnostic assessment of acute pancreatitis. In: Jarnagin, William R. (Ed.), Blumgart's surgery of the liver, biliary tract, and pancreas, 6th ed. Elsevier

Leese T, Shaw D, Holliday M (1988) Prognostic markers in acute pancreatitis: can pancreatic necrosis be predicted? Ann R Coll Surg Engl 70(4):227–232. PMID: 2458063

Macintyre S, Samols D, Dailey P (1994) Two carboxylesterases bind C-reactive protein within the endoplasmic reticulum and regulate its secretion during the acute phase response. J Biol Chem 269(39):24496–24503. PMID: 7929114

Makela JT, Heinonen E, Heikki K et al (2007) Computed tomography severity index and C-reactive protein values in predicting mortality in emergency and intensive care unit for patients with acute pancreatitis. Am J Surg 1994(1):30–34. https://doi.org/10.1016/j.amsurg.2006.08.089

Markanday A (2015) Acute phase reactants in infections: evidence based review and a guide for clinicians. Open Forum Infect Dis 2(3) Oxford University Press. https://doi.org/10.1093/ofid/ofv098

Martinez L, Johnson CD et al (2006) Obesity is a definitive risk factor for pancreatic necrosis in acute pancreatitis; an updated meta-analysis. Pancreatology 6:206–209. https://doi.org/10.1159/000092104. PMID: 16549939

Matull WR, Pereira SP, O'Donohue JW (2006) Biochemical markers of acute pancreatitis. J Clin Pathol 59(4):340–344. https://doi.org/10.1136/jcp.2002.002923

Mayer AD, McMahon MJ, Bowen M, Cooper EH (1984) C reactive protein: an aid to assessment and monitoring of acute pancreatitis. J Clin Pathol 37:207–211. https://doi.org/10.1136/jcp.37.2.207

Mayumi T, Inui K, Maetani I et al (2012) Validity of the urinary Trypsinogen-2 test in the diagnosis of acute pancreatitis. Pancreas 41(6):869–875. https://doi.org/10.1097/MPA.0b013e3182480ab7

Meher S, Mishra TS, Sasmal PK, Rath S, Sharma R, Routh B et al (2015) Role of biomarkers in diagnosis and prognostic evaluation of acute pancreatitis. J Biomarkers. https://doi.org/10.1155/2015/519534. PMID: 26345247

Moran RA, Garcia-Rayado G, de la Iglesia-García D et al (2018) Influence of age, body mass index and comorbidity on major outcomes in acute pancreatitis, a prospective nation-wide multicentre study. United Eur Gastroenterol J 6(10):1508–1518. https://doi.org/10.1177/2050640618798155. PMID: 30574321

Mortele KJ, Wiesner W, Intriere L et al (2004) A modified CT severity index for evaluating acute pancreatitis: improved correlation with patient outcome. AJR 183:1261–1265. https://doi.org/10.2214/ajr.183.5.1831261. PMID: 15505289

Mounzer R, Langmead CJ, Wu BU et al (2012) Comparison of existing clinical scoring systems to predict persistent organ failure in patients with acute pancreatitis. Gastroenterology 142:1476–1482. https://doi.org/10.1053/j.gastro.2012.03.005. PMID: 22425589

Mourad MM, Evans RPT, Kalindindi V, Navaratnam R, Dvorkin L, Bramhall SR (2017) Prophylactic antibiotics in acute pancreatitis: endless debate. Ann R Coll Surg Engl 99:107–112. https://doi.org/10.1308/rcsann.2016.0355. PMID: 27917667

Munigoti SP, Rees A (2011) Hypertriglyceridemia, LPL deficiency and pancreatitis. Br J Diab Vasc Dis 11(3):107–112. https://doi.org/10.1177/1474651411413192

Nawaz H, Mounzer R, Yadav D et al (2013) Revised Atlanta and determinant-based classification: application in a prospective cohort of acute pancreatitis patients. Am J Gastroenterol 108:1911–1917. https://doi.org/10.1038/ajg.2013.348. PMID: 24126632

Papachristou GI, Papachristou DJ, Avula H et al (2006) Obesity increases the severity of acute pancreatitis: performance of APACHE O score and co relation with the inflammatory response. Pancreatology 6:279–285. https://doi.org/10.1159/000092689. PMID: 16636600

Papachristou GI, Muddana V, Yadav D et al (2010) Comparison of BISAP, Ranson's, APACHE-II, and CTSI scores in predicting organ failure, complications, and mortality in acute pancreatitis. Am J Gastroenterol 105(2):435–441. https://doi.org/10.1038/ajg.2009.622. PMID: 19891954

Pepys MB, Gideon MH (2003) C-reactive protein: a critical update. J Clin Invest 111:1805–1811. https://doi.org/10.1172/JCI18921

Petrov MS, van Santvoort HC, Besselink MG, van der Heijden GJ, Windsor JA, Gooszen HG (2008) Enteral nutrition and the risk of mortality and infectious complications in patients with severe acute pancreatitis: a meta-analysis of randomized trials. Arch Surg 143(11):1111–1117. https://doi.org/10.1001/archsurg.143.11.1111. PMID: 19015471

Pongprasobchai S, Jianjaroonwong V et al (2010) Erythrocyte sedimentation rate and C-reactive protein for the prediction of severity of acute pancreatitis. Pancreas 39(8):1226–1230. https://doi.org/10.1097/MPA.obo13e3181deb33e

Raghu MG, Wig JD, Kochhar R et al (2007) Lung complications in acute pancreatitis. JOP 8 (2):177–185

Ranson JHC, Rifkind KM, Roses DF et al (1974) Prognostic signs and the role of operative management in acute pancreatitis. Surg Gynecol Obstet 139:69. PMID: 4834279

Ranson JH (1982) Etiological and prognostic factorsin human acute pancreatitis: a review. Am J Gastroenterol 77(9):633–638. PMID: 7051819

Rau B, Steinbach G, Gansauge F, Mayer JM, Grünert A, Beger HG (1997) The potential role of procalcitonin and interleukin 8 in the prediction of infected necrosis in acute pancreatitis. Gut 41:832–840. https://doi.org/10.1136/gut.41.6.832. PMID: 9462219

Rau B, Steinbach G, Baumgart K, Gansauge F et al (2000) Serum amyloid A versus C-Reactive protein in acute pancreatitis: clinical value of an alternative acute-phase reactant. Crit Care Med 28(3):736–742. https://doi.org/10.1097/00003246-200003000-00022. PMID: 10752923

Sakorafas GH, Tsiotou AG (2000) Etiology and pathogenesis of acute pancreatitis: current concepts. J Clin Gastroenterol 30(4):343–356

Saluja A, Dudeja V, Dawra R, Sah RP (2019) Early intra-acinar events in pathogenesis of acute pancreatitis. Gastroenterology 156:1979–1993. https://doi.org/10.1053/j.gastro.2019.01.268

Sarr MG, Nagorney DM, Mucha P Jr et al (1991) Acute necrotizing pancreatitis: management by planned, staged pancreatic necrosectomy/debridement and delayed primary wound closure over drains. Br J Surg 78:576–581. https://doi.org/10.1002/bjs.1800780518. PMID: 2059810

Tenner Scott, Steinberg William (2010) Acute pancreatitis (Ch 58: pp 1338-1362). In: Feldman M, Friedman LS, Brandt LJ (Eds.), Sleisenger and Fordtran's gastrointestinal and liver disease, 10th ed. Elsevier (Saunders)

Sharma A, Muddana V, Lamb J, Greer J, Papachristou GI, Whitcomb DC (2009) Low serum adiponectin levels are associated with systemic organ failure in acute pancreatitis. Pancreas 38:907–912. https://doi.org/10.1097/MPA.0b013e3181b65bbe. PMID: 19696691

Singh VK, Wu BU, Bollen TL et al (2009) A prospective evaluation of the bedside index for severity in acute pancreatitis score in assessing mortality and intermediate markers of severity in acute pancreatitis. Am J Gastroenterol 104:966–971. https://doi.org/10.1038/ajg.2009.28. PMID: 19293787

Somani P, Sunkara T, Sharma M (2017) Role of endoscopic ultrasound in idiopathic pancreatitis. World J Gastroenterol 23(38):6952–6961. https://doi.org/10.3748/wjg.v23.i38.6952. PMID: 29097868

Spanier B, Bruno MJ, Dijkgraaf MG (2013) Incidence and mortality of acute and chronic pancreatitis in The Netherlands: a nationwide record-linked cohort study for the years 1995–2005. World J Gastroenterol 19:3018–3026. https://doi.org/10.3748/wjg.v19.i20.3018. PMID: 23716981

Sproston NR, Ashworth JA (2018) Role of C-reactive protein at sites of inflammation and infection. Front Immunol 9:754. https://doi.org/10.3389/fimmu.2018.00754

Sternby H, Hartman H, Johansen D et al (2017) IL6 and CRP are superior in early differentiation of mild and non-mild acute pancreatitis. *Pancreatology* 17(4):550–554. https://doi.org/10.1016/j.pan.2017.05.392

Steinman WM, Nauck MA, Zinman B et al (2014) LEADER 3—lipase and amylase activity in subjects with Type 2 Diabetes: baseline data over 9000 subjects in the leader trial. Pancreas (41):1223–1231. https://doi.org/10.1097/MPA.0000000000000229

Stirling AD, Moran NR, Michael EK, Ridgway PF, Conlon KC (2017) The predictive value of C-Reactive Protein (CRP) in acute pancreatitis—is interval changes in CRP an additional indicator of severity? HPB 19:874–880. https://doi.org/10.1016/j.hpb.2017.06.001

Tenner S, Dubner H, Steinberg W (1994) Predicting gallstone pancreatitis with laboratory parameters: a meta-analysis. Am J Gastroenterol 89(10):1863–1866. PMID: 7942684

Tenner S, Baillie J, DeWitt J, Vege SS (2013) American College of Gastroenterology guideline: management of acute pancreatitis. Am J Gastroenterol. https://doi.org/10.1038/ajg.2013.218

Thomsen HS (2006) Nephrogenic systemic fibrosis: a serious late adverse reaction to gadodiamide. Eur Radiol 16:2619–2621. https://doi.org/10.1007/s00330-006-0495-8

Tilg H, Moschen AR (2006) Adipocytokines: mediators linking adipose tissue, inflammation and immunity. Nat Rev Immunol 6:772–783. https://doi.org/10.1038/nri1937. PMID: 16998510

Vasudevan S, Goswami P, Sonika U, Thakur B, Sreenivas V, Saraya A (2018) Comparison of various scoring systems and biochemical markers in predicting the outcome in acute pancreatitis. Pancreas 47(1):65–71. https://doi.org/10.1097/MPA.0000000000000957

Van Acker GJ, Saluja AK, Bhagat L et al (2002) Cathepsin B inhibition prevents trypsinogen activation and reduces pancreatitis severity. Am J Physiol Gastrointest Liver Physiol 283(3): G794–G800. https://doi.org/10.1152/ajpgi.00363.2001. PMID: 12181196

Vila JJ (2010) Endoscopic ultrasonography and idiopathic acute pancreatitis. World J Gastrointest Endosc 2(4):107–111. https://doi.org/10.4253/wjge.v2.i4.107. PMID: 21160725

Whitcomb DC (2006) Acute pancreatitis; acute pancreatitis. N Engl J Med 354:2142–2150. https://doi.org/10.1056/NEJMcp054958. PMID: 16707751

Whitcomb DC (2017) Epidemiology and genetics of pancreatitis. In: Adams DB, Cotton PB, Zyromski NJ, Windsor J (eds) Pancreatitis: medical and surgical management, 1st edn. John Wiley and Sons

Windsor JA (2008) Assessment of the severity of acute pancreatitis: no room for complacency. Pancreatology 8(2):105–109. https://doi.org/10.1159/000123604

Working Group IAP/APA (2013) Acute Pancreatitis Guideline: IAP/APA evidence-based guidelines for the management of acute pancreatitis. Pancreatology 13:e1–e15. https://doi.org/10.1016/j.pan.2013.07.063. PMID: 24054878

Wu BU, Johannes RS, Sun X, Tabak Y, Conwell DL, Banks PA (2008) The early prediction of mortality in acute pancreatitis: a large population-based study. Gut 57(12):1698–1703. https://doi.org/10.1136/gut.2008.152702. PMID: 18519429

Wu BU, Johannes RS, Sun X, Tabak Y, Conwell DL, Banks PA (2009) Early changes in blood urea nitrogen predict mortality in acute pancreatitis. Gastroenterology 137:129–135. https://doi.org/10.1053/j.gastro.2009.03.056. PMID: 19344722

Wu BU, Hwang JQ, Gardner TH, Repas K, DeLee R, Yu S, Smith B, Banks PA, Conwell DL (2011) Lactated Ringer's solution reduces systemic inflammation compared with saline in patients with acute pancreatitis. Clin Gastroenterol Hepatol 9:710–717. https://doi.org/10.1016/j.cgh.2011.04.026. PMID: 21645639

Wu BU, Butler RK, Chen W (2019) Factors associated with opioid use in patients hospitalized for acute pancreatitis. JAMA Netw Open 2(4):e191827. https://doi.org/10.1001/jamanetworkopen.2019.1827

Yadav D, Lowenfels AB (2013) The epidemiology of pancreatitis and pancreatic cancer. Gastroenterology 144:1252–1261. https://doi.org/10.1053/j.gastro.2013.01.068. PMID: 23622135

Yokoe M, Takada T, Mayumi T, Yoshida M, Isaji S, Wada K, Itoi T et al (2015) Japanese guidelines for the management of acute pancreatitis: Japanese Guidelines 2015. J Hepatobiliary Pancreat Sci 22:405–432. https://doi.org/10.1002/jhbp.259. PMID: 25973947

Young JY, Chen J, Phillips ARJ, Widsor JA, Petrov MS (2014) Predictors of severe and critical acute pancreatitis: a systematic review. Dig Liver Dis 46:446–451. https://doi.org/10.1016/j.dld.2014.01.158

Role of SNP in CRP and Biology of Cancer

6

Rishav Dasgupta and Shyamasree Ghosh

Contents

6.1 Introduction 177
6.1.1 History 177
6.1.2 Structure 177
6.1.3 Function 177
6.2 Regulation of CRP 180
6.2.1 Positive Regulation of CRP 181
6.2.2 Negative Regulation 182
6.3 Forms of CRP 183
6.3.1 pCRP 183
6.3.2 mCRP 183
6.3.3 CRP and Association with Diseases 184
6.4 The Role of CRP in Cancer 185
6.4.1 Cancer and CRP 186
6.4.2 CRP as a Prognostic Marker in Different Cancer 187
6.5 Single Nucleotide Polymorphisms and CRP 187
6.6 Discussion 188
References 189

Abstract

C-reactive protein (CRP) has gained significant traction in the medical world due to the discovery of its role in identifying and categorizing cancer. Researchers have worked with CRP in reference to detecting cardiovascular disease and inflammatory infections in the past, and this expansion into cancer has shown

R. Dasgupta
University of Illinois Urbana Champaign, Urbana, IL, USA

S. Ghosh (✉)
School of Biological Sciences, NISER, Bhubaneswar, Orissa, India

W. Ansar, S. Ghosh (eds.), *Clinical Significance of C-reactive Protein*,
https://doi.org/10.1007/978-981-15-6787-2_6

that CRPs often play a significant role in the risk, prognosis, and diagnosis of cancer. Single nucleotide polymorphisms (SNPs) in these CRPs have also been seen to have an active role in tumorogenesis and studying certain SNPs has shown increased or decreased risk for certain types of cancer. This chapter includes different SNPs of CRPs and their relationship to diseases ranging from coronary heart disease to breast cancer. While the data for certain SNPs from different experiments can conflict to whether or not it increases the risk of tumorigenesis, it also has shown consistent results for other cases. This method of analysis helps us build a foundation for a novel type of cancer diagnosis, one that can even predict the chance of certain diseases in an individual long before the symptoms may start to manifest.

Keywords

CRP · SNP · Cancer · Risk · Prognostic · Diagnostic · Marker

Abbreviations

AD	Alzheimer's disease
AMRD	Age-related macular degeneration
APP	Acute phase protein
C/EBP	CCAAT-enhancer-binding protein
CRP	C-reactive protein
CVD	Cardiovascular disease
hs-CRP	High-sensitivity CRP
IL	Interleukin
LDL	Low-density lipoprotein
mCRP	Monomeric C-reactive
NF-kB	Nuclear factor kappa
NO	Nitric oxide
pCRP	Pentametric isoform
PD	Parkinson's disease
PRR	Pattern recognition receptor
S. pneumoniae	*Streptococcus pneumoniae*
SAP	Serum amyloid P-competent
SLE	Systemic lupus erythematosus
SNP	Single nucleotide polymorphism
STAT3	Signal transducer and activator of transcription 3
T2D	Type 2 diabetes
TNF	Tumor necrosis factor

6.1 Introduction

C-reactive protein (CRP) is a protein of homopentametric structure, which is an acute phase protein (APP), mediator of host defense which may be undetectable in normal healthy individuals but is known to increase up to 10,000 times its original value (Pepys and Hirschfield 2003; Sproston and Ashworth 2018; Ansar and Ghosh 2013, 2016; Ansar et al. 2009) during inflammation due to infection or injury.

As a pattern recognition receptor (PRR), it is synthesized, for the most part, in the liver. However, it is also known to be synthesized extrahepatically from smooth muscle cells and macrophages, leukocytes, adipocytes, neuronal cells, renal cells, and respiratory epithelial cells.

6.1.1 History

CRP was identified as Fraction C by Tillett and Frances in 1930 as a serum component present in serum of acutely ill patients reacting with a specific (*S. pneumoniae*) extract. Since then large-scale experimental work has been conducted that has elucidated the role of CRP in biology and disease in human and other animals (Pepys and Hirschfield 2003; Tillett and Francis 1930).

6.1.2 Structure

The protein CRP secreted as 206 amino acid 23 κDa, non-glycosylated monomer polypeptide (Fig. 6.1), synthesized by the human hepatocytes and is coded by the human *crp* gene on chromosome 1q23, with two exons and one intron (Ansar and Ghosh 2013, 2016; Ansar et al. 2009; Lei et al. 1985; Woo et al. 1985; Black et al. 2004a, b; Thompson et al. 1999). The monomers in non-covalent association to one another form the homopentameric ring structure which makes it a member of the pentraxin family (Fig. 6.1).

6.1.3 Function

CRP is an ancient highly conserved pattern recognition molecule, being a member of the pentraxin family of proteins in the innate immune system and functions as an early defense system against infectious agents and pathogens. It can bind to phosphocholine (PC) on microorganisms and can bind to both extrinsic organisms including bacteria, fungus, parasite, and plant components and intrinsic components of damaged cell membranes, chromatin, histones, and apoptotic cells and activates the classical complement pathway and binds immunoglobulin receptors on phagocytes. CRP binds to pathogens, damaged tissue, nuclear antigens in a calcium-dependent manner. It can act as opsonins to pathogens.

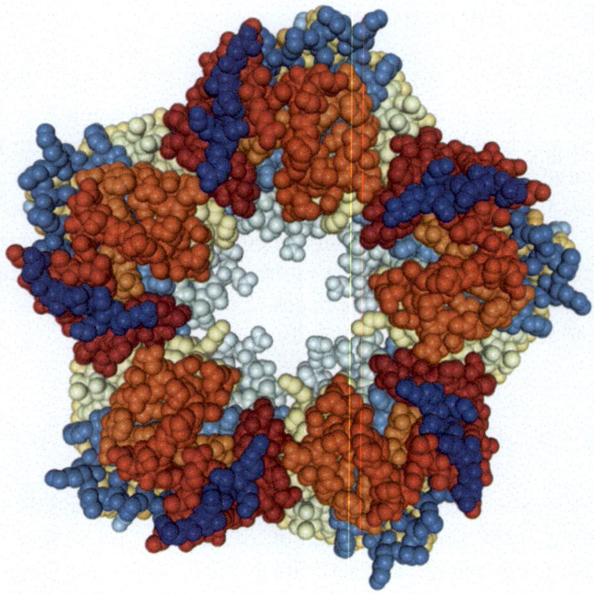

Fig. 6.1 CRP 3D structure, Image of 1GNH Human C-reactive protein from PDB (Shrive et al. 1996; Burley et al. 2019)

It is known to play a role in complement binding to pathogens and damaged cells, activation of complement receptors and thereby enable opsonin-mediated phagocytosis by macrophages. CRP-mediated phagocytosis could also be mediated directly through receptors for the IgGFc or FcγRs or CRP-specific receptor. CRP-mediated innate immune responses lead to increased pro-inflammatory signal and activation of the humoral, adaptive immune system.

Secreted by the liver in response to inflammatory cytokines, CRP levels are known to increase sharply in response to sudden stress, trauma, inflammation, injury, and infection and decrease also sharply with the removal of the condition.

Thus, quantitative estimation of CRP is indicative of inflammatory condition of the body. CRP has been implicated in different types of acute phase responses (APR, Fig. 6.2) and in a number of diseases and is reported a biomarker in different diseases (Frank and Hargreaves 2003) important for diagnosis, prognosis, and risk assessment of disease (Fig. 6.3).

Single nucleotide polymorphisms (SNP) in CRP have been associated with baseline CRP levels and affect the function (Szalai et al. 2005; Young et al. 2008). In this chapter with a brief introduction on CRP, we move on to describe the association of SNP in CRP and the alteration in structure and biological function.

Thus, plasma CRP is a biomarker indicative of inflammatory status of patients. The CRP expression and secretion by hepatocytes, in response to inflammation is regulated by pro-inflammatory cytokines interleukin 6 (IL-6) and to a less by interleukin 1β (IL-1β) by the activation of signaling pathways including the STAT3, NF-κB, and C/EBP family members including C/EBPβ and C/EBPα. STAT3 and Rel proteins on binding to the proximal CRP promoter lead to increased C/EBP binding leading to overexpression of CRP.

Table 6.1 The role of SNPs and CRP in disease

SNP type	Role of interaction	Associated disorder	Citation
SNP on CRP (rs3093059)	Higher circulating CRP in patients with history of family depression, elevated levels of chronic periodontitis among South Indians, susceptibility of hemorrhagic stroke in men	Clinical depression, periodontitis, stroke	Yibulaiyin et al. (2017)
Polymorphism on CRP 1846C > T (rs2808630)	Lymphatic invasion and adverse prognoses in endometrial cancer, risk of developing infantile sepsis, increased chance of lung cancer, prediction of response to chemotherapy	Endometrial cancer, infantile sepsis, lung cancer, chemotherapy (general)	Kito et al. (2015), Zhang et al. (2015), Wu et al. (2018)
CRP gene (rs1205)	Not related to the risk of developing colorectal cancer, possibility of inherited depression, colorectal adenoma risk, risk of tumorigenesis	Colorectal cancer, colorectal adenoma, depression, colorectal cancer	Zhang et al. (2005), Fang and Yu (2017), Su et al. (2014a, b), Wang et al. (2018)
CRP Gene rs1260326	Increased risk of colorectal cancer, increase in risk of fibrosis	Colorectal cancer, non-alcoholic fatty liver disease	Wang et al. (2019a, b), Hudert et al. (2019)
CRP Gene rs1044498	Increased chance of pediatric prevalence of non-alcoholic fatty liver disease, risk of coronary heart disease	Non-alcoholic fatty liver disease, heart diseases and disorders	Wang et al. (2019a, b), Hudert et al. (2019), Di et al. (2018)
CRP promoter SNP position 286	Recurrent mutations	Cancer	Wang et al. (2014)
Genetic polymorphisms of CRP	Late age-related macular degeneration		Cipriani et al. (2017)
−757 T > C polymorphism of C-reactive protein	Chronic periodontitis of South Indian population		Selvaraj et al. (2019)

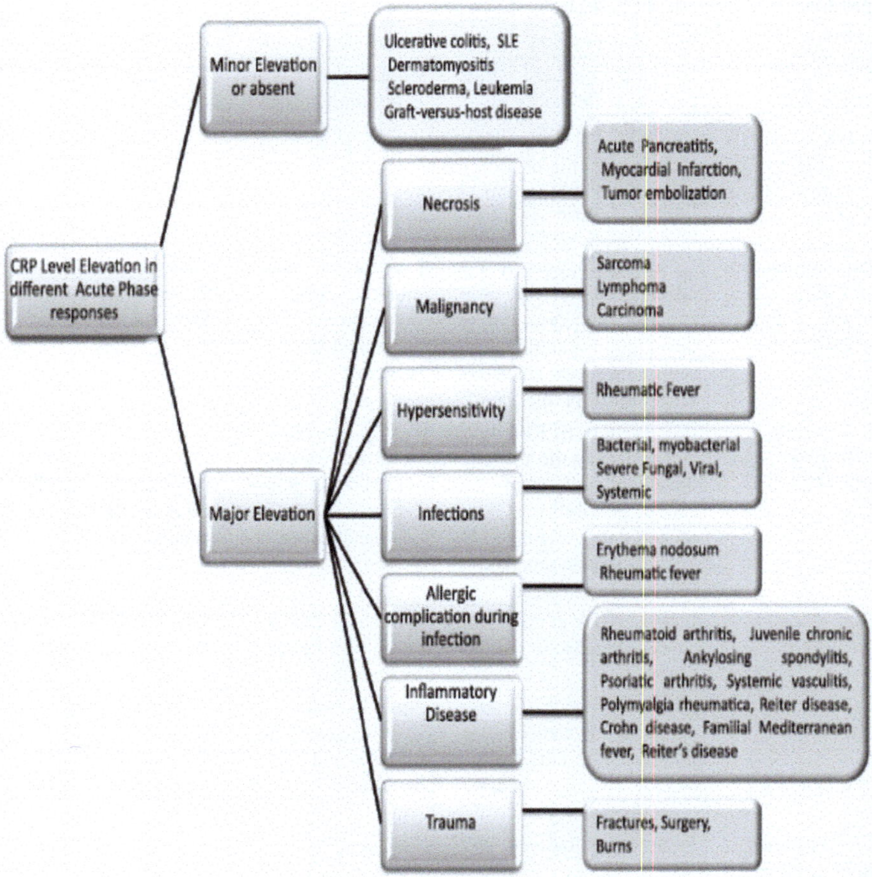

Fig. 6.2 CRP in acute phase response (APR) and disorders. Image and legend adapted from Ansar and Ghosh (2013) with permission

6.2 Regulation of CRP

CRP is predominantly hepatically synthesized and released in the serum (Kushner et al. 2006). Cytokines including IL-1β, IL-6, IL-17, TGF-β, and TNF-α are reported to regulate the transcription of the CRP (Zhang et al. 1996a, b; Patel et al. 2007). STAT3 has been reported to play a role in activation of CRP by IL-6 (Zhang et al. 1996a, b) and IL-17 has been reported to stimulate CRP synthesis in hepatocytes and smooth muscle cells mediated by p38 MAPK and ERK1/2-dependent NF-kappa B (NF-κB) and C/EBP beta activation pathways (Patel et al. 2007).

CRP is known to be positively and/or negatively regulated in response to stimuli of injury or infection or inflammation.

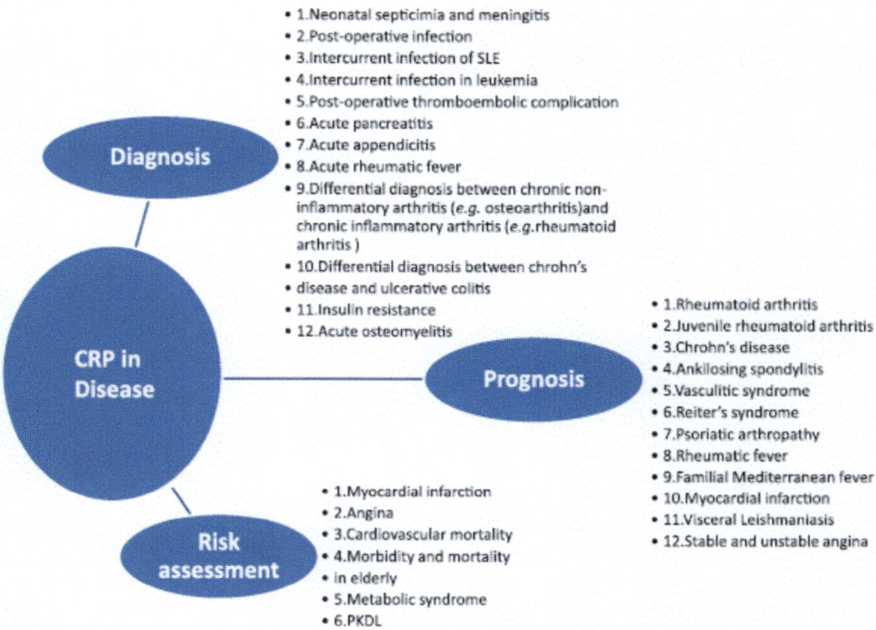

Fig. 6.3 CRP as a diagnostic, prognostic marker. Figure adapted from Ansar and Ghosh (2013) with permission

6.2.1 Positive Regulation of CRP

Inflammation is generally accompanied by hepatocytes in response to cytokines by the synthesis of acute phase proteins (APPs). In mice, this role is played mostly by the serum amyloid P-competent (SAP) and by CRP in a minor way (Young et al. 2008; Ochrietor et al. 2000; Fig. 6.4). Positive regulation is a cytokine-dependent process that regulates CRP production in humans (6–7). Studies from the transcription sites in the CRP gene promoter and the SAP gene promoter reveal that STAT3 and C/EBPβ is consistently associated with CRP gene expression in humans while C/EBPβ expression bears more correlation with SAP gene expression in mice (Young et al. 2008; Ochrietor et al. 2000). This is confirmed by another study which demonstrates that C/EBPβ binding to CRP promoters is associated with the synthesis of CRP. C/EBPβ was also found to increase CRP accumulation in these studies (Ochrietor et al. 2000). First 4000 bases of the CRP gene promoter shows that there are 13 potential κB sites in this gene and that transcription factor NF-κB can synergistically add to the effects of STAT3 and C/EBPβ; however, in-depth roles of NF-κB stimulating CRP gene transcription needs to be further explored. (Young et al. 2008; Ochrietor et al. 2000; Black et al. 2004a, b; Zhang et al. 1996a, b; Cha-Molstad et al. 2007; Agarwal et al. 2003; Buie et al. 2017).

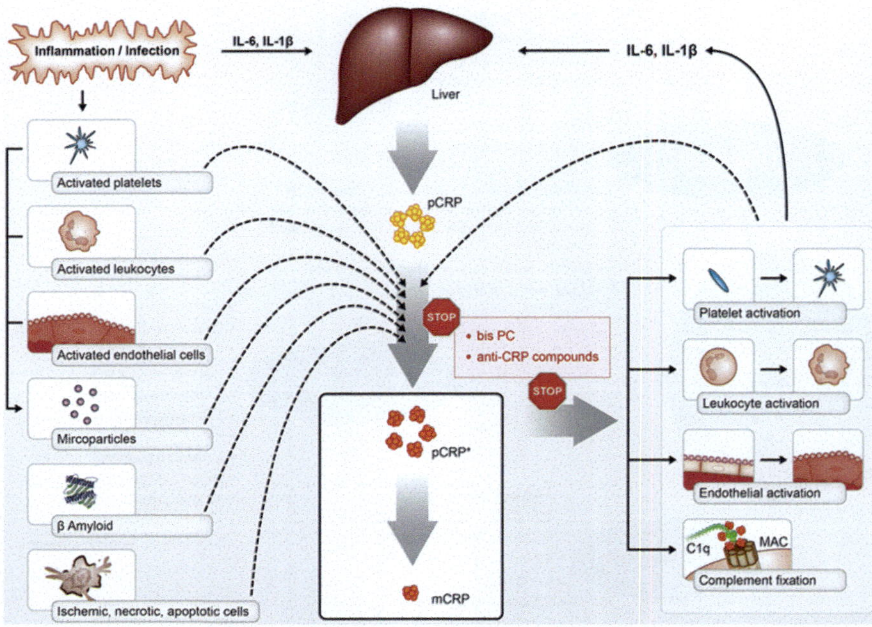

Fig. 6.4 pCRP is synthesized in liver as post inflammation and infection response, under influence of interleukin 6 (IL-6) and IL-1β cytokines. pCRP can be dissociated into pCRP* and monomeric CRP (mCRP) on activated platelets, leukocytes, endothelial cells, dying cells, and beta-amyloid surfaces. pCRP* and mCRP can activate platelets, leukocytes, endothelial cells and complement by C1q binding revealing pro-inflammatory role. bisPC can inhibit pCRP dissociation blocking mCRP effector binding. Image and legend adapted from open access article (McFadyen et al. 2018)

6.2.2 Negative Regulation

On the other hand, interferon-α, statins, and nitric oxide (NO) are known to suppress the induction of CRP expression by pro-inflammatory cytokines, leading to low serum CRP levels in viral infections and lupus.

Systemic lupus erythematosus (SLE) is an autoimmune disease and is a known factor for increasing risk which can lead to endothelial dysfunction. Interferon-α is known to cause symptoms as severe as impaired blood vessel dilatation and can negatively regulate NO production by repressing endothelial NO synthase activity. It is also known that CRP production is limited by IFN-α; thus, it has the same effect on CRP and NO production. By blocking such anti-atherogenic pathways, IFN-α promotes the development of cardiovascular disease (CVD) during SLE. In this case, CRP is seen directly correlated with the chance of cardiovascular disease, where the higher the CRP, the lower the risk of endothelial dysfunction during SLE (Buie et al. 2017).

Cardiovascular disease (CVD) is the leading cause of mortality in the worldwide adult population. A key pathological component of this is atherosclerosis and CRP happens to be a fundamental inflammatory molecule in this process. A study done in Venezuela poses interesting results, some of which are different from those presented by the previous study. In this case, which studies the progression of heart diseases in regular adult patients, CRP is seen to favor the establishment of a chronic inflammatory state and thus potentiating atherosclerosis. mCRP drives most of these effects and thus high levels of CRP can be seen to increase the risk of cardiovascular disease (Buie et al. 2017; Salazar et al. 2014a, b, c). The discrepancy between the two aforementioned cases displays how CRP levels can play different roles in different contexts and hence is not fixed to have a singular function.

6.3 Forms of CRP

While we have generalized all CRP into one category so far in this chapter, CRP is not only available in one form in the human body. CRP is known to have two conformational forms: pentametric isoform (pCRP) and monomeric isoform (mCRP, Fig. 6.5), each of which has unique features. Studies show that pCRP is susceptible to dissociations, which in turn because smaller mCRP units. In local expression, mCRP mRNA is found in high quantities, especially in inflammatory cells. This being said, mCRP is hardly found in circulation, which suggests that is used specifically for local expression (Thiele et al. 2014; Boncler et al. 2019). Besides these two forms of CRP, another form of CRP or glycosylated CRP exists (Fig. 6.6). Glycosylation is a post-translational protein modification and has been associated with several physiological diseases. The glycosylation of serum proteins can be seen primarily in inflammatory diseases. While glycosylation of CRP is not non-existent, it is not very deeply studied due to its rarity (Boncler et al. 2019)

6.3.1 pCRP

Of the two different conformational forms, native pentameric isoform (pCRP) and the monomeric isoform (mCRP) reveals distinct antigenic, electrophoretic, and biological features of which pCRP is the stable predominant form detected in serum (Salazar et al. 2014a, b, c).

6.3.2 mCRP

Monomeric CRP is scarcely found in circulation. mCRP mRNA has been detected in U937 macrophages of atherosclerotic lesions and their expression has been associated with diseases including atherosclerotic lesions of diabetics, systemic inflammation, local dissociation of pCRP into mCRP on apoptotic cells activated platelets in atherosclerotic plaques (Salazar et al. 2014a, b, c).

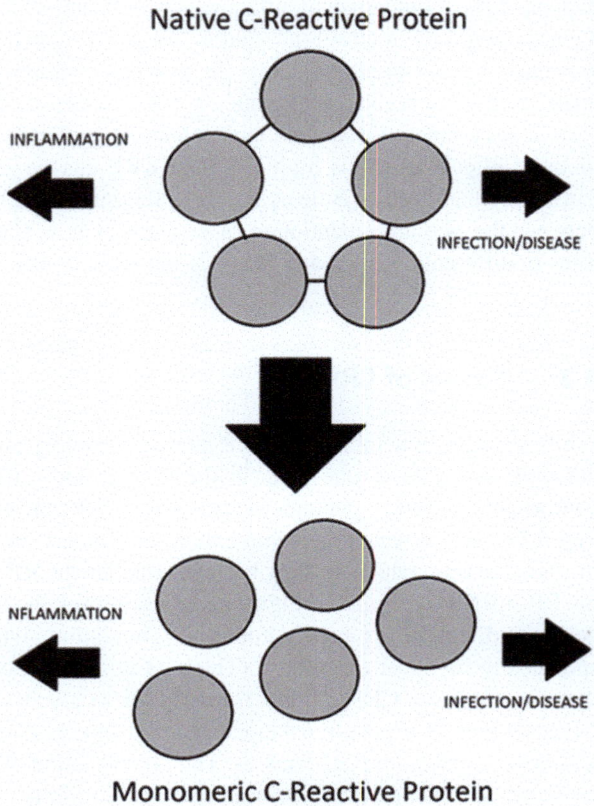

Fig. 6.5 Native and monomeric CRP in inflammation, infection, and disease. Figure and legend adapted from Open access article (Sproston and Ashworth 2018)

6.3.3 CRP and Association with Diseases

CRP has mostly been implicated as markers indicating infections or cardiovascular disease (CVD). Elevated CRP levels have been associated in CVD including atherosclerosis, chronic heart failure, type 2 diabetes (T2D) and insulin resistance, age-related macular degeneration (ARMD) leading to visual impairment and blindness, hemorrhagic stroke and brain injury, neurodegenerative disorder including Alzheimer's disease (AD), Parkinson's Disease (PD), dementia (Luan and Yao 2018); however, recently there is more evidence pointing to the fact that it may also have a role in the production of cytokines, particularly tumor necrosis factor-α. Although the correlation between cardiovascular disease (CVD) and CRP higher, reports are suggesting a role of CRP in cancer. Studies are now showing that elevated levels of CRP at certain stages of cancer development have the ability to predict the prognosis of certain cases during the diagnosis phase as well as predicting the chance of mortality in certain situations (Allin and Nordestgaard 2011; Lee et al. 2011).

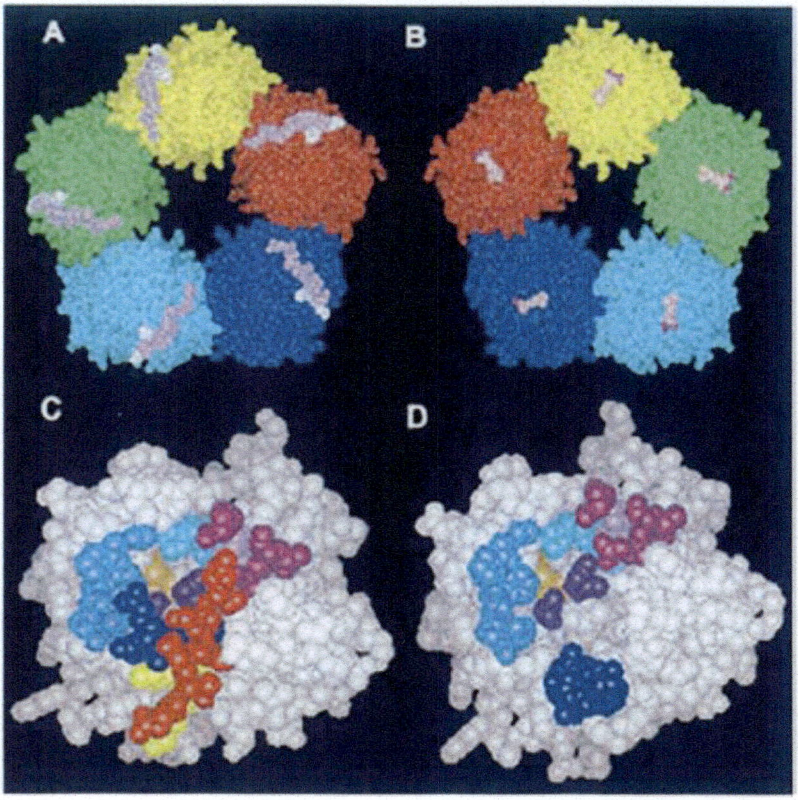

Fig. 6.6 Molecular modeling studies on CRP reveals (**a**) pentraxin structure of human CRP 1B09, (**b**) modeled CRP. (**a**) Structural motifs not found in the CRP isolated from diseased individuals shown in light pink (amino acids 1–6, QTDMSR) and white (amino acids 189–191, ALK). (**b**) back view (180° rotation along the X-axis), revealing calcium (light pink) and phosphorylcholine (PC) (deep pink) bound to the CRP. (**c**, **d**) Single subunit of CRP. (**c**) cleft on the structure of human CRP (1GNH, original X-ray crystallographic structure). (**d**) cleft on the modeled structure with two fragments missing. Colors indicate the different parts of cleft: deep pink and cyan: two rims of the cleft; deep blue and light blue; two sides of the cleft, in the floor toward the open end: cyan and light pink; two asparagine residues located centrally and colored purple; beneath the cleft floor, disulfide bridge in gold and red (amino acids 1–6, QTDMSR) constitutes the other side wall and yellow (amino acids 189–191, ALK) segments are the motifs that are absent in the modified variants of the CRP purified from patients. Image and legend adapted from Das et al. (2003)

6.4 The Role of CRP in Cancer

Cancer is life-threatening disease, affecting millions of life across the globe (Fig. 6.7) Elevated levels of high sensitivity C-reactive protein (hs-CRP) has been associated with diseases ranging from diabetes to arthritis and even various forms of cancer. In a normal, healthy human body, hs-CRP usually maintains levels that are too low to detect. Thus, the detection of this protein generally indicates the presence of an abnormality. We discuss the association of CRP in cancer.

6.4.1 Cancer and CRP

There are two schools of thought associated with CRP and its role with cancer. One states that elevated hs-CRP is related to the presence of cancer, while another states that inflammation and elevated CRP may have a part in carcinogenesis. The latter view believes that oxidative damage from inflammation cause cancer by inactivating tumor suppressor genes. (Lee et al. 2011). In order to understand which of the two theories is closest to reality, a group of Korean researchers studied incoming cancer patients at a hospital. Results of the study did not rule out the possibility that elevated CRP levels may be due to latent cancers. That being said, the serum hs-CRP was found to be positively associated with cancer risk. Thus, it can be said that low-grade systemic inflammation or chronic inflammation may increase the chances of carcinogenesis (Allin and Nordestgaard 2011; Lee et al. 2011).

It has also been suggested that elevated hs-CRP levels in cancer patients may have a correlation to their mortality rates. In order to further investigate this claim, another study was conducted in Korea among hospital patients over an extended period of time. Patients at Seoul National University Hospital who had been screened for CRP between 1995 and 2006 were used in the research. Patients were followed up with for the mortality baseline examination until the end of 2008. Many patients in the directory were excluded from study due to incomplete or inconclusive data, but the final study was done on 33,556 individuals. The mean follow-up period of the study was 9.42 years and in this time 506 deaths from cancer were recorded. It

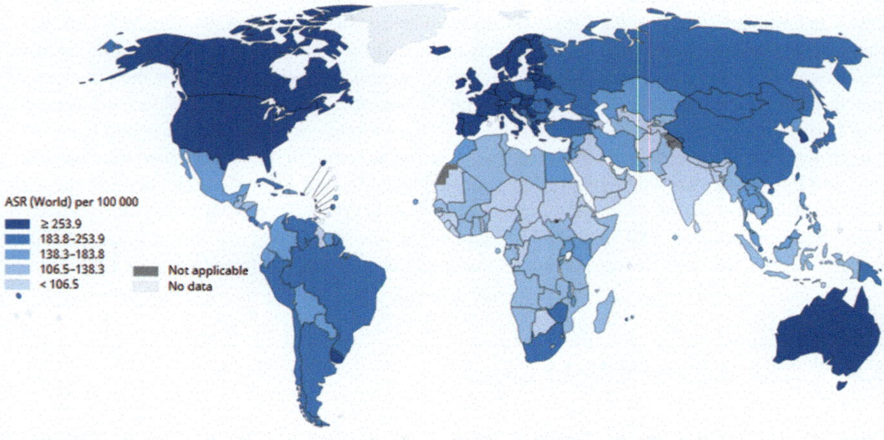

Fig. 6.7 Estimated age-standardized incidence rates (World) in 2018, all cancers, both sexes, all ages, powered by Globocon 2018 World Health, IARC, 2019, Cancer Today, Cancer whole world, map and statistics. Reprinted with permission from Data source: GLOBOCAN 2018 Graph production: IARC (http://gco.iarc.fr/today) (Ferley et al. 2018)

is important to note that diseases such as diabetes and elevated triglyceride also resulted in higher hs-CRP serum. The study revealed that elevated hs-CRP in cancer-free individuals increases the risk of cancer mortality in men, but not conclusively in women. Additionally, this elevated CRP level sees an even higher risk of mortality particularly in lung and colorectal cancers (Ko et al. 2012). The study could not conclusively state if preexisting inflammation led to carcinogenesis, but it did state this as a likely possibility.

6.4.2 CRP as a Prognostic Marker in Different Cancer

One of the most widely used applications of evaluating levels of CRP is in cancer prognosis. Studies have suggested that in patients with solid cancers, higher circulating levels of CRP can lead to a poor prognosis. It has also been seen that healthy patients with higher levels of CRP have a possible increased risk of cancer of various types. This association may be the result of reverse causality, causality, or confounding. (Allin and Nordestgaard 2011) CRP has been implicated as prognostic marker in different cancer. In the case of nasopharyngeal carcinoma, baseline CRP has been determined to be a prognostic factor. In a study conducted in Guangzhou, China, patients were divided into three groups—those whose baseline CRPs are normal, those whose CRPs are always raised and those whose baselines were elevated but never normalized. By studying these patients over time and applying mathematical methods, it was seen that those with an elevated baseline CRP level had a lower chance of survival in comparison to those with normal CRP levels. Elevated CRP levels during treatment were also not seen to be positive indicators. (Chen et al. 2019) In another study, the CRP to albumin ratio was identified as a prognostic marker. In this case, non-metastatic anal squamous cell carcinoma was analyzed. In this study as well, CRP was linked with an unfavorable outcome in the cancerous cases. It is important to note, however, this the reason for this common outcome may differ from case to case. (Martin et al. 2019; Wang et al. 2019a, b).

6.5 Single Nucleotide Polymorphisms and CRP

SNP are places within the genome where the kind of nucleotide can be different among a wide population. Generally speaking, there are only two alleles at an SNP locus. There are a number of SNP sites, and their quantity is in the millions. SNP sites and certain nucleotide substitutions can lead to many changes in a person's protein expression. If an SNP can possibly have this effect, it is called a coding SNP (Daiger et al. 2013).

CRPs may also play an active role in tumorigenesis if single nucleotide polymorphisms (SNP) are present (Su et al. 2014a, b). In the following sections, the role of CRP in cancer diagnosis, prognosis, and case will be discussed alongside the role of SNPs in CRP and their interactions with cancer.

Single nucleotide polymorphisms (SNPs) are cases in which one fundamental building block of a DNA sequence is replaced. SNPs are quite natural and most of them have little to no effect of the functioning of the human body. That being said, SNPs that do have an effect on human health may do in drastic and unexpected ways. Consider cases of inherited depression, for example. While it has been known for a long time that a history of depression in the family increases the likelihood of depression in a patient, the cause behind this could not be identified. However, a recent study conducted in China shows that a certain SNP on CRP may have an association with this occurrence. The study showed that patients with a family history of depression had higher CRP serum levels. Additionally, two out of the five variations in the CRP gene were seen to have a significant association with depression (Ancelin et al. 2015). As the study itself acknowledges, there are many inconsistencies in part of the research and its repeatability. In the case of other diseases, such as colorectal cancers, different papers have found contradictory results for whether rs1205 on CRP has any risk-based interactions.

6.6 Discussion

CRP has been reported in different cancers including association of risk, elevated levels in breast cancer (52) CRP levels>3 mg/L has been reported as a diagnostic marker in invasive breast cancer (30) and elevated CRP levels have been associated with poor survival in many malignant tumors, including breast cancer, soft tissue sarcoma, prostate cancer, renal cell carcinoma, colorectal cancer, non-small-cell lung cancer, malignant lymphoma, pancreatic cancer, potential prognostic marker in gastric cancer (Balaji et al. 2015; Nakamura et al. 2012; Stark et al. 2009; Pierce et al. 2009; Karakiewicz et al. 2007; Redmond, 2003; Scott et al. 2002; Legouffe et al. 1998; Falconer et al. 1995; Lukaszewicz-Zając et al. 2011; Shimura et al. 2012), associated with risk in ovarian cancer (Jing et al. 2017). Recent studies have revealed that elevated CRP levels associated with poor progression and overall survival on cancer patients even under treatment with PD-1 inhibitors or anti-programmed cell death protein-1 (Livanainen et al. 2019) that find application in standard cancer treatments in different cancers. Thus, CRP has a profound role in cancer.

We discuss in this chapter the different genetic variation in CRP and their association in cancer. Circulating CRP concentrations have been associated with risk in cancer (Heikkilä et al. 2011). From a study from FINRISK 1992, FINRISK 1997, and Health 2000, it was found that individuals with one or two variant T alleles at rs1892534 revealed increased overall cancer risk, while individuals with one or two variant A alleles at rs1169300 or rs2464196 increased risk of lung cancer; however, CRP SNPs did not reveal any disease-specific association in colorectal, prostate, or breast cancer either conferring risk of the disease or the probability of developing cancer. Studies on the exact role of CRP and genetic variations are being studied by moist groups working on CRP. Although studies are indicative of role of

CRP in cancer but whether genetic variation forms a cause in cancer remains to be proven in future studies.

References

Agrawal A, Cha-Molstad H, Samols D, Kushner I (2003) Overexpressed nuclear factor-kappaB can participate in endogenous C-reactive protein induction, and enhances the effects of C/EBPbeta and signal transducer and activator of transcription-3. Immunology 108(4):539–547

Allin KH, Nordestgaard BG (2011) Elevated C-reactive protein in the diagnosis, prognosis, and cause of cancer. Crit Rev Clin Lab Sci 48(4):155–170

Ancelin ML, Farré A, Carrière I, Ritchie K, Chaudieu I, Ryan J (2015) C-reactive protein gene variants: independent association with late-life depression and circulating protein levels. Transl Psychiatry 5(1):e499. https://doi.org/10.1038/tp.2014.145

Ansar W, Ghosh S (2013) C-reactive protein and the biology of disease. Immunol Res 56 (1):131–142. (Review)

Ansar W, Ghosh S (2016) Biology of C reactive protein in health and disease. Springer

Ansar W, Mukhopadhyay S, Habib SKH, Basu S, Saha B, Sen AK, Mandal CN, Mandal C (2009) Disease-associated glycosylated molecular variants of human C-reactive protein activate complement-mediated hemolysis of erythrocytes in tuberculosis and Indian visceral leishmaniasis. Glycoconj J 26:1151

Balaji S, Asegaonkar SB, Asegaonkar BN, Takalkar UV, Suresh Advani S, Thorat AP (2015) C-reactive protein and breast cancer: new insights from old molecule. Int J Breast Cancer:145647. 6 pages. https://doi.org/10.1155/2015/145647

Black S, Kushner I, Samols D (2004a) C-reactive protein. J Biol Chem 279:48487–48490

Black S, Kushner I, Samols D (2004b) C-reactive protein J. Biol Chem 279:48487

Boncler M, Wu Y, Watala C (2019) The Multiple faces of C-reactive protein—physiological and pathophysiological implications in cardiovascular disease. Molecules 24(11):2062

Buie JJ, Renaud LL, Muise-Helmericks R, Oates JC (2017) IFN-α Negatively regulates the expression of endothelial nitric oxide synthase and nitric oxide production: implications for systemic lupus erythematosus. J Immunol 199(6):1979–1988

Burley SK, Berman HM et al (2019) RCSB Protein Data Bank: biological macromolecular structures enabling research and education in fundamental biology, biomedicine, biotechnology and energy. Nucleic Acids Res 47:D464–D474

Cha-Molstad H, Young DP, Kushner I, Samols D (2007) The interaction of C-Rel with C/EBPbeta enhances C/EBPbeta binding to the C-reactive protein gene promoter. Mol Immunol 44 (11):2933–2942

Chen RW, Zhou Y, Yuan YJ, Zhang Q, He SS, Chen Y, Ren YF (2019) Effect of CRP and kinetics of CRP in prognosis of nasopharyngeal carcinoma. Front Oncol 9:89. https://www.frontiersin.org/articles/10.3389/fonc.2019.00089/full

Cipriani V, Hogg RE, Sofat R, Moore AT, Webster AR, Yates JRW, Fletcher AE, European Eye (EUREYE) Study Group (2017) Association of C-reactive protein genetic polymorphisms with late age-related macular degeneration. JAMA Ophthalmol 135(9):909

Stephen P. Daiger, Lori S. Sullivan, Sara J. Bowne, Chapter 31—Genetic mechanisms of retinal disease. In: Stephen J. Ryan, SriniVas R. Sadda, David R. Hinton, Andrew P. Schachat, SriniVas R. Sadda, C.P. Wilkinson, Peter Wiedemann (Eds.), Retina, 5th ed., WB Saunders, 2013, pp 624–634, ISBN 9781455707379

Das T, Sen A, Kemph T, Pramanik SR, Mandal C, Mandal C (2003) Induction of glycosylation in human C-reactive protein under different pathological conditions. Biochem J:373

Di JY, Dai ML, Zhang ZX (2018) ENPP1 K121Q (rs1044498 C > A) genetic polymorphism confers a high risk of susceptibility to coronary heart disease. Medicine 97(27)

Falconer JS, Fearon KC, Ross JA, Elton R, Wigmore SJ, Garden OJ, Carter DC (1995) Acute-phase protein response and survival duration of patients with pancreatic cancer. Cancer 75 (8):2077–2082

Fang D, Yu Y (2017) C-reactive protein Gene rs1205 polymorphism is not associated with the risk of colorectal cancer. Biosci Rep 37(4)

Ferlay J, Ervik M, Lam F, Colombet M, Mery L, Piñeros M, Znaor A, Soerjomataram I, Bray F (2018) Global cancer observatory: cancer today. International Agency for Research on Cancer, Lyon, France. Available from https://gco.iarc.fr/today, accessed January 11, 2020

Frank R, Hargreaves R (2003) Clinical biomarkers in drug discovery and development Nat. Rev Drug Disc 2:566

Heikkilä K, Silander K, Salomaa V, Jousilahti P, Koskinen S, Pukkala E, Perola M (2011) C-reactive protein-associated genetic variants and cancer risk: findings from FINRISK 1992, FINRISK 1997 and Health 2000 studies. Eur J Cancer 47(3):404–412

Hudert CA, Selinski S, Rudolph B, Bläker H, Loddenkemper C, Thielhorn R, Berndt N, Golka K, Cadenas C, Reinders J, Henning S, Bufler P, Jansen PLM, Holzhütter HG, Meierhofer D, Hengstler JG, Wiegand S (2019) Genetic determinants of steatosis and fibrosis progression in paediatric non-alcoholic fatty liver disease. Liver Int 39:540–556

Jing L, Xuedan J, Zhongfu Y, Haifeng Q, Ruixia G (2017) C-reactive protein and risk of ovarian cancer: a systematic review and meta-analysis. Medicine 96(34):e7822

Karakiewicz PI, Hutterer GC, Trinh QD, Jeldres C, Perrotte P, Gallina A, Tostain J, Patard JJ (2007) C-reactive protein is an informative predictor of renal cell carcinoma-specific mortality: a European study of 313 patients. Cancer 110(6):1241–1247

Kito M et al (2015) CRP 1846C>T Genetic polymorphism is associated with lymph node metastasis and/or severe lymphatic invasion in endometrial cancer. Tohoku J Exp Med 237 (1):25–30

Ko YJ, Kwon YM, Kim KH, Choi HC, Chun SH, Yoon HJ, Goh E, Cho B, Park M (2012) High-sensitivity C-reactive protein levels and cancer mortality. Cancer Epidemiol Biomark Prev 21 (11):2076–2086

Kushner I, Rzewnicki D, Samols D (2006) What does minor elevation of C-reactive protein signify? Am J Med 119(2):166.e17–166.e28

Lee S et al (2011) High-sensitivity C-reactive protein and cancer. J Epidemiol 21(3):161–168

Legouffe E, Rodriguez C, Picot MC, Richard B, Klein B, Rossi JF, Commes T (1998 Oct) C-reactive protein serum level is a valuable and simple prognostic marker in non Hodgkin's lymphoma. Leuk Lymphoma 31(3–4):351–357

Lei KJ, Liu T, Zon G, Soravia E, Liu TY, Goldman ND (1985) Genomic DNA sequence for human C-reactive protein. J Biol Chem 260:13377

Livanainen S, Ahvonen J, Knuuttila A, Tiainen S, Koivunen JP, Elevated CRP (2019) Levels indicate poor progression-free and overall survival on cancer patients treated with PD-1 inhibitors. ESMO Open 4:e000531. https://doi.org/10.1136/esmoopen-2019-000531

Luan Y-Y, Yao Y-M (2018) The clinical significance and potential role of C-reactive protein in chronic inflammatory and neurodegenerative diseases. Front Immunol 9:1302

Lukaszewicz-Zając M, Mroczko B, Gryko M, Kędra B, Szmitkowski M (2011) Comparison between clinical significance of serum proinflammatory proteins (IL-6 and CRP) and classic tumor markers (CEA and CA 19-9) in gastric cancer. Clin Exp Med 11(2):89–96. https://doi.org/10.1007/s10238-010-0114-5. Epub 2010 Oct 12

Martin D, Rödel F, Balermpas P, Winkelmann R, Fokas E, Rödel C (2019) C-reactive protein-to-albumin ratio as prognostic marker for anal squamous cell carcinoma treated with chemoradiotherapy. Front Oncol 9:1200. https://doi.org/10.3389/fonc.2019.01200

McFadyen JD, Kiefer J, Braig D, Loseff-Silver J, Potempa LA, Eisenhardt SU, Peter K (2018) Dissociation of C-reactive protein localizes and amplifies inflammation: evidence for a direct biological role of C-reactive protein and its conformational changes. Front Immunol 9:1351. https://doi.org/10.3389/fimmu.2018.01351

Nakamura T, Matsumine A, Matsubara T, Asanuma K, Uchida A, Sudo A (2012) Clinical significance of pretreatment serum C-reactive protein level in soft tissue sarcoma. Cancer 118 (4):1055–1061

Ochrietor JD, Harrison KA, Zahedi K, Mortensen RF (2000) Role of STAT3 and C/EBP in cytokine-dependent expression of the mouse serum amyloid P-component (SAP) and C-reactive protein (CRP) genes. Cytokine 12(7):888–899

Patel DN, King CA, Bailey SR, Holt JW, Venkatachalam K, Agrawal A, Valente AJ, Chandrasekar B (2007) Interleukin-17 stimulates C-reactive protein expression in hepatocytes and smooth muscle cells via p38 MAPK and ERK1/2-dependent NFkappaB and C/EBPbeta activation. J Biol Chem 282:27229–27238

Pepys MB, Hirschfield GM (2003) C-reactive protein: a critical update. J Clin Invest 111 (12):1805–1812

Pierce BL, Ballard-Barbash R, Bernstein L, Baumgartner RN, Neuhouser ML, Wener MH, Baumgartner KB, Gilliland FD, Sorensen BE, McTiernan A, Ulrich CM (2009) Elevated biomarkers of inflammation are associated with reduced survival among breast cancer patients. J Clin Oncol 27(21):3437–3444. https://doi.org/10.1200/JCO.2008.18.9068. Epub 2009 May 26

Redmond HP (2003) Systemic inflammatory response predicts survival following curative resection of colorectal cancer. Br J Surg 90(7):889. author reply 889

Salazar J, Martínez MS, Chávez-Castillo M, Núñez V, Añez R, Torres Y, Toledo A, Chacín M, Silva C, Pacheco E, Rojas J (2014a) C-reactive protein: an in-depth look into structure, function, and regulation. Int Sch Res Notices:653045. 11 pages Review Article

Salazar J, Martínez MS, Chávez-Castillo M, Núñez V, Añez R, Torres Y, Toledo A, Chacín M, Silva C, Pacheco E, Rojas J, Bermúdez V (2014b) C-reactive protein: an in-depth look into structure, function, and regulation. Int Sch Res 2014:653045. 11 p. https://doi.org/10.1155/2014/653045

Salazar J, Martínez MS, Chávez-Castillo M, Núñez V, Añez R, Torres Y, Toledo A, Chacín M, Silva C, Pacheco E, Rojas J (2014c) C-reactive protein: an in-depth look into structure, function, and regulation. Int Sch Res

Scott HR, McMillan DC, Forrest LM, Brown DJ, McArdle CS, Milroy R (2002) The systemic inflammatory response, weight loss, performance status and survival in patients with inoperable non-small cell lung cancer. Br J Cancer 87(3):264–267

Selvaraj SM, Christina JJ, Gurumoorthy S, Jayaraman BG, Vellaichamy A (2019) Association of – 757T > C Polymorphism of C-reactive protein gene with chronic periodontitis of South Indian Population. Inflamm Res 68(5):347–349

Shimura T, Kitagawa M, Yamada T, Ebi M, Mizoshita T, Tanida S, Kataoka H, Kamiya T, Joh T (2012) C-reactive protein is a potential prognostic factor for metastatic gastric cancer. Anticancer Res 32(2):491–496

Shrive AK, Cheetham GM, Holden D et al (1996) Three dimensional structure of human C-reactive protein. Nat Struct Biol 3(4):346–354

Sproston NR, Ashworth JJ (2018) Role of C-reactive protein at sites of inflammation and infection. Front Immunol 9:754

Stark JR, Li H, Kraft P et al (2009) Circulating prediagnostic interleukin-6 and C-reactive protein and prostate cancer incidence and mortality. Int J Cancer 124(11):2683–2689. https://doi.org/10.1002/ijc.24241

Su HX, Zhou HH, Wang MY et al (2014a) Mutations of C-reactive protein (CRP) -286 SNP, APC and p53 in colorectal cancer: implication for a CRP-Wnt crosstalk. PLoS One 9(7):e102418. Published 2014 Jul 15. https://doi.org/10.1371/journal.pone.0102418

Su H-X et al (2014b) Mutations of C-reactive protein (CRP) -286 SNP, APC and p53 in colorectal cancer: implication for a CRP-Wnt crosstalk. PLoS One 9(7)

Szalai AJ, Wu J, Lange EM, McCrory MA, Langefeld CD, Williams A, Zakharkin SO, George V, Allison DB, Cooper GS, Xie F, Fan Z, Edberg JC, Kimberly RP (2005) Single-nucleotide polymorphisms in the C-reactive protein (CRP) gene promoter that affect transcription factor binding, alter transcriptional activity, and associate with differences in baseline serum CRP level. J Mol Med 83(6):440–447

Thiele JR, Habersberger J, Braig D, Schmidt Y, Goerendt K, Maurer V, Bannasch H, Scheichl A, Woollard KJ, von Dobschütz E, Kolodgie F, Virmani R, Stark GB, Peter K, Eisenhardt SU (2014) Dissociation of pentameric to monomeric C-reactive protein localizes and aggravates inflammation: in vivo proof of a powerful proinflammatory mechanism and a new anti-inflammatory strategy. Circulation 130(1):35–50

Thompson D, Pepys MB, Wood SP (1999) The physiological structure of human C-reactive protein and its complex with phosphocholine. Structure 7:169

Tillett WS, Francis T (1930) Serological reactions in Pneumonia with a non-protein somatic fraction of Pneumococcus. J Exp Med 52(4):561–571

Wang MY, Zhou HH, Zhang SC et al (2014) Recurrent mutations at C-reactive protein gene promoter SNP position −286 in human cancers. Cell Res 24(4):505–508. https://doi.org/10.1038/cr.2014.7

Wang S, Zhong H, Lu M, Song G, Zhang X, Lin M, Yang S, Qian M (2018) Higher serum C reactive protein determined C Reactive protein single-nucleotide polymorphisms are involved in inherited depression. Psychiatry Invest 15(8):824–828

Wang X, Dai JY, Albanes D, Arndt V, Berndt SI, Bézieau S, Brenner H, Buchanan DD, Butterbach K, Caan B, Casey G, Campbell PT, Chan AT, Chen Z, Chang-Claude J, Cotterchio M, Easton DF, Giles GG, Giovannucci E, Grady WM, Hoffmeister M, Hopper JL, Hsu L, Jenkins MA, Joshi AD, Lampe JW, Larsson SC, Lejbkowicz F, Li L, Lindblom A, Marchand LL, Martin V, Milne RJ, Moreno V, Newcomb PA, Offitt K, Ogino S, Pharoah PDP, Pinchev M, Potter JD, Rennert HS, Rennert G, Saliba W, Schafmayer C, Schoen RE, Schrotz-King P, Slattery ML, Song M, Stegmaier C, Weinstein SJ, Wolk A, Woods MO, Wu AH, Gruber SB, Peters U, White E (2019a) Mendelian randomization analysis of C-reactive protein on colorectal cancer risk. Int J Epidemiol 48(3):767–780

Wang X, Liu S, Zhao X, Fang E, Zhao X (2019b) The value of C-reactive protein as an independent prognostic indicator for disease-specific survival in patients with soft tissue sarcoma: a meta-analysis. PLoS One 14(7):e0219215

Woo P, Korenberg JR, Whitehead AS (1985) Characterization of genomic and complementary DNA sequence of human C-reactive protein, and comparison with the complementary DNA sequence of serum amyloid P component. J Biol Chem 260:13384

Wu Q, Nie J, Wu F, Zou X, Chen F (2018) Association of single-nucleotide polymorphisms of C-reactive protein gene with susceptibility to infantile sepsis in Southern China. Med Sci Monit 24:590–595

Yibulaiyin H, Sun H, Yang Y (2017) Depression is associated with CRP SNPs in patients with family history. Transl Neurosci 8:201–206. https://doi.org/10.1515/tnsci-2017-0027. Published 2017 Dec 29

Young DP, Kushner I, Samols D (2008) Binding of C/EBPβ to the C-Reactive Protein (CRP) Promoter in Hep3B cells is associated with transcription of CRP mRNA. J Immunol 181 (4):2420–2427

Zhang D, Sun M, Samols D (1996a) Kushner (I) STAT3 participates in transcriptional activation of the C-reactive protein gene by Interleukin-6 (*). J Biol Chem 271:9503–9509

Zhang D, Sun M, Samols D, Kushner I (1996b) STAT3 participates in transcriptional activation of the C-reactive protein gene by interleukin-6. J Biol Chem 271:9503–9509

Zhang SM, Buring JE, Lee IM, Cook NR, Ridker PM (2005) C-reactive protein levels are not associated with increased risk for colorectal cancer in women. Obstet Gynecol 105(6):1480

Zhang S, Thakur A, Liang Y, Wang T, Gao L, Yang T, Li Y, Geng T, Jin T, Chen T, Liu JJ, Chen M (2015) Polymorphisms. In C-reactive ProteinandGlypican-5 are associated with lung cancer risk andgartrokine-1influences cisplatin-based chemotherapy response in a Chinese Han population. Dis Markers 2015:1–8

Role of C-Reactive Protein in Tropical Infectious Diseases

7

Junaid Jibran Jawed

Contents

7.1 Introduction 195
7.2 Visceral Leishmaniasis 197
7.2.1 Epidemiology 200
7.2.2 CRP Binding Transforms Parasite 200
7.2.3 CRP Facilitates Parasite Entry 201
7.2.4 CRP in VL Prognosis 201
7.3 Tuberculosis 202
7.3.1 Epidemiology 203
7.3.2 CRP in Pulmonary Tuberculosis 204
7.3.3 CRP in Musculoskeletal Tuberculosis 205
7.4 Malaria 205
7.4.1 Epidemiology 207
7.4.2 CRP in Malaria 207
7.4.3 CRP in Cerebral Malaria 208
7.5 Discussion 209
References 209

Abstract

Tropical diseases are the deadly infectious diseases which are caused by protozoan parasites, and some pathogenic species of bacteria mostly prevalent in tropical and subtropical regions of the world. These diseases mostly affect the population having weaker socioeconomic condition, poor hygiene, and health condition and are highly fatal if left untreated. Visceral leishmaniasis (VL), malaria, and tuberculosis are among the notable contributor of highest mortality

J. J. Jawed (✉)
School of Biotechnology, Presidency University-New campus, Kolkata, West Bengal, India
e-mail: junaid.dbs@presiuniv.ac.in

W. Ansar, S. Ghosh (eds.), *Clinical Significance of C-reactive Protein*,
https://doi.org/10.1007/978-981-15-6787-2_7

due to infections globally. The occurrence of HIV with any of the former infections leads to more serious consequences in the developing countries. Therefore, it is necessary to understand the possible mechanism of their eradication and understanding the prognosis of the disease. In most of these infections, the common event is the increased expression of serum acute phase protein (s) which indicates the onset of infection. Increased expression of acute phase proteins is employed as a possible diagnostic solutions, among which the C-reactive protein (CRP) is the important one. CRP is the first pattern recognition receptors identified till date and it is expressed by the liver cells in response to the various factors secreted by macrophages. In case of VL, during infection macrophages induce high serum CRP to initiate complement cascade and phagocytosis, but *L. donovani* utilizes CRP and its receptor to facilitate its entry into macrophage. While in tuberculosis and malaria, the increased serum level of CRP correlates the severity of the disease. Taken together, this chapter highlights the importance of CRP as an important acute phase protein in the most fatal tropical infectious diseases.

Keywords

C-reactive protein · Tropical diseases · Neglected tropical diseases · Infectious diseases · Visceral leishmaniasis · Tuberculosis · Malaria · Prognosis

Abbreviation

CD 4	Cluster of differentiation 4
CM	Cerebral malaria
CRP	C-reactive protein
Fcγ R	Fc gamma receptors
GIPL	Glycosylinositol phospholipid
gp63	Surface glycoprotein 63
GSPL	Glycosphingophospholipid
HIV	Human immunodeficiency viruses
IFN γ	Interferon gamma
IL10	Interleukin 10
IL-1β	Interleukin 1 beta
IL-6	Interleukin 6
LPG	Lipophosphoglycan
MDR	Multidrug resistant
MHC	Major histocompatibility complex
NTD	Neglected tropical diseases
RBC	Red blood cells
SSG	Sodium stibogluconate
TB	Tuberculosis
TGF	Tumor growth factor
TGF-β	Transforming growth factor beta

TNF-α	Tumor necrosis factor alpha
VL	Visceral leishmaniasis
WHO	World Health Organization

Highlights

- The importance of CRP in tropical infectious diseases.
- Visceral leishmaniasis pathology and clinical significance of CRP.
- Pulmonary and musculoskeletal tuberculosis and symptoms.
- Role of CRP in prognosis and disease progression of tuberculosis.
- Malaria and its different forms including cerebral malaria.
- Clinical significance of CRP in malaria and cerebral malaria.

7.1 Introduction

Vertebrate physiological system is made up of highly complicated and diverse processes, chemical pathways, cells, tissues, and organs which secrete various soluble factors to maintain the homeostasis of the body (Song et al. 2020). Any disturbances to the normal physiological process trigger immune response commonly known as inflammation. Inflammation can be triggered by allergen, pathogen, cell damage, and can be stress induced (Díaz et al. 2020) resulting in the elevation of several soluble serum proteins collectively known as acute phase proteins. Acute phase proteins are certain group of proteins the expression of which frequently fluctuate with the onset of inflammation. The expression of acute phase protein can increase and decrease depending upon the nature of soluble factor and type of inflammation (Lüthje et al. 2020) and mostly used as indicator for the diagnosis purpose. In various studies, monitoring the acute phase protein expressions is used as a parameter to understand the progression of infection, efficiency of therapies, development of resistance and also has been used to design vaccines (Janmohammadi et al. 2020). C-reactive protein (CRP) is an acute phase serum protein and a part of innate immune defense system. It binds with a number of surface expressed receptors/ligands of various microbes, activate defensive responses to succumb the immunological fate of the microbe (Sproston and Ashworth 2018). The surface receptors include phosphocholine and its derivative, parasite expressed lipophosphoglycan (LPG), etc. and trigger strong immune response (Olivier et al. 2005).

CRP is primarily produced by liver cells in response to inflammatory cytokines like IL-6, IL-1β, and TNF-α, where the main source of the cells are macrophages and also from smooth muscle cells and endothelial cells (Boras et al. 2014). The biophysical methods like X-ray crystallography had made it possible to understand the detailed structure of human CRP. Studies showed that CRP is a pentameric protein arranged in annular ring shape in which one surface is occupied by the binding ligands like phosphocholine, whereas the other is exposed for the attachment to the C1q complement protein (Thompson et al. 1999). The name CRP came from the fact that initially it was identified as a soluble serum protein which can react with the antibodies of C-polysaccharide of *pneumococcus*.

CRP expression rapidly rises with infection and inflammation and increases the secretions of inflammatory cytokines, activates complement cascade to clear pathogen or often increased the phagocytosis (Mihlan et al. 2011). It is also involved in clearing apoptotic cells, prevent autoimmune reactions (Gershov et al. 2000). Monitoring the expression of CRP has also significance contribution in predicting the problems of cardiovascular diseases, autoimmunity, renal malfunctioning and can be utilized in predicting the design of an effective vaccine. The increased expression of CRP is mostly used as the parameter to monitor the infection stage and therapeutic efficacies; however, in many infections it has been found that CRP assists the parasite entry into the host system and somehow involved in pathogenic response (Olivier et al. 2005). Infectious diseases like malaria, leishmaniasis, and tuberculosis (TB) trigger strong inflammatory response due to which various serum proteins like interferons, interleukins, cytokines, and chemokines expression rises. The pathogen of these deadly diseases try all possible methods to survive in the hostile environment of the host and thereby make the host immune compromised. However, in many infectious diseases with the onset and progression of the infection the serum CRP expression starts elevating which provides a positive signal for the progression of the infection. The elevated expression of the CRP in many cases helps in the clearance of the parasite from the host system by mounting complement cascade-mediated destruction. Perhaps in many situations, the elevated serum CRP expression is utilized by the parasite for its speedy entry into the host cells and the onset of disease symptoms. In this chapter, the detailed role of the CRP and its importance as a noninvasive and quick prognosis marker has been highlighted in important tropical infectious diseases which include visceral leishmaniasis (VL), tuberculosis, and malaria. Malaria and leishmaniasis are infections of protozoan parasite which have almost similar mode of infections and also vector-borne disease and both results into anemic situation in the susceptible host (Ansar et al. 2009a, b). The major differences between these two infections are their target tissues and types of symptoms. In VL, the affected individuals show enlarged liver and spleen, darkening of skin accompanied by fever, whereas the fever in case of malaria is accompanied by chilling sensation due to cyclic destruction of RBC. Both these events in VL and malaria increase serum CRP concentration but have different fate in each disease. Tuberculosis is a bacterial infection which has different mode of infections and target tissues. This is the most fatal infection and responsible for highest mortality in infectious disease category. The pathogen of tuberculosis can affect both pulmonary and extra-pulmonary tissue which is also known as musculoskeletal tuberculosis commonly called bone TB. In tuberculosis, the bacterial infection leads to increased expression of CRP which is used as an indicator for staging of the infection and as an important prognosis value. Different studies also pointed out controversial role of CRP in these diseases that are the major contributors to the global fatality due to infection and hence their diagnostics advantage based on CRP expression and the mechanism of action is matter of extensive research.

7.2 Visceral Leishmaniasis

Leishmaniasis is an infectious disease caused by protozoan parasite of genus *Leishmania* mostly affecting the tropical and subtropical countries and marking 1.5–2 million new cases each year and making 350 million people at risk causing around 70,000 deaths of each year (Torres-Guerrero et al. 2017). The disease is mainly effecting the population having weaker socioeconomic condition of several countries and hence also considered as neglected tropical disease (NTD). Due to the involvement of different species of *Leishmania* (Fig. 7.1), the clinical symptoms of the disease are different and are named as cutaneous, mucocutaneous, diffuse cutaneous, and visceral forms (Fig. 7.1) (Piscopo and Mallia 2007; Murray 2002; Desjeux 2004).

Post-kala-azar dermal leishmaniasis (PKDL) is a condition when *Leishmania donovani* invades skin cells, resides, and develops there and manifests as dermal lesion. Some of the kala-azar cases manifests PKDL after a few years of treatment. Actually, it is believed that PKDL develops after resolution of visceral leishmaniasis. The time interval to the development of PKDL is variable. Recently, it is believed that PKDL may appear without passing through visceral stage (Piscopo and Mallia 2007; Murray 2002; Desjeux 2004).

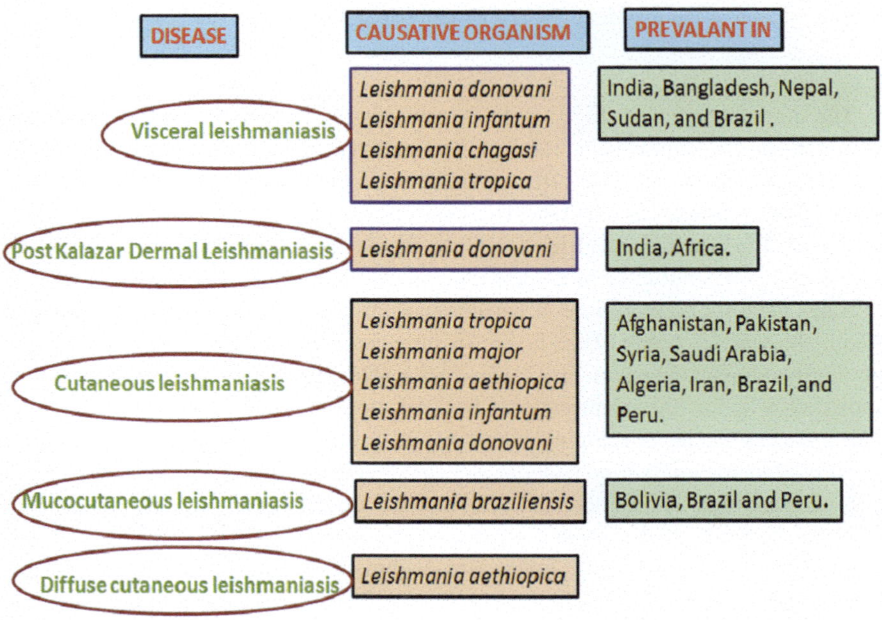

Fig. 7.1 Organisms involved in Leishmaniasis with their geographical prevalence. There are at least 20 species of Leishmania. Each species has a disease specific host response. Organism differs by geographical prevalence also. PKDL develops after resolution of VL.Leishmaniasis is prevalent in 60 countries worldwide (Piscopo and Mallia 2007; Murray 2002; Desjeux 2004)

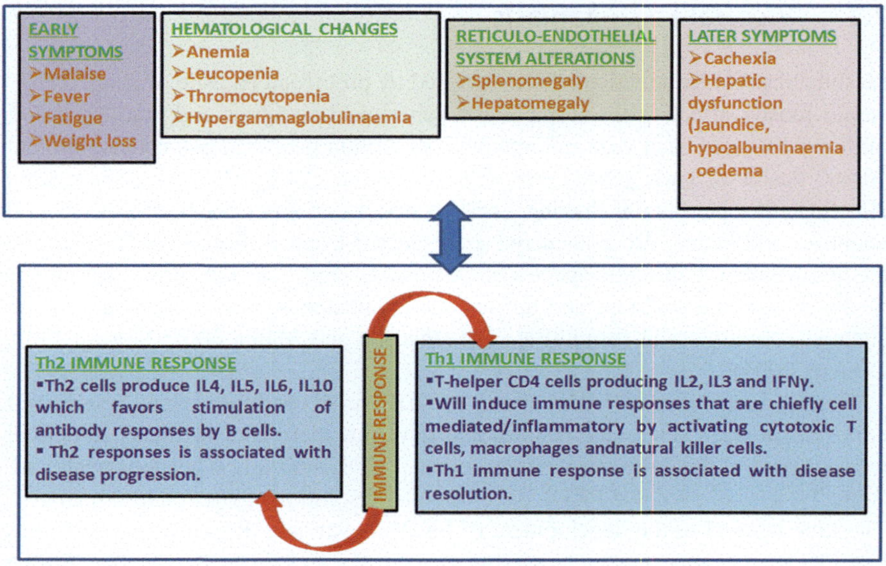

Fig. 7.2 immune responses and clinical presentation in VL. Features of VL include high grade fever, hepatosplenomegaly, weight loss and abdominal discomfort with no rigors. In Indian VL disease, hyperpigmentation of skin is a also a clinical feature. Other features were also noted. Immune response is mostly cell mediated (Paul and Bagga 2019; Piscopo and Mallia 2007)

The immune response (Fig. 7.2) to *Leishmania* infection is mostly cell mediated. The parasite is exclusively intracellular, mainly residing in macrophages as replicating amastigotes. The effect of infection will depend on whether the host accumulates primarily a T-helper (Th)-1 or Th2 response to succumb the intruding organism. The parasite also tries to reduce the T cell responses of the host by the release of many soluble factors (Heinzel et al. 1989; Piscopo and Mallia 2007).

It has been found that visceral leishmaniasis is the most fatal one as the parasite infects the visceral organ of the host causing enlargement of liver and spleen in the susceptible host (Jawed et al. 2016, 2018). The disease is highly fatal if left untreated, whereas the occurrence of HIV-VL co-infections leads to more serious consequences in the developing countries. It is the second largest parasite killer globally after malaria yet considered as neglected tropical infection. Children aged 1–4 years are regarded as more susceptible to the disease. As the protective immune response is mainly cell-mediated immunity that effect in subclinical infection and sometimes spontaneous cure in most cases. Failure of immunity in children results in illness, like in every case of VL; there are nearly 30 subclinical infections. Malnutrition and HIV co-infection also predispose the children to this infection (Paul and Bagga 2019).

Although modern therapeutic approach has been tested in experimental model successfully, the effective human vaccine is yet under trial (Olivier et al. 2005). The importance of several serum proteins has been highlighted in order to implement a

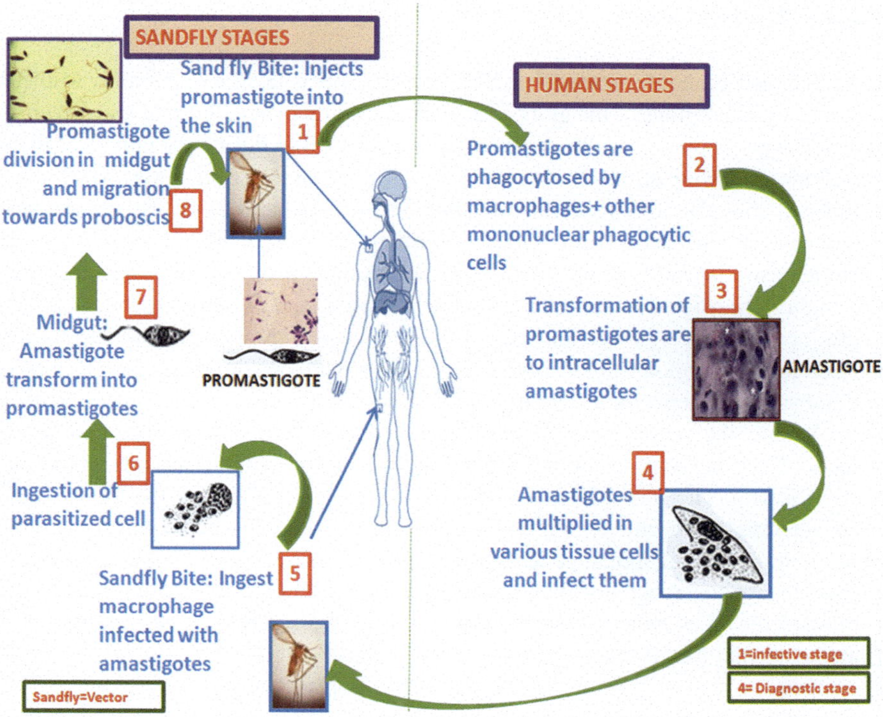

Fig. 7.3 Lifecycle of *Leishmania* sp. The lifecycle of Leishmania goes through two host (Human and Sandfly)with promastigote and amastigote stage. The infective stage is marked (cdc.gov)

successful immune-therapeutic or diagnostic approach among which CRP is the important one. CRP is the first pattern recognition receptors identified till date and it is expressed by the liver cells in response to the various factors secreted by macrophages (Newling et al. 2019). The protein is named as C-reactive because of its property that enables it to bind with the C-polysaccharide of the pneumococcus (Singh et al. 1999).

Leishmania parasite is transmitted by the bite of female phlebotomine sandflies. The sandflies inject the infective stage, promastigotes, during blood meals (1). Promastigotes that reach the puncture wound are phagocytized by macrophages (2) and transform into amastigotes (3). Amastigotes multiply in infected cells and affect different tissues, depending in part on the *Leishmania* species (4). This originates the clinical manifestations of leishmaniasis. Sandflies become infected during blood meals on an infected host when they ingest macrophages infected with amastigotes (5 and 6). In the sandfly's midgut, the parasites differentiate into promastigotes (7), which multiply and migrate to the proboscis (8) (Fig. 7.3).

7.2.1 Epidemiology

Leishmaniasis is an infectious disease of protozoan parasite affecting mainly tropical and subtropical regions of the globe. The disease is prevalent in drought and poverty-afflicted region and spreads by the bite of infected sandfly during the course of their blood meal (Jawed et al. 2016). WHO (World Health Organization) 2017 data confirmed that the seven countries which include India, Kenya, Brazil, Sudan, Ethiopia, Somalia, and South Sudan contribute to 94% of all the new cases of leishmaniasis. In 2018, India was among the leading nations which had reported the highest registered cases of visceral leishmaniasis. *Leishmania donovani* the causative agent of visceral leishmaniasis is a tropical protozoan parasite which needs temperate region and its hospitality for its growth, development, pathogenesis, and transmission hence India and its neighboring countries are among the vulnerable one. Among the major affected territories in India most of the infections were reported from the state of Bihar, Uttar Pradesh, West Bengal, Assam, and some part of Sikkim. The state of Bihar alone contributes around 70% of the VL cases in India whereas the specific region includes Purnia, Araria, and Kishanganj along with their vicinities. Of the Indian patients, 40% of the VL patients had already developed resistance against most effective antimonial drugs whereas the phenomena still continued for other effective drugs. India along with its neighboring countries like Bangladesh and Nepal shares the 67% of the global burden of VL which is a matter of concern.

In general, leishamnaisis is a collective name of disease caused by the protozoa belonging to the genus of *Leishmania*, perhaps the VL is considered as the most fatal one prevalent mostly in African, South American, South Asian, and some other regions of the world. At present, it is endemic in 80 countries globally. Around 30,000 cases registered each year for VL and in each case the morbidity and mortality are very high due to drug resistance, negligence in diagnosis, and unresponsive treatment. 20,000 deaths due to VL are reported each year. Conditions like high population, famine, drought, etc. mostly favor the spread of disease.

7.2.2 CRP Binding Transforms Parasite

Leishmania donovani is the protozoan parasite which has several glycoprotein and glycolipid moieties on its cell surface which include lipophosphoglycan (LPG), gp63, glycosphingophospho lipid (GSPL), glycosylinositol phospholipid (GIPL), etc. (Olivier et al. 2005). These surface molecules have several functions and advantages for the parasite. The onset of VL takes place when the infected sandfly bites the mammalian host for its blood meal, during which the promastigote (with flagella) form of the parasite enters into the circulation and from there it attacks the liver, spleen, and bone marrow of the susceptible host. During the course of infection, the parasite changes its surface molecules and transforms from promastigotes to amastigotes (without flagella) of morphological stage (Olivier et al. 2005). The most virulent form of the parasite is metacyclic promastigote stage which infects the host macrophages and starts the onset of infection (Jawed et al. 2019). The factors which are responsible for the transformation of the parasite include pH, temperature, and specific serum protein among which CRP have shown to be playing an important role. It was reported that CRP binds to the specific disaccharide residue of LPG expressed by the promastigotes form of the parasite and helps in its transformation from promastigotes to amastigotes (Mbuchi et al. 2006). However, *L. major* or other strain having deficient LPG or the disaccharide unit of LPG showed less or no binding to CRP. It was also evidenced that direct opsonization of *Leishmania* LPG with CRP increases its uptake by human macrophages (Culley et al. 1996).

7.2.3 CRP Facilitates Parasite Entry

The parasite *Leishmania* is an obligate intracellular protozoan; therefore, their survival and pathogenesis starts once they evade innate immune response and enter into the hostile environment of the macrophages (Parveen et al. 2018). In order to do so, the surface molecules of the parasite combine with several serum proteins which opsonize it and make it available for the attachment to the different receptors present on the surface of macrophages. During their uptake, the macrophages cause less harm to the parasite and thus parasite gets survival advantage (Wright and Silverstein 1983). CRP has found to be one of the important opsonization factors which combines with LPG residues of the *Leishmania* and helps its uptake to the macrophage. The increased concentration of CRP causes more opsonization of the parasite and aids in combining with the specific receptor known as FcγR receptors including the subtypes FcγR I and FcγRIIa of the phagocytes and thereby facilitates the entry of the parasite (Bodman-Smith et al. 2002). Although CRP is responsible for promoting Th-1 inflammatory environment to the host system by inducing IFNγ, TNF-α, and other pro-inflammatory cytokines which provide strong leishmanicidal effect (Mortensen and Zhong 2000), but due to the facilitated entry of the parasite, the parasite downregulates these effect by its LPG and increased synthesis of TGF-β and IL-10.

Thus, it was found that CRP is a critical acute phase protein which although required as an innate defense mechanism, in case of leishmaniasis it helps in the transformation of the parasite, facilitates its entry into the hostile environment of the macrophages. It was also evidenced that the expression of CRP can be used as a parameter to monitor the progress of infection, disease prognosis, and emergence of drug resistance during treatment.

7.2.4 CRP in VL Prognosis

The serum level of CRP is helpful in monitoring the therapeutic progression and prognosis value. Since initially most of the treatment progression were monitored through invasive techniques like splenic and bone marrow aspirations. The successful progress of the treatment can be monitored with the estimation of decreased CRP expression (Singh et al. 1999), and it can be analyzed through simple spectrophotometric techniques of serum samples. The study also provided the insight to understand the possibility of resistance development, in which the raised serum CRP level after sodium stibogluconate treatment every 10 days predicts the non-functioning of the drug and higher parasite burden. Whereas for changing the type of drug from SSG to any other like amphotericin B, pentamidine restored the normal serum CRP level (Singh et al. 1999). Another study further clarified the notion of increased serum CRP level in SSG-treated patient proves the unresponsiveness and development of resistance varieties against drug (Ansari et al. 2007). Therefore, the expression of CRP was proved to be a direct correlation with the parasite burden and diseases status.

7.3 Tuberculosis

Tuberculosis is an air-borne infectious disease caused by bacterial pathogen *Mycobacterium tuberculosis*. The pathogen is highly infectious and after successful infection the symptoms of lung damage are associated with severe cough, chest pain, and fever arises which in most of the cases leads to fatal end of the host if remain untreated (Fogel 2015). Every year 1.3 million deaths occur worldwide out of 10.4 million cases reported each year and estimated one quarter of the world's population were infected by tuberculosis pathogen by 2018 (Pezzella 2019). As per the World Health Organization report, due to the highest rate of mortality the disease is considered as the number one cause of the highest death rate under infectious disease category. The disease is the most ancient one and the anthropological studies showed that it was also prevalent 4000 years ago among the Egyptian's population (Stead 1997).

Tuberculosis infection onset takes place by inhaling the aerosols contaminated with live *Mycobacterium tuberculosis* where they enter into the pulmonary cavity. After reaching the alveolar passage, they are engulfed by the resident macrophages, pneumocytes, and dendritic cells (Delogu et al. 2013). The process of phagocytosis takes place with the help of various receptors which facilitate the attachment and entry of the pathogen to the resident cells which include Toll-like receptors, mannose-binding receptors, complement receptors, etc. During the process of infection, the host membrane cholesterol is also found to be helpful in facilitating the entry of the bacteria (Miner et al. 2009). Once entered, it resides in the endocytes of the host and inhibits intracellular calcium which delays the phagosomal maturation and thereby presentation to CD4+ T cells by MHC-II (Sakai et al. 2014). At this point, when the pathogen takes over the control of host system, these infected cells secrete various chemokines which attract other inactivated immune cells like monocytes, neutrophils, and lymphocytes. This assembly helps in the spread of the bacteria and ultimately the aggregate immune cells form a granulomatous lesion or granuloma (Silva Miranda et al. 2012). The granuloma gradually liquefies by the action of bacterial secreted molecules and the lesion becomes highly efficient

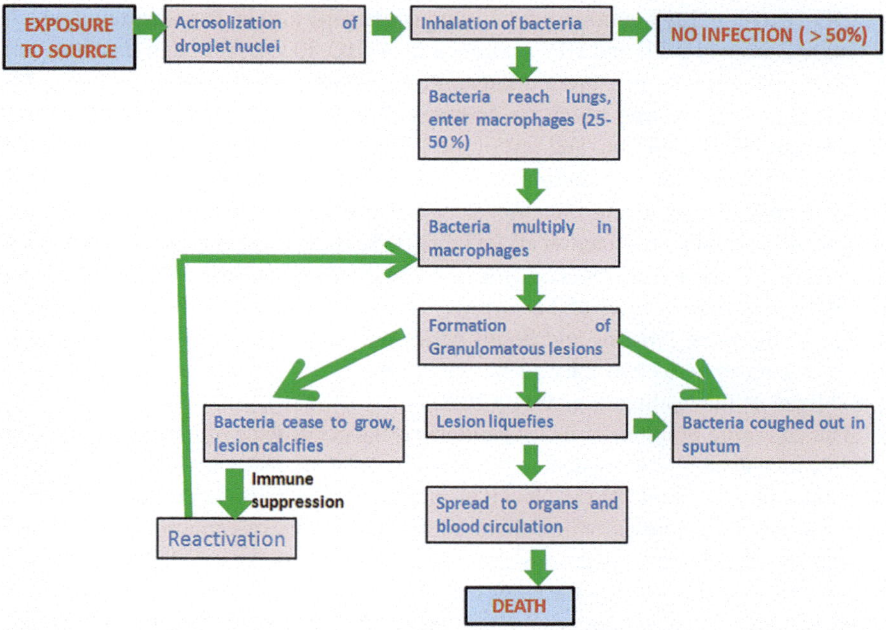

Fig. 7.4 Pathogenesis of tuberculosis. Schematic presentation of pathogenesis of tuberculosis

medium for the growth of upcoming bacteria. After this, the pathogen spreads throughout lungs and the onset of active tuberculosis takes place; at this stage, the person becomes infectious.

The site of infection is not only limited to the pulmonary tissues, but the disease can also effect the skeletal tissues which is known as musculoskeletal tuberculosis or extra-pulmonary tuberculosis (Leonard and Blumberg 2017; Gorse et al. 1983). During this condition, the musculoskeletal tissues of the body which include joints, long bones, and spines get affected by the pathogen (Pigrau-Serrallach and Rodríguez-Pardo 2013). Joint and spine tuberculosis is now emerging as a serious threat as the treatment cost is very high due to surgical procedures and the suffering of the patients is unbearable. Emerging drug resistance is a common phenomenon associated with tuberculosis which is the main reason for the failure of the medication and the area of extensive research. The detailed pathogenesis of tuberculosis is shown in Fig. 7.4.

7.3.1 Epidemiology

Tuberculosis is a fatal infectious disease caused by acid-fast bacteria belonging to the genus of *Mycobacteria*. Globally, TB is considered as one of the top 10 causes of death whereas the leading cause of death is due to single infection of a pathogen

(WHO 2020). In 2018, around 1.5 million people died of this disease among which 251,000 people had TB and HIV co-infection (WHO 2020). Although treatable and curable, data showed that around ten million people got infected with active TB pathogen in 2018, among which the numbers of men were 5.7 million, women were infected around 3.2 million, and rest of the 1.1 million were the numbers of affected children. Among this, 1.1 million affected children's data showed that 205,000 children died due to severe infection which also includes the numbers of children having TB and HIV co-infection (WHO 2020). Studies suggested that in 2018, 87% of the new TB cases were reported from the countries which were considered among the 30 highest burden cases of TB formerly. India, Bangladesh, Pakistan, Indonesia, China, Philippines, Nigeria, and South Africa are among the top 8 countries which together contribute to two third of the global burden of TB among which India is considered as the leading one (WHO 2020).

The emerging problems with tuberculosis is the occurrence of multidrug resistant strain which even found resistant to the first-line drug of tuberculosis like rifampicin. The multidrug resistant TB (MDR-TB) is a severe threat to the vulnerable countries and also for the countries which share higher percentage of economically affected population. WHO recent data suggested that 484,000 cases of resistant tuberculosis arises among which 78% are the cases of MDR-TB. A sharp decline of mortality due to tuberculosis had been noticed from 2000 to 2018 which is accounted for saving the life of around 58 million people. Even the cases of tuberculosis have been found to be decreased by 2% globally, but a constant check in the incidence of MDR-TB is required. The sustainable development has targeted to end the TB endemic by 2030.

7.3.2 CRP in Pulmonary Tuberculosis

Pulmonary tuberculosis is the major infection caused by *Mycobacterium tuberculosis* which affects the lung tissues whereas other forms comprise of musculoskeletal tuberculosis where the bones and spinal tissues are damaged (Gorse et al. 1983). Infection of pulmonary tuberculosis is found to increase the expression of serum CRP level. Studies have revealed that serum CRP level is higher in the infected individuals, and the levels are gradually found to be slowing down upon successful therapy (Schleicher et al. 2005). This study also confirmed the direct correlation between the serum CRP level and therapeutic efficacy. Monitoring the expression of CRP also confirmed to be useful in case of patients suffering from HIV co-infection. It was found that point-of-care CRP-based tuberculosis screening revealed progress of therapy in patients having HIV infection and undergoing retroviral therapy (Yoon et al. 2017). Although CRP level cannot be used as a rule-out test for active tuberculosis, the bacterial load can be confirmed by its expression. Several studies revealed the fact that the CRP screening in stored serum specimen also showed 2–6-fold higher specificity (Lawn et al. 2013). Overall, it was found that among the other biomarkers to be as the monitoring parameters for the antituberculosis therapy, CRP is important and its change in expression gives a significance prognosis value for successful therapy in more than 90% population (Yoon et al. 2017).

7.3.3 CRP in Musculoskeletal Tuberculosis

Musculoskeletal tuberculosis is another form of active tuberculosis where the joints, bones, and spines of the patient are severely affected by *Mycobacterium tuberculosis* infection and is 10% of all the extra-pulmonary tuberculosis as per the database of US tuberculosis cases (Leonard and Blumberg 2017). The suffering of the patient is very high in this form of tuberculosis as the most common symptoms of the disease are severe pain associated with fluid accumulation. Spinal tuberculosis is predominant of all musculoskeletal tuberculosis and comprises around 50% of the extra-pulmonary tuberculosis, where the vertebral bone and the spinal tissues are affected. Percivall Pott was the first person to describe the occurrence of spinal tuberculosis and therefore the disease is often called Pott's disease after his name (Gorse et al. 1983). In most of the cases, surgical removal of the vertebral bone left is the only option for the patients. Studies revealed that the CRP levels were higher in patients having active spinal tuberculosis, the level of which gradually falls after surgical removal of the affected vertebra. Since more than six months, the follow-up studies showed that the expression of CRP was continuously decreasing (Cao et al. 2018) giving rise to the indication of restoration of healthy physiology and decreased parasite burden. Although the detailed mechanism of CRP in the progression of extra-pulmonary tuberculosis still needs much study, the correlation of CRP with the active phase of disease gives enough of prognosis importance and valuable information in noninvasive diagnosis for the spinal tuberculosis.

7.4 Malaria

Malaria (literal meaning bad air) is a vector-borne infectious disease caused by the spread of protozoan parasite *Plasmodium* in the mammalian host and is transmitted by the bite of infected mosquitoes (Tangpukdee et al. 2009). The spread of infection takes place by the bite of infected female Anopheles mosquito, the saliva of which harbors the sporozoites form of the parasite *Plasmodium* (Fig. 7.5). After entering into the blood stream, the sporozoites travel through the blood vessels to the liver where the asexual reproduction of the parasite takes place in the hepatocytes (Yap et al. 2019). The asexual cycle of the parasite is also known as schizogony where the sporozoites produce thousands of other morphological form of the parasites known as merozoites (Yap et al. 2019). These merozoites then start infecting the red blood cells and produce 8–24 new merozoites per cell after each round of infection and the infection continues to spread. The RBC bursts and releases these merozoites where the cell debris trigger strong immune responses causing appearance of chilling fever after every round of cycle which is the most common symptoms of malaria (Mohandas and An 2012). The other symptoms of the disease are weakness, headache, and vomiting and usually appears between tenth and fifteenth days of infection. The rest of the merozoites transformed into immature gametocytes which can give rise to male and female gametes (Yap et al. 2019). These gametocytes enter into the mosquito gut and form male and female gamete during next round of

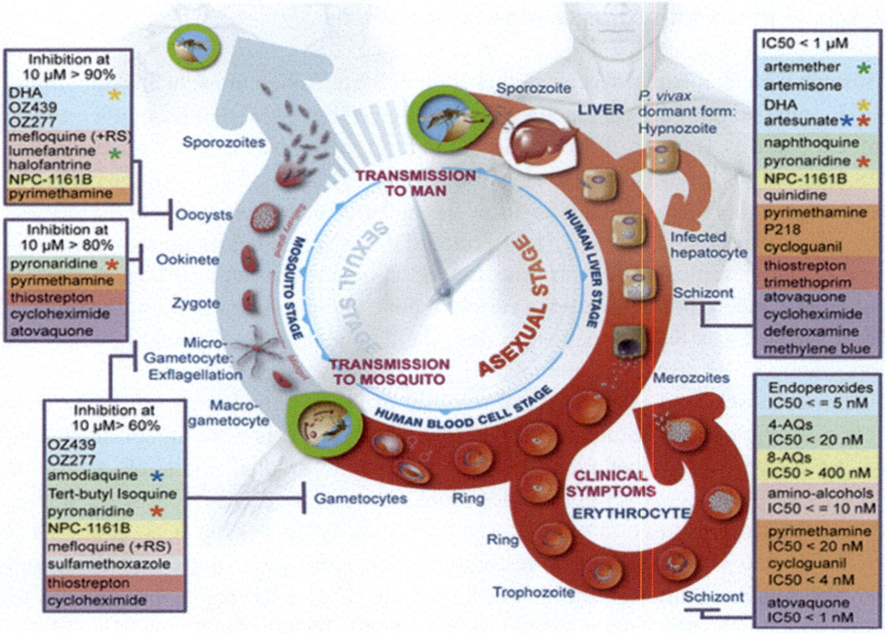

Fig. 7.5 Summary of the activity of the most widely used antimalarials throughout the life cycle of *Plasmodium*. The three main phases, i.e., liver stage, blood stage, and vector stage, of the life cycle of *Plasmodium* are shown. The two key entry points leading to transmission of the parasites from vector to host and from host to vector are indicated (green circles). Parasite forms specific to each stage are highlighted and drugs identified as inhibitors of development of these forms are listed in boxes and coloured as described in Fig. 7.5. Stars highlight components of the main artemisinin combination therapies: green, coartem; red, pyramax; orange, eurartesim; blue, ASAQ. Image adapted from Delves M, Plouffe D, Scheurer C, Meister S, Wittlin S, Winzeler EA, et al. (2012). The Activities of Current Antimalarial Drugs on the Life Cycle Stages of *Plasmodium*: A Comparative Study with Human and Rodent Parasites. PLoS Med 9(2): e1001169. https://doi.org/10.1371/journal.pmed.1001169. Under Creative commons Liscence

mosquito bite in the infected host, and these gametes fuse to form ookinete which is a fertile and motile zygote of the parasite (Aly et al. 2009). These ookinetes then develop into sporozoites which enter into the salivary gland of the mosquitoes and with next mosquito bite they infect the healthy host and in this way the spread of the infection continues (Aly et al. 2009). The different species of *Plasmodium* are *P. falciparum, P. ovale, P. vivax, and P. malariae* which can infect human where *P. falciparum* is the most deadly one causing cerebral malaria.

The disease is mostly prevalent in the tropical and subtropical countries and mostly endemic in major continents like Africa, Asia, Central and Latin America. It is one of the reasons for high fatality in infection diseases worldwide. Every year around 220 million new cases registered for this disease among which 1–2 million deaths have been reported (Lamb et al. 2006). The usage of different antimalarials with different life cycle point is mentioned in Fig. 7.5.

7.4.1 Epidemiology

Malaria is another deadly infectious disease caused by the bite of infected mosquitoes which transmits the protozoan parasite *Plasmodium* into the blood stream of the infected host. The disease is preventable and curable but still counted among the leading of death due to single infection globally. The parasite mostly evolves in tropical and subtropical regions globally. In 2018, an estimated 228 million cases of malaria were reported globally which resulted in 405,000 deaths including men, women, and children (WHO). Among children falling under the age group of 5 years are the most vulnerable one and accounted for 272,000 (around 67%) death due to malaria globally. African region is the most vulnerable among all other countries in sharing the highest proportion of malaria-affected individuals which accounts for 93% of the patients and that resulted into 94% of the deaths due to the same. Six counties held accounted for sharing half for more than half of the total global burden of malaria which includes Nigeria, Uganda, Democratic Republic of Congo, Mozambique, Cote d'Ivoire, and Niger among which Nigeria shares the highest burden (WHO 2020). Since the different species of *Plasmodium* are involved in spreading varieties of pathogenesis which include *P. falciparum, P. malariae. P.ovale, and P. vivax,* the most fatal one is *P. falciparum.* As per data, it was found that P. falciparum causes 99.7% cases of malaria in WHO designated African region, 50% in Southeast Asian region, 71% in and around Eastern Mediterranean, and 65% in Western Pacific region. *P. vivax* is dominant among American population accounting for 75% of the incidence of malaria (WHO). Development of resistance is another common phenomenon associated with this disease as found in other similar infectious diseases.

7.4.2 CRP in Malaria

CRP helps to activate the immune system and has been found to be an effective protein, the level of which gets increased after microbial pathogen infection. From very long time, the levels of CRP have been used to monitor the extent of infection in several diseases and therefore the same concept holds in case of malaria. As malaria is a parasite infectious disease of blood, it is very likely that the infection induces certain changes in the expression pattern of several serum acute phase proteins and so the level of CRP too gets affected. It has been found that during *Plasmodium* infection the serum level of CRP gets significantly increased marking the onset and progress of infection, and during the ongoing therapy the level gradually falls. Being a molecule of the immune system, the fluctuation in the expression of CRP is used as immune marker to monitor the prognosis of the malaria. Studies on African population of malaria-affected individuals revealed that CRP levels were perfectly correlated with the parasite burden and clinical complications in case of malaria (Paul et al. 2012). Even in the study performed in Eastern Indian scenario, it has been revealed that CRP indeed is useful as a biomarker for the assessment of the severity of malaria and to use this as a tool for follow-up of patient for the therapeutic

Table 7.1 CRP in Malaria prognosis

Major research findings	References
1. Elevated level of CRP can be useful in predicting *P. falciparum* disease severity even in absence of clinic manifestation like fever	Bhardwaj et al. (2019)
2. Study in patients having increased CRP levels confirmed that they had more than an eight-fold likelihood for parasitemia considered after correction for other parameters	Sarfo et al. (2018)
3. High serum-CRP yielded a high sensitivity, accuracy and specificity for high-grade, moderate and low malaria, respectively, therefore, may serve as an effective supplementary diagnostic and prognostic biomarker for *Plasmodium* parasite infection	Addai-Mensah et al. (2019)
4. Study performed in Tanzanian children who were in the endemic area of malaria showed that CRP associated with increased parasite densities provides an objective measure of malaria-specific morbidity	Hurt et al. (1994)
5. After the malarial treatment If level of CRP and PCT fail to return to normal, it indicates that the patient may suffer from malaria recrudescence	Li et al. (2019)
6. CRP is an effective biomarker in assessing malaria severity in patient from eastern India which also corroborated the similar finding performed in African population	Paul et al. (2012)
7. CRP and procalcitonin are not sufficiently accurate for diagnosing invasive bacterial infections in the population of hospitalized children with complicated severe acute malnutrition	Page et al. (2014)
8. The combination of travel history, fever prior to blood sampling, and CRP serum levels above or below 10.8 mg/L upon hospital admission, best discriminated between malaria patients and control persons	Stauga et al. (2013)

progress (Paul et al. 2012). Although recent advancement in case of CRP and malaria, correlation studies performed in sub-Saharan African population showed that complete reliability of the malarial severity and CRP expression cannot always be used to predict parasitemia (Sarfo et al. 2018). Although the data showed more than half population have increased level of serum CRP and showing parasite positive indication (Sarfo et al. 2018). Prognostic role of CRP in malaria is evaluated in Table 7.1.

7.4.3 CRP in Cerebral Malaria

Cerebral malaria (CM) is the most severe form of malaria caused by *P. falciparum*. The fatality rate of CM is more than any other of its forms as it affects mainly the brain and thus the entire central nervous system. During the infection, the parasite infects the blood cells causing blockage of the blood vessels connected to the brain (Lamb et al. 2006). As a result of infection, swelling occurs to various regions of the brain resulting in fatal end of the affected patient. The disease mainly affects the children and immune-compromised individuals. It was found that serum CRP level was increased with the increased parasite burden in case of CM (Haghighi 1969), showing a direct correlation of the disease exacerbation and expression of CRP

(Naik and Voller 1984). Other studies further put light into the similar findings that CRP is required for the progression of the disease as genetically knockout CRP mice showed less severe CM complexity than that of the wild type. It was also evidenced that CRP knockout mice showed higher survival rate than WT mice during experimental cerebral malaria (Szalai et al. 2014). Taken together these studies provided the correlation between CRP and cerebral malaria and also the evasion of immune system by the pathogen through taking advantage of the CRP to beat immune surveillance.

7.5 Discussion

CRP is one of the important acute phase proteins, a soluble pattern recognition receptor, and an important biomarker of the immune system to study the extent of microbial infection and metabolic disease. It mainly combines with the molecular pattern found on the surface of the pathogen and induces strong immune response to destroy them as a result of which its expression becomes high even in almost all infections. Frequent fluctuation in the expression of CRP also indicates that occurrence of resistance or studies have also confirmed that serum CRP expression is useful in vaccine design and implementations. Yet the importance of the CRP lies in fact that the increased expression of the CRP also promotes increased parasite burden and helps pathogens for their entry into the hostile environment of the host. Although more studies are needed to establish the mechanism of *Mycobacterium tuberculosis* (*M. tuberculosis*) and *Plasmodium* survival and entry using the serum increased CRP expression, enough studies have been performed to establish the role of CRP in the progression of infectious diseases like leishmaniasis, tuberculosis, and malaria. CRP combined with *Leishmania* surface expressed LPG molecule promotes its entry to the macrophages and thereby assists in infection. Similar phenomena have been observed in case of cerebral malaria where the CRP knockout strain shows more survival advantages. However, the significance of CRP relies upon its convenient usage and easy to identify molecular marker to study the progress of microbial infection, to monitor therapeutic efficacies, and to understand the occurrence of drug resistance in the patient.

References

Addai-Mensah O, Annani-Akollor ME, Fondjo LA, Anto EO, Gyamfi D, Sallah L, Agama D, Djabatey R, Owiredu EW (2019) High-sensitivity C-reactive protein: a potential ancillary biomarker for malaria diagnosis and morbidity. Dis Markers 2019:1408031. https://doi.org/10.1155/2019/1408031

Aly AS, Vaughan AM, Kappe SH (2009) Malaria parasite development in the mosquito and infection of the mammalian host. Annu Rev Microbiol 63:195–221

Ansari NA, Sharma P, Salotra P (2007) Circulating nitric oxide and C-reactive protein levels in Indian kala azar patients: correlation with clinical outcome. Clin Immunol 122(3):343–348. Epub 2007 Jan 9

Ansar W, Habib SK, Roy S, Mandal CN, Mandal C (2009a) Unraveling the C-reactive protein complement-cascade in destruction of red blood cells: potential pathological implications in Plasmodium falciparum malaria. Cell Physiol Biochem 23(1–3):175–190. https://doi.org/10.1159/000204106. https://www.who.int/news-room/fact-sheets/detail/tuberculosis

Ansar W, Mukhopadhyay S, Habib SK, Basu S, Saha B, Sen AK, Mandal CN, Mandal C (2009b) Disease-associated glycosylated molecular variants of human C-reactive protein activate complement-mediated hemolysis of erythrocytes in tuberculosis and Indian visceral leishmaniasis. Gycoconj J 26(9):1151–1169. https://doi.org/10.1007/s10719-009-9236-y

Bhardwaj N, Ahmed MZ, Sharma S, Nayak A, Anvikar AR, Pande V (2019) C-reactive protein as a prognostic marker of Plasmodiumfalciparum malaria severity. J Vector Borne Dis 56 (2):122–126. https://doi.org/10.4103/0972-9062.263727

Bodman-Smith KB, Mbuchi M, Culley FJ, Bates PA, Raynes JG (2002) C-reactive protein-mediated phagocytosis of Leishmania donovani promastigotes does not alter parasite survival or macrophage responses. Parasite Immunol 24(9–10):447–454

Boras E, Slevin M, Alexander MY, Aljohi A, Gilmore W, Ashworth J, Krupinski J, Potempa LA, Al Abdulkareem I, Elobeid A, Matou-Nasri S (2014) Monomeric C-reactive protein and Notch-3 co-operatively increase angiogenesis through PI3K signalling pathway. Cytokine 69 (2):165–179

Cao G, Rao J, Cai Y, Wang C, Liao W, Chen T, Qin J, Yuan H, Wang P (2018) Analysis of treatment and prognosis of 863 patients with spinal tuberculosis in Guizhou Province. Biomed Res Int 2018:3265735

Culley FJ, Harris RA, Kaye PM, McAdam KP, Raynes JG (1996) C-reactive protein binds to a novel ligand on Leishmania donovani and increases uptake into human macrophages. J Immunol 156(12):4691–4696

Delogu G, Sali M, Fadda G (2013) The biology of mycobacterium tuberculosis infection. Mediterr J Hematol Infect Dis 5(1):e2013070

Desjeux P (2004) Leishmaniasis: current situation and new perspectives. Comp Immunol Microbiol Infect Dis 27:305–318

Díaz CM, Bullon B, Ruiz-Salmerón RJ, Fernández-Riejos P, Fernández-Palacín A, Battino M, Cordero MD, Quiles JL, Varela-López A, Bullón P (2020) Molecular inflammation and oxidative stress are shared mechanisms involved in both myocardial infarction and periodontitis. J Periodontal Res. https://doi.org/10.1111/jre.12739

Fogel N (2015) Tuberculosis: a disease without boundaries. Tuberculosis (Edinb) 95(5):527–531

Gershov D, Kim S, Brot N, Elkon KB (2000) C-Reactive protein binds to apoptotic cells, protects the cells from assembly of the terminal complement components, and sustains an anti-inflammatory innate immune response: implications for systemic autoimmunity. J Exp Med 192(9):1353–1364. Erratum in: J Exp Med 2001 Jun 18;193(12):1439

Gorse GJ, Pais MJ, Kusske JA, Cesario TC (1983) Tuberculous spondylitis. A report of six cases and a review of the literature. Medicine (Baltimore) 62(3):178–193

Haghighi L (1969) C-reactive protein in malaria. J Clin Pathol 22(4):430–432

Heinzel FP, Sadick MD, Holaday BJ et al (1989) Reciprocal expression of interferon gamma or interleukin 4 during the resolution or progression of murine leishmaniasis. Evidence for expansion of distinct helper T cell subsets. J Exp Med 169:59–72

Hurt N, Smith T, Teuscher T, Tanner M (1994) Do high levels of C-reactive protein in Tanzanian children indicate malaria morbidity. Clin Diagn Lab Immunol 1(4):437–444

Li Y, Chen L, Fang W, Chen H (2019) Application value of procalcitonin, C-reactive protein and interleukin-6 in the evaluation of traumatic shock. Exp Ther Med 17(6):4586–4592. https://doi.org/10.3892/etm.2019.7492

Janmohammadi A, Sheikhi N, Nazarpak HH, Nikbakht BG (2020) Effects of vaccination on acute-phase protein response in broiler chicken. PLoS One 15(2):e0229009

Jawed JJ, Majumder S, Bandyopadhyay S, Biswas S, Parveen S, Majumdar S (2016) SLA-PGN-primed dendritic cell-based vaccination induces Th17-mediated protective immunity against

experimental visceral leishmaniasis: a crucial role of PKCβ. Pathog Dis 74(5). https://doi.org/10.1093/femspd/ftw041. pii: ftw041; Epub 2016 May 4
Jawed JJ, Banerjee S, Bandyopadhyay S, Parveen S, Chowdhury BP, Saini P, Majumdar S (2018) Immunomodulatory effect of Arabinosylated lipoarabinomannan restrict the progression of visceral leishmaniasis through NOD2 inflammatory pathway: functional regulation of T cell subsets. Biomed Pharmacother 106:724–732
Jawed JJ, Dutta S, Majumdar S (2019) Functional aspects of T cell diversity in visceral leishmaniasis. Biomed Pharmacother 117:109098
Lamb TJ, Brown DE, Potocnik AJ, Langhorne J (2006) Insights into the immunopathogenesis of malaria using mouse models. Expert Rev Mol Med 8(6):1–22
Lawn SD, Kerkhoff AD, Vogt M, Wood R (2013) Diagnostic and prognostic value of serum C-reactive protein for screening for HIV-associated tuberculosis. Int J Tuberc Lung Dis 17 (5):636–643
Leonard MK, Blumberg HM (2017) Musculoskeletal tuberculosis. Microbiol Spectr 5(2). https://doi.org/10.1128/microbiolspec.TNMI7-0046-2017
Lüthje FL, Blirup-Plum SA, Møller NS, Heegaard PMH, Jensen HE, Kirketerp-Møller K, Gottlieb H, Skovgaard K, Jensen LK (2020) The host response to bacterial bone infection involves a local upregulation of several acute phase proteins. Immunobiology 15:151914. https://doi.org/10.1016/j.imbio.2020.151914
Mbuchi M, Bates PA, Ilg T, Coe JE, Raynes JG (2006) C-reactive protein initiates transformation of Leishmania donovani and L. mexicana through binding to lipophosphoglycan. Mol Biochem Parasitol 146(2):259–264
Mihlan M, Blom AM, Kupreishvili K, Lauer N, Stelzner K, Bergström F, Niessen HW, Zipfel PF (2011) Monomeric C-reactive protein modulates classic complement activation on necrotic cells. FASEB J 25(12):4198–4210
Miner MD, Chang JC, Pandey AK, Sassetti CM, Sherman DR (2009) Role of cholesterol in Mycobacterium tuberculosis infection. Indian J Exp Biol 47(6):407–411
Mohandas N, An X (2012) Malaria and human red blood cells. Med Microbiol Immunol 201 (4):593–598
Mortensen RF, Zhong W (2000) Regulation of phagocytic leukocyte activities by C-reactive protein. J Leukoc Biol 67(4):495–500
Murray HW (2002) Kala-azar—progress against a neglected disease. N Engl J Med 22:1793–1794
Naik P, Voller A (1984) Serum C-reactive protein levels and falciparum malaria. Trans R Soc Trop Med Hyg 78(6):812–813
Newling M, Sritharan L, van der Ham AJ, Hoepel W, Fiechter RH, de Boer L, Zaat SAJ, Bisoendial RJ, Baeten DLP, Everts B, den Dunnen J (2019) C-Reactive protein promotes inflammation through FcγR-induced glycolytic reprogramming of human macrophages. J Immunol 203 (1):225–235
Olivier M, Gregory DJ, Forget G (2005) Subversion mechanisms by which Leishmania parasites can escape the host immune response: a signaling point of view. Clin Microbiol Rev 18 (2):293–305
Page AL, de Rekeneire N, Sayadi S, Aberrane S, Janssens AC, Dehoux M, Baron E (2014) Diagnostic and prognostic value of procalcitonin and C-reactive protein in malnourished children. Pediatrics 133(2):e363–e370. https://doi.org/10.1542/peds.2013-2112
Parveen S, Chowdhury AR, Jawed JJ, Majumdar SB, Saha B, Majumdar S (2018) Immunomodulation of dual specificity phosphatase 4 during visceral leishmaniasis. Microbes Infect 20(2):111–121
Paul VK, Bagga A (2019) Ghai essential pediatrics, 9th edn. CBS Publishers and Distributors, New Delhi
Paul R, Sinha PK, Bhattacharya R, Banerjee AK, Raychaudhuri P, Mondal J (2012) Study of C reactive protein as a prognostic marker in malaria from Eastern India. Adv Biomed Res 1:41
Pezzella AT (2019) History of pulmonary tuberculosis. Thorac Surg Clin 29(1):1–17. https://doi.org/10.1016/j.thorsurg.2018.09.002

Pigrau-Serrallach C, Rodríguez-Pardo D (2013) Bone and joint tuberculosis. Eur Spine J 22(Suppl 4):556–566
Piscopo TV, Mallia AC (2007) Leishmaniasis. Postgrad Med J 83(976):649–657
Sakai S, Mayer-Barber KD, Barber DL (2014) Defining features of protective CD4 T cell responses to Mycobacterium tuberculosis. Curr Opin Immunol 29:137–142
Sarfo BO, Hahn A, Schwarz NG, Jaeger A, Sarpong N, Marks F, Adu-Sarkodie Y, Tamminga T, May J (2018) The usefulness of C-reactive protein in predicting malaria parasitemia in a sub-Saharan African region. PLoS One 13(8):e0201693. https://doi.org/10.1371/journal.pone.0201693
Schleicher GK, Herbert V, Brink A, Martin S, Maraj R, Galpin JS, Feldman C (2005) Procalcitonin and C-reactive protein levels in HIV-positive subjects with tuberculosis and pneumonia. Eur Respir J 25(4):688–692
Silva Miranda M, Breiman A, Allain S, Deknuydt F, Altare F (2012) The tuberculous granuloma: an unsuccessful host defence mechanism providing a safety shelter for the bacteria? Clin Dev Immunol 2012:139127
Singh UK, Patwari AK, Sinha RK, Kumar R (1999) Prognostic value of serum C-reactive protein in kala-azar. J Trop Pediatr 45(4):226–228. https://doi.org/10.1093/tropej/45.4.226
Song X, Sun X, Oh SF, Wu M, Zhang Y, Zheng W, Geva-Zatorsky N, Jupp R, Mathis D, Benoist C, Kasper DL (2020) Microbial bile acid metabolites modulate gut RORγ(+) regulatory T cell homeostasis. Nature 577(7790):410–415
Sproston NR, Ashworth JJ (2018) Role of C-reactive protein at sites of inflammation and infection. Front Immunol 9:754
Stauga S, Hahn A, Brattig NW, Fischer-Herr J, Baldus S, Burchard GD, Cramer JP (2013) Clinical relevance of different biomarkers in imported plasmodium falciparum malaria in adults: a case control study. Malar J 12:246. https://doi.org/10.1186/1475-2875-12-246
Stead WW (1997) The origin and erratic global spread of tuberculosis. How the past explains the present and is the key to the future. Clin Chest Med 18(1):65–77
Szalai AJ, Barnum SR, Ramos TN (2014) Deletion of C-reactive protein ameliorates experimental cerebral malaria? Trans R Soc Trop Med Hyg 108(9):591–593
Tangpukdee N, Duangdee C, Wilairatana P, Krudsood S (2009) Malaria diagnosis: a brief review. Korean J Parasitol 47(2):93–102
Thompson D, Pepys MB, Wood SP (1999) The physiological structure of human C-reactive protein and its complex with phosphocholine. Structure 7(2):169–177
Torres-Guerrero E, Quintanilla-Cedillo MR, Ruiz-Esmenjaud J, Arenas R (2017) Leishmaniasis: a review. F1000Res 6:750
https://www.who.int/news-room/fact-sheets/detail/tuberculosis
Wright SD, Silverstein SC (1983) Receptors for C3b and C3bi promote phagocytosis but not the release of toxic oxygen from human phagocytes. J Exp Med 158(6):2016–2023
Yap XZ, Lundie RJ, Feng G, Pooley J, Beeson JG, O'Keeffe M (2019) Different life cycle stages of Plasmodium falciparum induce contrasting responses in dendritic cells. Front Immunol 10:32
Yoon C, Semitala FC, Atuhumuza E, Katende J, Mwebe S, Asege L, Armstrong DT, Andama AO, Dowdy DW, Davis JL, Huang L, Kamya M, Cattamanchi A (2017) Point-of-care C-reactive protein-based tuberculosis screening for people living with HIV: a diagnostic accuracy study. Lancet Infect Dis 17(12):1285–1292

Acute Respiratory Ailments in Pediatric Age Group and Role of CRP in Diagnosis and Management

8

Chandra Shekhar Das

Contents

8.1 Introduction 215
8.2 Epidemiology 216
8.3 Spectrum of Respiratory Illness or Infections 218
8.4 Risk Factors 220
8.5 Pathophysiology of Upper Respiratory Tract Infections 223
8.6 Role of CRP in Upper Respiratory Tract Infections 228
8.7 Pathophysiology of Lower Respiratory Tract Infections 230
8.8 Role of CRP in Lower Respiratory Tract Infections 233
8.9 Management of Acute Respiratory Tract Infections 236
8.10 Clinical Study Design 238
8.10.1 Materials and Methods 238
8.10.2 Results and Interpretation 242
8.11 Discussion on the Clinical Study 243
References 243

Abstract

Respiratory diseases account for nearly 25% of all pediatric consultations. Acute respiratory infections (ARI) in infant, children, and adolescent age group are common incidence in India and worldwide. ARI need special care and knowledge from physicians to diagnose the exact pathology. It is sometimes very challenging to cure. In the intrauterine (fetal) life, gaseous exchange of oxygen and carbon-di-oxide does not occur in lungs as the placenta helps in exchange process. After birth hypoxia, temperature fluctuations, hypercapnia, and sensitivities of chemoreceptor play important role in breathing. With increasing age, there is expansion in lung

C. S. Das (✉)
Department of Respiratory Medicine, Nilratan Sarkar Medical College Hospital, Kolkata, West Bengal, India

Kolkata Municipal Corporation, Kolkata, West Bengal, India

W. Ansar, S. Ghosh (eds.), *Clinical Significance of C-reactive Protein*,
https://doi.org/10.1007/978-981-15-6787-2_8

volume, multiplication of alveoli and vessels for better and improved lung ventilation. The patients can be presented with cough (irritation of pharynx, larynx, trachea, bronchi, and pleura), rattling (due to excessive secretion in trachea-bronchial pathway), wheezing (audible whistling sound), stridor (upper respiratory obstruction by hoarseness, retraction of chest), tachypnea (abnormally rapid respiration), and dyspnea (difficult breathing). As the children cannot expectorate smoothly, there are always higher chances of lower respiratory tract infections like bronchiolitis, pneumonia, lung abscess, etc. These infections have to be diagnosed properly with blood investigations: pulmonary functions test (PFT), bronchoscopy, and imaging (X-ray or CT scan) techniques. Arterial blood gas analysis with supportive and definitive line of management is always crucial. Serial estimation of C-reactive protein (CRP) can guide the true recovery or deteriorating phases of infections in cumulative disease conditions apart from viable signs and symptoms.

Keywords

Acute respiratory infections · Pneumonia · Under-five children · Pediatric respiratory illness · Asthma · CRP · Bronchitis · RSV · Coronavirus

Abbreviation

ADA	Adenosine deaminase
AIDS	Acquired immune deficiency syndrome
AOM	Acute otitis media
ARI	Acute respiratory infections
BMI	Basal Metabolic Rate
bpm	Breaths/minute
CAP	Community-acquired pneumonia
CBC	Complete blood count
CBNAAT	Cartridge-based nucleic acid amplification test
CHERG	Child Health Epidemiology Reference Group
COVID-19	Coronavirus disease 2019
CRP	C-reactive protein
CT scan	Computed tomography scan
ED	Emergency department
ESR	Erythrocyte sedimentation rate
GAPP	Global Action Plan for Pneumonia
GBD	Global Burden of Diseases
H1N1	Influenza type A virus
Hib	Haemophilus influenzae type b
HIV	Human immunodeficiency virus
HMPV	Human metapneumovirus
Hs-CRP	High-sensitivity C-reactive protein
ICU	Intensive care unit
IL-6	Interleukin-6
ILI	Influenza-like illness

IV	Intravenous
LBW	Low birth weight
LMIC	Lower and middle income countries
LRTI	Lower respiratory tract infection
MDR	Multidrug resistant
NIS	National Immunization Schedule
OPD	Outpatient department
PCOS	Polycystic ovary syndrome
PFT	Pulmonary function test
RSV	Respiratory syncytial virus
RTI	Respiratory tract infection
RV	Respiratory virus
SARS-CoV-2	Severe Acute Respiratory Syndrome Coronavirus 2
TB	Tuberculosis
UNICEF	The United Nations Children's Fund
URTI	Upper respiratory tract infection
VAP	Ventilator-associated pneumonia
WBC	White blood cells
WHO	World Health Organization

Highlights

1. In pediatric age group, common upper respiratory tract infections (URTI) like rhinitis (common cold), sinusitis, ear infections, acute pharyngitis, acute tonsillitis, epiglottitis, and laryngitis are discussed.
2. The common lower respiratory tract infections (LRTIs) in children are pneumonia bronchiolitis and tuberculosis are discussed with pathogens, diagnosis, and risk factors.
3. Tuberculosis is an important disease with very high prevalence (as high as 40%) in developing countries like India. Asthma accounts for 10% of all the respiratory childhood consultations.
4. C-reactive protein (CRP) is widely used to detect bacterial infection in children. CRP level along with other clinical findings is crucial in taking important decision regarding management like de-escalation of antibiotics. Role of CRP in URTIs and LRTIs are discussed. Management strategies of acute respiratory infections (ARIs) are discussed.
5. A clinical study is included to correlate the prognostic role of CRP with RTIs and related antibiotic dosage.

8.1 Introduction

Acute and chronic respiratory infections and subsequent diseases represent a global public health burden because of their increasing incidence and severity worldwide (Aaronson et al. 1955). This can be featured to several factors: (1) the significant spike in the occurrence of early allergen sensitization in childhood leading to asthma; (2) the frequent reappearance of viral infections usually associated with pediatric population (Box 8.1); and (3) the increased survival of awfully preterm and fragile children born with bronchopulmonary dysplasia. All these factors supplement to the increased risk of acute expressions/symptoms becoming chronic. The persistent lung function deterioration thus leads to the development of chronic respiratory diseases in adulthood (Cutrera et al. 2017). Acute respiratory infections (ARIs) contribute to foremost disease associated mortality and morbidity among children under 5 years. Most of these deaths are due to bronchiolitis and pneumonia. Emergence of new microbial pathogens, re-emergence of previously controlled disease (s), widespread antibiotic usage and subsequent antibiotic resistance, and suboptimal coverage by immunization even after many novel efforts are major issues responsible for high incidence of acute respiratory infections. Low-cost interventions like hand washing, breastfeeding, availability of quick and feasible array of diagnostic assays, introduction of some new vaccines, and country-wise National Immunization Schedule (NIS) may reduce the burden of ARI by some extent. Epidemiological data on the incidence of different respiratory diseases are very scarce. The admissions of children with acute respiratory diseases on hospital or emergency department are becoming a routine phenomenon (Cutrera et al. 2017).

8.2 Epidemiology

In ARI incidence, Southeast Asia stands first in number (Wardlaw et al. 2006). Southeast Asia together with sub-Saharan African countries account for more than 80% of all incidences of ARI (UNICEF 2008). In India, due to pneumonia, more than 4 lakh deaths occur every year. Death from pneumonia account for 13–16% of all deaths in the hospital pediatric wards (Jain et al. 2001; Vashishtha 2010).

Box 8.1: Grouping of Pediatric Population Based on Age

GROUP	AGE
1. Neonate	Birth to 28 days
2. Post neonate	29 days to < 1 year
3. Newborn	First 4 weeks after birth
4. Infant	Birth to < 12 months
5. Toddler	I year to 36 months
6. Preschool child	37-72 months
7. School age child	73 months-12 years
8. Adolescence	
i. Early	10-13 years
ii. Middle	14-16 years
iii. Late	17-20 years

From 2000 to 2003, it is recorded that 10.6 million deaths annually occur in children under 5 years (Sharma et al. 2013). ARI alone accounted for 19% of children deaths and which is just over 2 million deaths. Bronchiolitis and pneumonia are the leading causes of death mostly caused by viruses (Bryce et al. 2005; Hart and Cuevas 2007). It is the most common infection in children in all age groups. Among the upper and lower respiratory tract infections, if we consider the milder form of the disease(s), in a country like India, it consists of the bulk of the cases reported from all socioeconomic status.

Sharma et al. (2013) found out that overall prevalence of acute respiratory infections (ARI) was 27% (Sharma et al. 2013). ARI was noticed more among low social class (79.3%), illiterate mothers (37.8%), those living in kutcha houses (52.6%), overcrowded houses (63.7%), use of smoky fuel for cooking (67.4%), inadequate cross ventilation (70.4%), with history of parental smoking (55.6%), low birth weight children (54.8%), and malnourished children (57.8%). Rural children (62.2%) were more affected than urban children.

Kuldeep et al. found that the overall prevalence of ARI was 32% (130/406) (Temani et al. 2016). Winter season, illiterate mother, >2 under-five children at home, overcrowding, smoker in house, family member suffering from cough and cold in last month, smoky chulhas, low birth weight (LBW), partial immunization, inappropriate breastfeeding were significant risk factors for ARI in children in India. No association was found between prevalence of ARI and age, sex, religion of child, geographic location of house in terms of main road, place of birth (home or hospital), and birth order of the child.

Pneumonia was ranked worldwide as the single largest killer of post-neonatal children in 2015. Pneumonia is attributable to nearly 15.5% of all deaths in children below 5 years of age. It is responsible for the deaths of around 900,000 children every year and is one of the most frequent causes of health facility consultation. The main cause of this disease in Southeast Asia and sub-Saharan Africa is low economic profile of population. In Bhutan, in spite of free traditional and modern medicine, the major public health challenge is acute respiratory infection (ARI) and represented 15% of the deaths in under-five children. Administration of *Haemophilus influenzae* type b (Hib) vaccine, pneumococcal conjugate vaccine, and childhood immunization schedule (from January 2019) were introduced to reduce ARI (Jullien et al. 2020; Liu et al. 2016; UNICEF 2018; Walker et al. 2013).

Common cold occurs round the year, but the incidence is greatest from the early fall until the late spring, reflecting seasonal predominance of viral pathogen. The recent incidence and outbreak of coronavirus infection in China and other countries started from early fall to late spring only. Young children have an average of 6–8 colds per year (Kliegman et al. 2020).

One of the leading under-five causes of mortality was pneumonia. In southern Asia and in sub-Saharan Africa, the leading cause of under-five deaths was pneumonia and preterm birth complications (Liu et al. 2016).

In 2010, estimated 120 million episodes of pneumonia (14 million of which progressed to severe episodes) in children below 5 years were noted of which

1.3 million of pneumonia led to death. *Streptococcus pneumoniae* (18·3%) is the most common cause of vaccine-preventable severe pneumonia (Walker et al. 2013).

In 2001, the Child Health Epidemiology Reference Group (CHERG), World Health Organization (WHO), and UNICEF estimated that pneumonia was the foremost cause of child mortality. This contributed to the initiation of Global Action Plan for Pneumonia (GAPP), a global effort (Rudan et al. 2013).

Williams et al. (2002) suggested that throughout the world 1.9 million children died from ARI in 2000, of which 70% reported from Africa and Southeast Asia (Williams et al. 2002).

The total number of pneumonia deaths of 1–59 months children for the year 2008 for 122 countries (with low (<85%) or no coverage of death certification) was estimated to be 1.18 M (calculated by regression model), which represented 23.27% of all 1–59 months children deaths (Theodoratou et al. 2011).

During 2000 and 2003, out of 73% of the 10.6 million yearly deaths reported in children younger than age 5 years, pneumonia accounted 19%, neonatal pneumonia or sepsis (10%), and asphyxia at birth (8%) (Bryce et al. 2005).

Study in 2008 showed that there are about 156 million new episodes of childhood pneumonia each year worldwide, of which 151 million episodes are reported in the developing world. Most incident cases occur in India (43 million), China (21 million), and Pakistan (10 million), with high numbers in Bangladesh, Indonesia, and also in Nigeria (6 million each). Pneumonia is responsible for nearly 19% of all pediatric deaths in children aged <5 years. More than 70% of the worldwide deaths of childhood pneumonia takes place in Southeast Asia and sub-Saharan Africa. The main pathogens associated with childhood pneumonia are respiratory syncytial viruses, *Streptococcus pneumoniae* and *Haemophilus influenzae*.

8.3 Spectrum of Respiratory Illness or Infections

Respiration is the act of breathing which consists of inspiration that means inhaling the air and expiration that is exhaling the air. Respiratory tract consists of organs which help in the action are nose, sinuses air pockets around the nose, mouth cavity, throat, larynx, trachea and its branches, bronchus and its divisions, and lungs. Depending upon the positions in the anatomy, the respiratory tract is divided into upper and lower respiratory tract. Upper respiratory tract mainly consists of nose, nasal sinuses, ethmoid, sphenoid, maxillary, and frontal. Lower respiratory tract consists of airway below the vocal cords, consists mainly trachea, bronchus, bronchiole, and lung parenchyma.

All infections of the upper (URI) and lower (ALRI) respiratory tract comes under the purview of ARI. Most of the infective respiratory viruses succumb throughout the respiratory tract without any representative clinical manifestations (Hart and Cuevas 2007).

For ARI, other allied factors are environmental (indoor air pollution, humidity), family upbringing (poverty, overcrowding, birth order, access to medical care), medical access (malnutrition, HIV/AIDS, prematurity, diarrheal disease, measles,

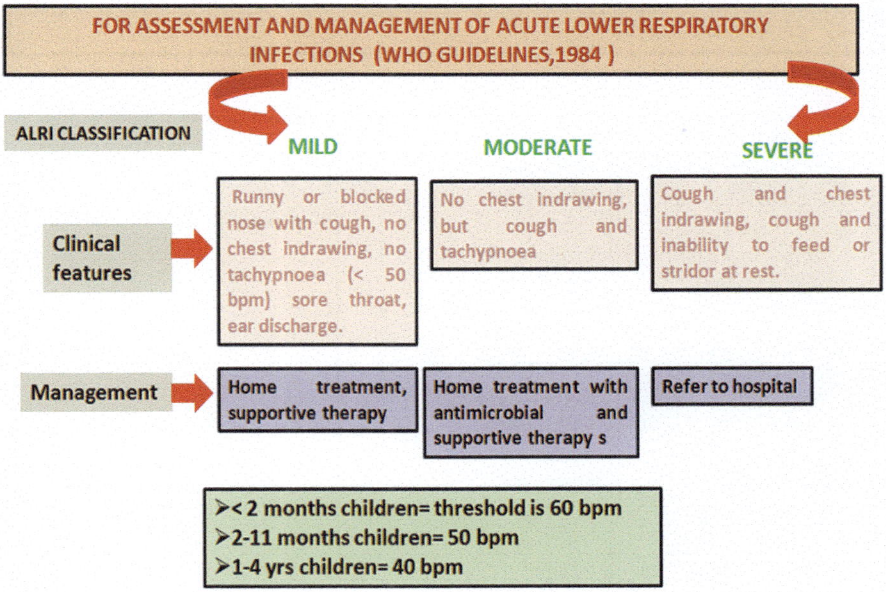

Fig. 8.1 The most widely clinical definition of ALRI was promulgated by the WHO. It assesses three principal signs to resolve severity namely rapidity of breathing, chest indrawing and inability to feed. From this point ALRI is grouped as mild, moderate or severe. Classification determines therapeutic interventions (Hart and Cuevas 2007; Shann et al. 1984)

chronic lung disease, malaria, micronutrient deficiency), etc. (Hart and Cuevas 2007).

Each child in the first 2 years of life experiences nearly 8–9 episodes of URI per year. However in the progression from URI to ALRI, children more frequently develops acute tracheitis, laryngitis, bronchitis, bronchiolitis, and lung infections (pneumonia, tuberculosis, and empyema) (Hart and Cuevas 2007).

ALRI were assessed by WHO guidelines (Shann et al. 1984) (Fig. 8.1) and classified as mild, moderate, or severe based on the three principal signs of rapidity of breathing, chest in drawing, and the inability to feed. This categorization determines therapeutic interventions. The modified assessment utilizes different thresholds for tachypnea. The threshold is 60 breaths/min (bpm) for children <2 months; it is 50 bpm in those of 2–11 months and 40 bpm in those of 1–4 years. This improves greatly the sensitivity of case identification. Any infection that affects the airways below the epiglottis is defined as ALRI and includes clinical pneumonia. The chief bacterial pathogens of pneumonia are *Streptococcus pneumoniae and Haemophilus influenzae*. An increasing number of other pathogens (Fig. 8.2) causing clinical pneumonia or ARLI are enlisted. Vaccination agains most of these pathogens has not been enrooted (Hart and Cuevas 2007; Lanata et al. 2004).

Lower respiratory tract infections (LRI) is a broad term that includes bronchiolitis and pneumonia, as defined by the Global Burden of Diseases, Injuries, and Risk Factors Study (GBD) (Troeger et al. 2018; Jullien et al. 2020).

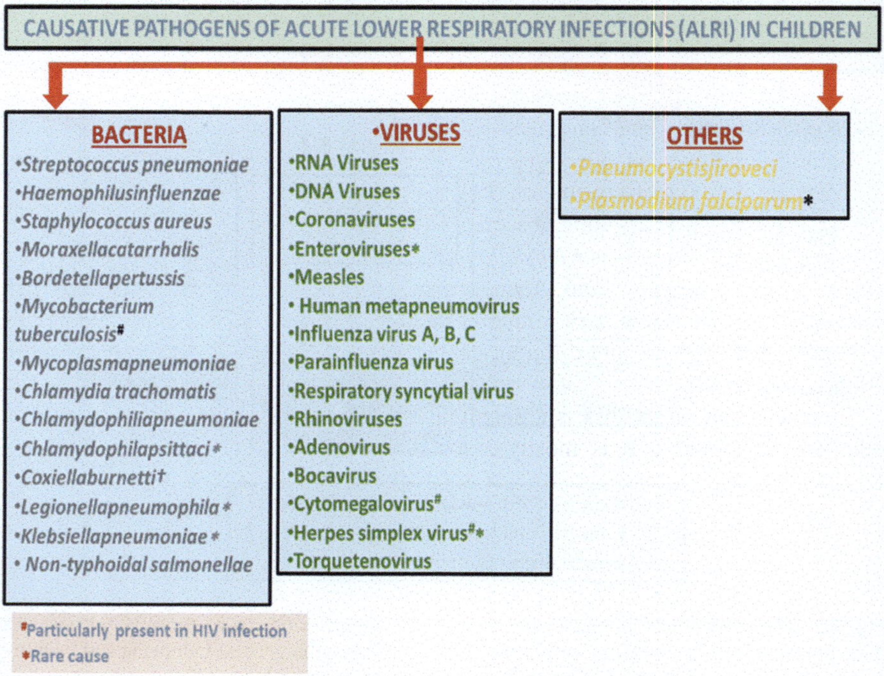

Fig. 8.2 The most common pathogens of ALRI were mentioned along with some pathogens which are rare pathogens and some are present in HIV infections

About 200 antigenically distinct viruses have been suggested as causes of epidemic or sporadic respiratory infections in infants and children. Since more than 30 years rhinovirus, enterovirus, corona virus, respiratory syncytial virus and parainfluenza virus were summed up with adenovirus, influenza and measles virus as primary causes of respiratory tract infections. This list of viable pathogens was extended in 2006 with the discovery of human bocavirus, metapneumovirus and polyomavirus. The family of herpesvirus was also included as pathogen for respiratory diseases in immune-compromised patient (Brouard et al. 2007).

8.4 Risk Factors

In India, some crucial risk factors of ARI are unsatisfactory extent of breastfeeding, malnutrition of the child, suboptimal coverage of immunization, low literacy, poor socioeconomic factors, poor quality cooking fuel, indoor air pollution, and others to increase the burden of ARI among children (Acharya et al. 2003; Savitha et al. 2007; Selvaraj et al. 2014).

Mostly children are presented to hospitals or primary health care before clinicians with acute cough which may be clinically diagnosed and treated accordingly. As children could not give a clear-cut history of their illness, so diagnosis based on visible symptoms is sometimes difficult. The causes of acute cough may be pulmonary or non-pulmonary. The treatment decisions of children having acute cough or difficulty in breathing is done on the basis of symptoms and other clinical evaluations (Boxes 8.2 and 8.3).

Box 8.2: Causes of Acute Cough*

➢<u>URTI:</u> Common cold, rhinitis, postnasal discharge during sinusitis (in older children), hypertropied tonsils and adenoids, laryngitis, pharyngitis and tracheobronchitis.
➢Whooping cough & measles.
➢Foreign body in air passages causing allergic responses.
➢Asthma & nasobronchial allergy.
➢Pneumonia, bronchiolitis, empyema and pulmonary suppuration.
➢Non pulmonary cause: Congenital Heart Failure.

*[Paul and Bagga,2019]

Children with atleast one atopic parent and atopic allergen sensitization within the first 2 years of childhood was the major risk factor for developing asthma at 5 years.

In sinusitis, viral infections predispose the child to bacterial sinusitis and bacterial middle ear infections. Acute rheumatic fever is one of the complications of tonsils and adenoids. Complication of croup occurred around 15% of viral croups. The highest risk in infants with acute bronchiolitis includes further respiratory compromise in the first 48–72 h after onset of dyspnea and cough, desperate illness, air hunger, respiratory acidosis, and apnea (Reuter et al. 2014).

In acute bronchiolitis, risk is higher for infants with young aged mother or with mothers who smoked during pregnancy. Infants with preexistent smaller airways and diminished lung functions are predisposed to this disease. Severe form of the disease encompasses 12 weeks, preterm birth and underlying comorbidity (like pulmonary, cardiovascular, neurologic, or immunologic disease) (Oymar et al. 2014). The most common complications of common cold are acute otitis media (AOM), sinusitis, and exacerbation of asthma along with indiscriminate use of antibiotics.

More than two episodes of either rhinovirus or respiratory syncytial virus infection or suffering, or a single lower respiratory tract infection increase the risk factor to nearly 5–7 folds more. These children are more susceptible in developing asthma in comparison to their atopic peers who were sensitized by allergen after 2 years.

Children, who were allergic at one year of age, are more susceptible in developing asthma at age of 14 years. Allergic sensitization was a prime risk factor below 2 years and becomes irrelevant in children >5 years (Kusel et al. 2007; Rubner et al. 2017; Cutrera et al. 2017).

Box 8.3: Treatment Decisions for Children (2 Months to 5 Years) with Difficulty in Breathing or Cough*

	Signs & symptoms	Type of disease/ Clinical condition	Therapeutic remedy
1.	COUGH OR COLD Without fast breathing No chest indrawing No indicators of severe illness	No Pneumonia	Remedies at home
2.	RESPIRATORY RATE RR/min — Age 60 or more — <2 months 50 or more — 2-12 months 40 or more — 12-60 months	Pneumonia	Cotrimoxazole (at Home)
3.	CHEST INDRAWING	Severe pneumonia	IV/IM penicillin (at Hospital)
4.	CYANOSIS Severe chest indrawing Inability to feed	Very severe pneumonia	IV Chloramphenicol (at Hospital)

* IV=Intravenous; IM=Intramuscular [Paul and Bagga,2019].

The risk factors of sinusitis are meningitis, epidural abscess, subdural empyema, and brain abscess (Wagenmann et al. 1992). Complications of pneumonia include the widespread bacterial infections in thoracic cavity, hematological spread, and bacteremia. Rare complications of pneumonia include suppurative arthritis, meningitis, osteomyelitis, and hematological spread of H. influenza or pneumococcal infection (Kliegman et al. 2020).

8.5 Pathophysiology of Upper Respiratory Tract Infections

Acute respiratory tract infections are not only affecting the respiratory tract but they also have systemic effects because of spreading of infection through blood (sepsis) or due to microbial toxins.

In pediatric age group, common upper respiratory ailments are rhinitis (common cold), sinusitis, ear infections, acute pharyngitis, acute tonsillitis, epiglottitis, and laryngitis. Ear infections and pharyngitis may cause complications like deafness and acute rheumatic fever.

Common cold is also known as rhinitis. It is mostly due to viral infection of the upper respiratory tract, children usually have runny nose (rhinorrhea), and nasal blockage are prominent. Systemic signs and symptoms such as fever, headache, body ache, fever, and lethargy are either mild or absent (Box 8.4).

Box 8.4: Common Cold and Its Pathogens

		COMMON SIGNS & SYMPTOMS
PATHOGENS*	FREQUENT#	
	Human rhinoviruses	Wheezing/bronchiolitis
	OCCASIONAL#	
	Coronaviruses	
	RSV	Brochiolitis in children < 2 y age
	Human metapneumovirus	Pneumonia and bronchiolitis
	UNCOMMON#	
	Influenza viruses	Influenza, pneumonia, croup
	Parainfluenza viruses	Croup, bronchiolitis
	Adenoviruses	Pharyngoconjunctival fever
	Enteroviruses	Herpangina

*Pathogens associated with common cold.
#Pathogens (relative frequency) includes all the frequent, occasional and uncommon viruses causing cold [Kleigman *et al*,. 2016].

Box 8.5: URTI Common Cold in Upper Respiratory Tract Infections*

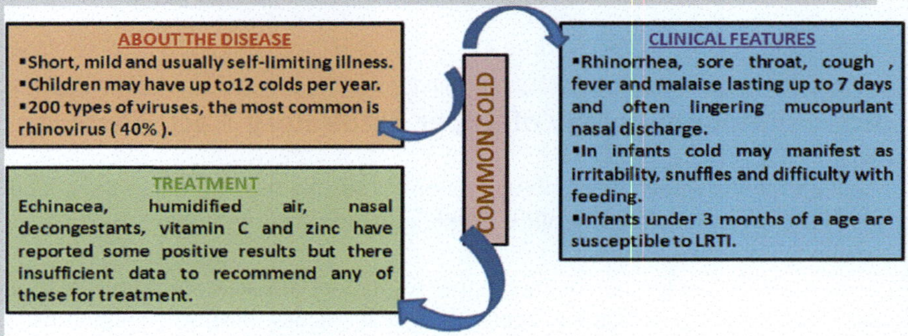

*[Hems and Henderson, Respiratory disorders. Forfar and Arneil' Textbook of Pediatrcs, 715,2008].

Viruses causing rhinitis are associated with symptoms of cough, wheezing, and fever. Most common causes of common cold are rhinoviruses. Common cold is frequently known as *infectious rhinitis (rhinosinusitis)* with the involvement of sinus mucosa. Nearly 200 types of human rhinoviruses are the most common pathogen of common cold. Rhinoviruses account for around 30% of URIs; respiratory syncytial viruses (RSVs), human metapnemoviruses, parainfluenza, and influenza viruses, and adenoviruses for around 35%; coronaviruses for 10%, and in <10% etiology cannot be ascertained. Enteroviruses are also the rare causes of common cold (Box 8.5).

These viral diseases show seasonal variation, in a tropical country like India. Rhinoviral infection is very common just before rainy season in June and July and also from October to February (Matthew et al. 2009).

Small children are more vulnerable to common cold, in India usually 4–6 episodes per year of common cold reported in children. However, incidence decreases as they grow up. Most of the common cold viruses are spread by aerosol, the small particles of biological fluid that form when we talk, sneeze, or cough. Viruses causing common cold are spread by direct hand contact, inhalation of air-borne aerosols from coughing or deposition of larger size aerosols deposited during sneezing on nasal or conjunctival mucosa. It has been shown that rhinovirus and RSV spread by direct contact also. Symptoms of virus infected common cold usually have throat irritation, pain in throat, runny nose, cough, fever, and malaise. Few viruses may also cause ear pain with infection of middle ear. Infections with adenoviruses and rhinoviruses result in development of serotype-specific immunity. Due to the presence of distinct serotype of each virus, repeated infections occur. The interaction of coronaviruses with host immune system is not clear-cut, but the distinct serotype variants of coronaviruses induce short-term protective immunity (Ulrich et al. 2020).

Routine blood studies are not recommended either useful for treating common cold, most of the time disease is self-limiting or remission occurred with minimal supportive treatment. Viral culture in secretion is costly and not cost worthy in a developing country like India. In nasal secretion, neutrophil predominance may be seen which does not mean bacterial superinfection should be remembered.

Sinusitis is one of the complications of common cold (Box 8.6). It is a common illness of childhood and adolescence. Both bacterial and viral pathogens cause acute sinusitis. The maxillary and ethmoidal sinuses are present in a child since birth but only ethmoidal sinuses contain air within the cavities of sinus which are well developed. Maxillary sinus is solid and does not contain cavity since birth, the maxillary sinus contains cavity and air after 4 years of age. The sphenoidal sinuses develop after 5 years and frontal sinuses develop after 8 years. These sinuses are linked with nasal cavity by small ducts which drain sterile secretion by ciliary cleaning movements. So sinusitis is essentially a complication in children who are generally 5–8 years or older. Common viral infection can give rise to secondary bacterial sinusitis in 2–4% of the patients. About 70–80% of etiologies of infection in sinusitis are due to three bacteria, *Streptococcus pneumoniae* (~30%), *Moraxella catarrhalis* (~20%), and nontypeable *Haemophilus influenzae* (~20%). Child with sinusitis presented with nasal stuffiness, purulent nasal discharge, runny nose, fever, and cough. *Staphylococcus aureus*, other streptococci and anaerobic bacteria are uncommon causes of sinusitis. Acute bacterial sinusitis can occur in any age with upper respiratory infections and allergic rhinitis. Viral infection produces a viral rhinosinusitis.

Box 8.6: URTI Sinusitis*

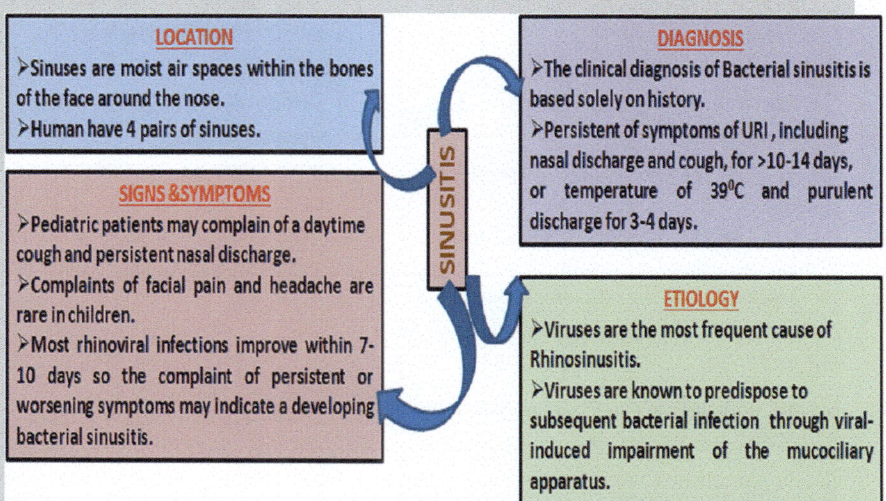

Acute Pharyngitis, the inflammation of pharynx, indeed the name is misleading but this is the actual upper airway tract infection where a child has typical throat pain or discomfort, classically known as sore throat. Common etiological bacterial agents are Group A beta hemolytic *Streptococci, Arcanobacterium haemolyticum, Fusibacterium necrophorum, Haemophilus influenzae, Streptococcus pneumoniae, Francisella tularensis*, rarely *corynebacterium diphtheriae, Neisseria,* etc. Viral infectious agents are adenovirus, coronavirus, cytomegalovirus, enterovirus, etc. (Kliegman et al. 2020).

Pharyngitis differs etiologically according to the season; in winter season, it is mostly due to viruses, children with all age groups are equally involved. Whereas bacterial infection, mainly streptococcus infection, is present throughout the year and it is more common in children from 3 to 14 years. In adolescent children, arcanobacterium may cause many cases of pharyngitis.

Box 8.7: URTI Pharyngitis*

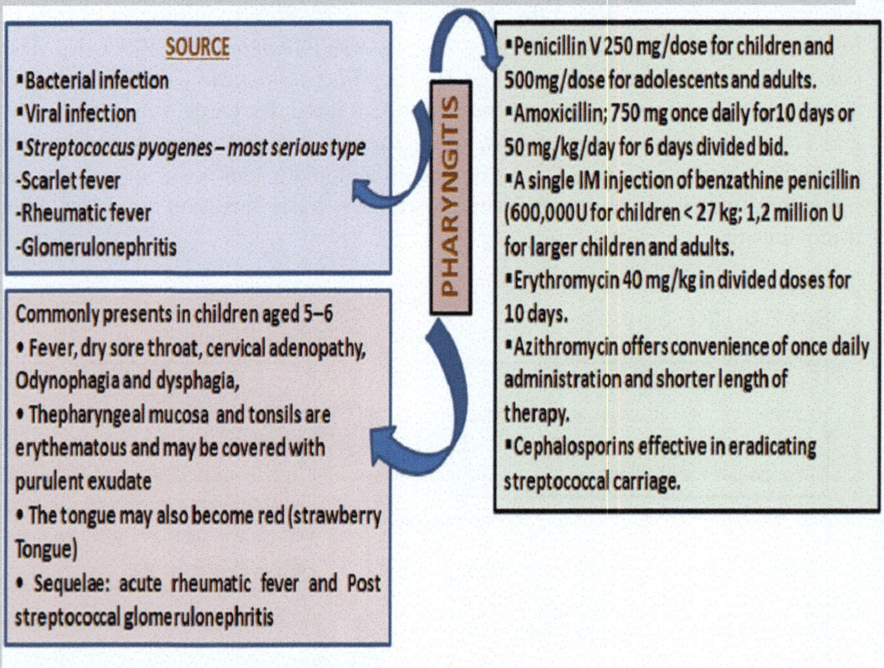

*[Pappas and Henley: Nelson textbook of Pediatrics, 2007].

Child with pharyngitis presented with fever, sometimes very high grade, erythema, edema, pain in throat which causes difficulty in deglutition, generalized malaise, sometimes vague abdominal symptoms also accompany these symptoms. Tonsil becomes enlarged, oral mucosa and palatal wall become congested. The lymph nodes drain oral cavity and tonsil becomes swollen and painful.

The viral pharyngitis is distinct and can be clinically diagnosed as they present more with nasal stuffiness, nasal discharge, conjunctivitis, vesicle in pharyngeal mucosa.

The lymphatic tissue around our opening of gullet and glottis, means around the oral cavity and nasopharynx is called Waldeyer's ring, palatine **tonsils** are part of this ring. Small collection of lymphoid tissue in posterior part of tongue is called lingual tonsil, also a part of this ring. Other members are scattered collection of tissue in posterior pharyngeal wall, the **adenoids**, a small collection between nasal septum and nasopharynx.

Common episodes of pharyngotonsillitis are caused by viruses. Tonsillitis can be acute or chronic infection. Acute tonsillitis, the most common bacterial cause is *Streptococci*, β hemolytic group A. Other uncommon are *Staphylococcus aureus*, *Neisseria*, and Corynebacterium. The chronic infection may be due to polymicrobial, both aerobic and anaerobic. In aerobic, *Haemophilus influenzae* is common whereas *Peptostreptococcus* is common among anaerobes as it is a normal commensal present in oral cavity.

Clinical manifestation of tonsillitis is fever, sometimes high grade and with malaise, bad breath may be present, pain in throat, difficulty in deglutition, enlarged neck glands may also present. In Chronic Tonsillitis, all symptoms may be present but more indolent together with may be foul smelling or cheesy discharge present.

Parapharyngeal and retropharyngeal abscess are formed due to infection in and around neck in the soft tissue. In neck, there are glands which are deeply situated; infection and inflammation are due to the draining organs; nodes infected can give rise to pus formation and abscess formation; depending upon the place of pus discharge the abscess is known as parapharyngeal or retropharyngeal abscess.

These are more common in 3–5 years of age group. A patient who has fever, neck pain, and stiffness will refuse to eat. In some cases, the child may be very ill with respiratory difficulty and stridor.

Bacteria which invade directly into the surrounding from infected tonsil can infect the deep tissues around the tonsil and form peritonsillar abscess. This is more common in the adolescent groups with history of tonsillitis. The child usually has fever, neck pain, and throat pain. *Streptococcus* and anaerobes bacteria of mouth are the most common offending agents.

Inflammation of the vocal cords and the structures inferior to the vocal cords are known as **laryngitis, laryngotracheitis,** and **laryngotracheobronchitis. Laryngotracheobronchitis** is also known as Croup. Croup is a heterogeneous group of acute and infectious reactions symptomized by brassy cough, hoarseness, respiratory distress, and inspiratory stridor. This is mainly a viral disease of children in which upper airway in and around glottis is affected, but if it does not resolve of its own within a week, superadded bacterial infection can occur. Croup affects larynx, trachea, and bronchi. The affected children usually have fever which may be low grade to very high grade in few instances, a typical barking cough, runny nose, and throat pain. Pathogens associated with croup are influenza A and B, respiratory syncytial virus, adenoviruses, and measles; rarely with *Mycoplasma pneumonia.*

Spasmodic croup is a variety of croup which is common in lower age group (1–3 years age), clinically similar to acute laryngotracheobronchitis, but mostly viral in origin. Also some psychological and allergic factors are associated with spasmodic croup. The occurrence of symptoms are often sudden and occur in late evening or nighttime where the child awakens with metallic cough, characteristic barking, noisy inspiration, respiratory distress with frightened and anxious behavior.

Acute Epiglottitis (supraglottitis) is an acute condition, life threatening sometimes as the child develop acute airway obstruction due to disease. The child has fever, prostration, sore throat, dyspnea, drooling of saliva, and difficulty in swallowing. The child may go into severe respiratory distress with stridor (harsh, high-pitched respiratory sound) suddenly, with cyanosis. Diagnosis is by visualization of a large and swollen epiglottis by laryngoscopy.

Acute laryngitis is mainly due to viral etiology (diphtheria is an exception), the child presented with fever, cough, pain in throat, and characteristic hoarseness. The illness is generally mild where respiratory distress is quite unusual. Bacterial laryngitis is uncommon and in past mostly due to diphtheria, which is rare now even in developing countries.

Bacterial Tracheitis is the acute infection (upper airway tract) of trachea mostly due to bacteria such as *Moraxella catarrhalis, Streptococcus pneumoniae, Haemophilus influenzae,* and some anaerobic organisms. The most common pathogen is *Staphylococcus aureus*. It takes place between 5 and 7 years of age, with a slight predominance toward male population. It is often referred to as bacterial complications of viral disease, not just a primary bacterial illness.

The children usually have high fever with expectoration (brassy cough), purulent airway secretions, they usually do not have difficulty in deglutition or drooling of saliva as the upper airway inflammation is relatively absent. The cough is typical, brassy in nature with different grade of respiratory difficulty.

8.6 Role of CRP in Upper Respiratory Tract Infections

In a study by Appenzeller et al. (2002) (van der Meer et al. 2005), 9205 cases of common cold studied, 5788 (62.9%; 3394 men and 2394 women) patients tested positive for respiratory virus (RV) (Appenzeller et al. 2002).

In RV-positive cases, the rate of abnormal CRP levels was 62.8% (3633 of 5788 cases). In cases of single RV infection, the rate of abnormal CRP levels was 61.9% (2855 of 4610 cases), whereas in cases of viral co-infection (two or more infections), the rate of abnormal CRP levels was 66.0% (778 of 1178 cases). Abnormal CRP levels were observed in 66.0% (695 of 1053 cases) of double-infection cases and in 66.4% (83 of 125 cases) of ≥triple-infection cases.

The differences in abnormal CRP rates between overall positive cases and cases of co-infection were not statistically significant ($P = 0.28$), and no significant differences were observed between single infection and ≥ double infections ($P = 0.48$ and $P = 0.37$, respectively). The average CRP levels were 2.80 ± 4.87 mg/dL in all RV-positive patients, 2.82 ± 4.94 mg/dL in patients with single infections, and 2.71 ± 4.56 mg/dL in patients with double infections.

But the fact can be appreciated that even if it is in abnormal range still it is quite low than bacterial infection.

Schalek et al. did a study between positive culture from middle meatus and sinuses with CRP level; they found that positive middle nasal meatus cultures were observed in 24 (30%) patients while the remaining 56 patients were culture negative. The most common bacteria cultured was *S. pneumoniae* (12 patients), followed by *H. influenzae*, *S. aureus*, and *M. catarrhalis* (six, four, and two patients, respectively) (Schalek et al. 2015).

CRP values ranged from 0 to 101 mg/L, ($\bar{x} = 23.70$; SD $\pm$ 24.74). The distribution of individual CRP levels was 0–20 mg/L in 46 patients (57.50%), 21–100 mg/L in 33 patients (21.25%), and in one patient the CRP value was greater than 100 mg/L (1.25%). The correlation coefficient was $r = 0.221$ ($P = 0.0498$). Therefore, there was a significant.

In a study by Andreeva et al. shows that in 179 subjects use of CRP reduces usage of antibiotics significantly without compromising recovery (Andreeva and Melbye 2014).

In a study by Calvino et al. a total of 149 patients were enrolled. The most frequent etiology found was group A streptococcus, present in 83 cases (55.7%) (Calviño et al. 2014). The highest CRP concentration was observed among patients with group C streptococcus infection, with a mean of 56.3 mg/L (95% confidence interval, 25.7–86.5 mg/L). For patients with group A streptococcus infection, the mean CRP value was 34.4 (95% confidence interval, 25.6–43.3 mg/L). So they conclude that CRP concentrations are not associated with group A streptococcus infection in patients with acute pharyngitis. The use of this point of care test is therefore not useful for distinguishing patients who require antibiotic therapy.

In a study by Melby et al., 41 subjects are studied; an etiological agent was established in 23 of the 41 included subjects. These were: influenza A, influenza B, rhinovirus, and other agents. Among the 15 patients examined on both the second and the third day of illness, the median CRP value increased from 7 to 10 mg/L, and the mean value was from 19 to 24 mg/L between day 2 and day 3. Peak CRP values were reached on days 2–4. Higher CRP values were found in those infected with influenza A and B than in the other subjects ($P < 0.001$). A CRP value >10 mg/L was found in 26 subjects during the first 7 days, compared to five subjects after 1 week (Melbye et al. 2004).

Evidence of a secondary infection with group A streptococci was found in two of these five subjects. The development of the symptoms of sore throat, fatigue, clamminess, and pain from muscles and joints followed a similar course as the CRP response, while stuffy nose, cough, sputum production, and dyspnea tended to persist after the CRP values had approached the normal range.

A moderately elevated CRP value (10–60 mg/L) is a common finding in viral upper respiratory tract infection, with a peak during days 2–4 of illness.

Moderately elevated CRP values cannot support a diagnosis of bacterial infection when the illness has lasted <7 days, but may indicate a complication of viral infection after a week.

In a retrospective study by Cetin et al., evaluated medical records of 12 pediatric (<18 years old) cases diagnosed with deep neck abscess or abscess containing suppurative lymphadenitis and treated with only medical therapy between 2010 and 2015 (Çetin et al. 2017). They found that the laboratory examinations resulted in a high count of white blood cell (WBC), CRP, and ESR before the beginning of the medical treatment. All of the patients underwent intravenous (iv) antibiotic therapy immediately. The mean values of the baseline (the first day of hospitalization and parenteral treatment) WBC, CRP, and ESR were 18,050/μL (10,600–25,200), 99.8 mg/L (9–190), and 73.1 mm/h (39–100), respectively. Subsequent laboratory tests which are performed with a 2 or 3 days' interval showed a declining trend in WBC, CRP, and ESR. The mean values of the last control (the day before the discharge from hospital) WBC, CRP, and ESR were 8166/μL (5200–12,000), 34.1 mg/L (2.1–101), and 35.3 mm/h (8–60), respectively, which proves that the infective markers are quite high in deep neck abscesses.

A moderately elevated CRP value (10–60 mg/dL) is very common in viral common cold, which peaks after 2–4 days of illness (van der Meer et al. 2005).

Since sinusitis is mostly the bacterial infection, the blood culture and specimen culture yield good results. High level of serum CRP has also been observed.

Like other etiological factors the diagnosis of pharyngitis is aided by the patients' history, and it has been shown in many studies that usage of antibiotics can be lowered with serum CRP level and chest radiograph. In case of viral etiology, the CRP level is generally low as compared with the bacterial etiology.

Diagnosis of tonsillitis and adenoid is essentially dependent on patient's history, different clinical assessment with direct visualization of the tissue with laryngoscopy with history. Culture from throat swab and CRP values are helpful in diagnosis of bacterial infection. At high CRP values, antibiotics should be started.

In parapharyngeal abscess, CT scan may be helpful in diagnosis and aspiration of pus and different routine culture is helpful in identifying the microbe(s). In peritonsillar abscess patients, CRP is generally high as the level of bacteremia is very high.

The diagnosis of laryngotracheobronchitis is essentially based on different clinical tests. Since most of them are viral, the CRP value is of little help in the diagnosis.

In acute epiglottitis, CRP level is normal to high depending upon the course of the disease since this disease developed very fast so CRP value nearly remains normal when the child is first presented in healthcare facilities. CRP level is often normal in acute laryngitis and diagnosis is by clinical examination of the child.

Diagnosis of bacteria tracheitis is usually done by clinical examination and patient's history. The CRP level is usually high due to bacterial preponderance.

8.7 Pathophysiology of Lower Respiratory Tract Infections

Acute bronchiolitis is caused by obstruction in the flow by internal narrowing of the airways. It is predominantly a viral disease where 50% of the incidence is caused by RSV. Other pathogens include adenovirus, rhinovirus, human metapneumovirus, human bocavirus, parainfluenza, and *Mycoplasma*. Bronchiolitis is rarely a bacterial cause. Huge numbers of children below 1 year are hospitalized with RSV infection. The disease is common among boys, in crowded population and in children who have not been breastfed. It is characterized by bronchiolar obstruction, edema, mucus, and cellular debris. Infants with acute bronchiolitis who are having respiratory distress (inability to take oral feeding, apnea, tachypnea, hypoxia) should be hospitalized.

Pneumonia is an infection which involves different pathological inflammatory process in lung parenchyma; the causes are varied from infection through microbes to non-infective causes also. It is the leading cause of death worldwide among children younger than age 5 year. Community-acquired pneumonia is actually the direct invasion of the lung by the bacterium present in environment. This can be attributed to both viruses and bacteria. Nearly one third of the cases are due to multiple infective organisms.

Streptococcus pneumoniae and respiratory syncytial virus are the most common offending agents. Apart from that *Streptococcus* pyogenes, *Staphylococcus aureus*, and *Haemophilus influenzae* type b are the major bacterial causes. Other common viral infective pathogens are influenza, parainfluenza, and adenovirus. In developing countries, *S. pneumoniae, H. influenza,* and *S. aureus* are the prime causes of hospitalization and deaths of children from bacterial pneumonia.

These pathogenic organisms directly invade the lower airway tract and destroy local immunity. The microbes proliferate and cause destruction of alveoli to form exudates in the alveolar cavity which give rise to the classical radiological impression of consolidation.

Classical pneumonia with classical radiological consolidation is becoming rare now particularly in pediatrics age group due to use of prophylactic antibiotics, especially in a country like India.

Clinically, the child may have history of other upper respiratory tract infection, few days prior to pneumonia. The child usually having fever, often high grade, cough, chest pain, in advanced stage respiratory difficulty is profound.

Abscess is due to infection or inflammation of certain region of lungs as a result of which the central part of necrosis of tissue occurs and a cavity is filled with material in radiograph. Children with **lung abscess** are mostly due to reflux disease, secondary infection with bacteria, developmental anomaly like trachea esophageal fistula, and pneumonia itself can cause abscess.

Bacterial lung abscess is due to fusobacterium, bacteroides, and other anaerobes which are present in reflux material. The child usually present with cough, fever, respiratory difficulty, and hemoptysis.

A lung abscess is an infectious pulmonary disease accompanied by pus-filled (or often air-fluid) cavity formation. Polymicrobial and multiple anaerobic bacteria are the causes of primary lung abscesses. The most common organism reported in lung abscess cases in Asia is *Klebsiella pneumonia, but now Staphylococcus aureus, Streptococcus pyogenes,* and *Actinomyces are some new pathogens that are included in the spectrum. Other factors include* unhygienic oral cavity, alcoholics with poor oral hygiene, and different dental diseases (Guo et al. 2019; Wang et al. 2005).

Exacerbation of asthma is a potentially serious complication of cold, but it is relatively uncommon. It is mostly associated with common cold. Since mostly asthma exacerbation are related to non-infective causes, discussion of which is beyond scope of this chapter; we can concentrate on infective causes of asthma. The treatment of common cold is not always used as preventive therapeutics in exacerbation of asthma. Indiscriminate use of antibiotics causes complications in common cold. **Childhood asthma** is a reversible airway disease which is caused by bronchial hypersensitivity due to various factors that result in airway flow limitation.

Most common infective cause of exacerbation of asthma is viral, mostly by RSV (respiratory syncytial virus); other viruses are adenovirus, parainfluenza, and rhinovirus. Bacterial infection is not so common with asthma and whenever present is due to secondary to lower body immunity that is due to the disease itself.

Child present with all symptoms of allergic upper respiratory tract involvement like rhinorrhea, fever, coughs with wheezy respiration. Viral prodrome may precede the acute exacerbation by days to weeks.

Tuberculosis is a disease caused by *Mycobacterium tuberculosis* complex. In India, its prevalence is 40%, that means 40% people are exposed in general population with these bacilli in some form or the other and they are tuberculin skin test positive.

Generally, in lower age group according to wallgrens calendar extra-pulmonary (mainly pleural effusion and meningitis) and disseminated tuberculosis are common, but tuberculosis is an enigma of a disease, in children the author witnessed all forms of tuberculosis in India. Surprisingly, the author successfully treated three cases of MDR tuberculosis, and the youngest is a female of 6 months of age with extra-pulmonary MDR. Tuberculosis spreads through inhalation route, via droplets. Each bout of cough can have as much as 3000 infectious droplets which contain 1000–10,000 bacili per microliter.

After ingestion, they are engulfed by the TH1 lymphocyte, which triggers an immune response which leads to the release of many mediators specifically the IFN-γ, which help minimum aggregation of more cells and formation of classic granuloma.

In children, the first response is often formation of a granuloma with draining lymph node called a primary complex. The allergic reaction to tubercular antigen causes pleural effusion whereas if body immune response is not adequate to contain the bacilli it spreads and causes dissemination of infection.

Diagnosis is often difficult in tuberculosis. In case of dissemination of tuberculosis, it is mainly clinical with chest radiograph showing military shadows or visibility of hepatosplenomegaly in abdominal ultrasonography. The meningeal tuberculosis is a severe form of tuberculosis in children. The child presents with high fever and signs of meningism. It is often complicated with stupor or coma and convulsions. TB meningitis is diagnosed by clinico-radiological methods. MRI of the brain is gold standard in diagnosis (Daniel et al. 2019). CSF examination is very helpful in establishing the diagnosis, which shows high protein, low sugar, lymphocytic predominance in cell type, and ADA more than 10 (Gupta et al. 2013).

Pulmonary tuberculosis is diagnosed radiologically by chest X ray and direct visualization of the bacilli by microscopy (always preferable). Now DNA study (GeneXpert/CBNAAT and truNAAT) helps in early diagnosis of drug-resistant tuberculosis mainly in case of Rifampicin-resistant tuberculosis.

Blood investigation may show a lymphocytic predominant response which lacks sensitivity. ESR has got no role in diagnosis.

8.8 Role of CRP in Lower Respiratory Tract Infections

Pneumonia is diagnosed by history, chest radiograph, and blood investigation, routine as well as culture. Blood count and CRP are quite high in bacterial pneumonias. CRP is also helpful in determining response and choices of antibiotics. Gonzalez et al. studied 557 subjects who are admitted with a diagnosis of pneumonia; the following microorganisms were found in 171 (30.7%): *Streptococcus pneumoniae* (highest percentage), *Pseudomonas aeruginosa Chlamydophila pneumoniae*, *Haemophilus influenzae*, *Legionella pneumophila*, influenza A, *Mycoplasma pneumoniae*, *Mycobacterium tuberculosis* and one for each of *Staphylococcus aureus*, *Moraxella catarrhalis*, *Escherichia coli*, and *Enterococcus faecium* (Ruiz-González et al. 2016).

Subjects with pneumonia had serum CRP concentrations (median 187 mg/L). In a study by Garcia et al., a cohort of 1222 patients with CAP was assessed. CRP levels were analyzed in 258 patients with a single etiological diagnosis. The mean CRP values in patients with pyogenic, atypical, viral, and *Legionella pneumophila pneumonia* were measured and noted as 16 mg/dL, 13 mg/dL, 14 mg/dL, and 25 mg/dL, respectively (Vázquez et al. 2003).

Serum CRP was not useful to distinguish between pneumococcal, chlamydial, and viral etiology in 193 pediatric patients with pneumonia in a prospective, population-based study (Heiskanen-Kosma and Korppi 2000).

Concerning the conclusion that higher CRP levels appear to predict severity of illness, Seppä et al. (2001) reported that a CRP level $\geq$ 100 mg/L is a marker independently associated with higher risk of death in patients with lower respiratory tract infections (Seppä et al. 2001).

Additionally, Hedlund observed the following: patients with higher CRP levels had longer duration of fever; patients had longer hospital stay; and fewer patients had recovered clinically or radiographically at 8 weeks of follow-up after discharge. These findings suggest that CRP could be used together with other clinical, radiographic, and laboratory findings for the risk stratification of patients with pneumonia (Hedlund 1995).

A study by Fujita et al. shows 109 patients with bronchial asthma, with or without attacks. Mean serum hs-CRP levels were significantly higher in patients without attacks (0.473 mg/L) and with attacks (0.908 mg/L) ($P < 0.001$ for both) than in controls (0.262 mg/L) (Fujita et al. 2007).

Serum hs-CRP levels were inversely correlated with forced expiratory volume in 1 s/forced vital capacity in asthmatic patients ($r = -0.4915$; $P < 0.01$). So they conclude, serum hs-CRP levels may be related to the state of asthma exacerbation and allergic inflammation.

In another study by Monadi et al., a total of 120 patients and 115 controls were studied. They found that median serum hs-CRP in asthma was higher than control ($P = 0.001$), so they conclude that serum hs-CRP in asthma is higher than healthy control (Monadi et al. 2016).

In a study by Kwas et al. (2015), 44 patients were treated for pulmonary tuberculosis. In this study, they saw patients with cavitary lung lesions more frequent with elevated CRP (16 cases *vs.* 3 cases) and often with an extensive and bilateral pulmonary involvement (15 cases *vs.* 3 cases). Sputum positive acid-fast bacilli found in sputum was more frequent in the group of patients with elevated CRP (22 cases *vs.* 6 cases) (Kwas et al. 2015).

Wilson et al. performed a study in an area of South Africa that has a very high prevalence of HIV infection (40%) and a very high annual incidence of active TB (1094 cases/100,000 population) (Wilson et al. 2011). Serum CRP levels were measured in 364 patients who were being evaluated in primary care clinics for possible TB; based on extensive clinical, imaging, and laboratory data collected over 8 weeks of follow-up, 37% of the patients were categorized as having confirmed TB, 31% as having possible TB, and 32% as not having TB. The proportion of patients with normal CRP levels in each group was 2%, 20%, and 59%, respectively. An elevated serum CRP level was found to have 98% sensitivity and 59% specificity for the diagnosis of TB; the positive predictive value was 74%, and the negative predictive value was 96%. Importantly, HIV status did not significantly influence median CRP levels in the three diagnostic categories, and among HIV-positive patients with confirmed TB, CRP levels were highest in those with advanced HIV disease.

The diagnostic value of CRP in a total of 207 cases was evaluated by different number of patients with exudative-fibrotic tuberculosis, tuberculous pneumonia, tuberculous effusion, cavitary tuberculosis, miliary tuberculosis, mycoplasmal pneumonia, bacterial pneumonia, bacterial pneumonia with effusion, lung abscess, *Strongyloides stercoralis* pneumonia, aspergilloma, and *Pneumocystis carinii* pneumonia. Depending on the type of pathogen and the severity of inflammation, CRP level varies in different disease state (Lin et al. 1990).

Patients at admission with CRP <100 mg/L level has reduced risk for 30-day mortality and complicated pneumonia (lung abscess, empyema, or complicated parapneumonic effusion). However, when the CRP level could not fall by 50% (or more) at day 4, then it leads to an increased risk of mortality (30-day), mechanical ventilation, and complicated pneumonia. Thus, CRP is an independent marker of severity in community-acquired pneumonia (Chalmers et al. 2008).

For community-acquired pneumonia and in ventilator-acquired pneumonia, the laboratory investigations include CRP tests. In asthma diagnosis, CRP is often low to normal even with viral infection as compared to the bacterial infection. However, hs-CRP is often found elevated in exacerbation as compared to non-infective stage.

In China, the Coronavirus disease 2019 (COVID-19) caused by the Severe Acute Respiratory Syndrome Coronavirus 2 (SARS-CoV-2) infection has tremendous health burden. The allergy status and clinical characteristic of patients infected with SARS-CoV-2 were investigated by the researchers. Significantly high level of CRP was recorded in 140 hospitalized COVID-19 patients (Zhang et al. 2020).

Marked elevated levels of CRP were most frequently found in pneumonia patients. However, in most patients with acute exacerbation of chronic obstructive bronchitis, acute asthma, and acute bronchitis the values of CRP were within the normal range. CRP values reflected sensitivities and specificities in the diagnosis of these diseases (Ritland and Melbye 1991).

CRP and BMI are potential predictors of development of asthma in Egyptian polycystic ovary syndrome (PCOS) patients. Asthmatic female patients with PCOS had higher CRP levels and IL-6 representing inflammatory condition. As a predictor of asthma, CRP had a specificity of 66% (Nasser et al. 2019).

However in a large study population of 1955 adolescent participants, no association between hs-CRP levels and any allergic (asthma, eczema, allergic rhinitis, food sensitization, aeroallergen sensitization, and any sensitization) sensitization was found (Yang et al. 2019).

Serum hs-CRP was lower in 79 infants with recurrent physician-diagnosed wheezing with recurrent respiratory syndrome. In early childhood wheezing, hs-CRP-mediated low-grade systemic inflammation is very rare (Määttä et al. 2017).

Research documentation revealed that hs-CRP can be a useful marker of airway inflammation in bronchial asthma in nonsmoking group without complications like hypertension heart disease, hyperlipidemia, infection or chronic obstructive pulmonary disease. Serum hs-CRP concentration significantly correlated with eosinophils and neutrophils count in the sputum. The serum hs-CRP level was higher in asthmatic patients as compared to healthy controls (Shimoda et al. 2015).

The mean serum hs-CRP levels were significantly higher in acute asthmatic patients as compared with healthy controls. Mean hs-CRP levels in patients were not correlated with indices of pulmonary function tests (forced expiratory volume in one second, forced vital capacity, and forced mid-expiratory flow). Thus, CRP is a useful diagnostic tool for detecting and monitoring inflammation in acute asthmatic patients (Razi et al. 2012).

Different studies were done to investigate the role of CRP in tuberculosis, it has been found that rise of CRP is different with different subset of patients as compared to mild localized disease, cavitary pulmonary disease, or bilateral diffuse infiltrative disease patients used to have higher CRP. In high CRP patients, chance of being sputum positive means chances of finding bacteria in sputum is also high.

High concentration of plasma CRP acts as an important predictor of pediatric active tuberculosis patients from healthy children in South Indian population (Albuquerque et al. 2019).

Children with active pulmonary tuberculosis and extra-pulmonary tuberculosis had significantly elevated levels of CRP, α-2 macroglobulin, hemoxygenase 1, and haptoglobin. CRP acts as a circulating biomarker (Kumar et al. 2013).

In a study by Bajaj et al., CRP levels were determined in 100 cases of tuberculosis and 30 age- and sex-matched children. Serial estimations, 1 and 3–6 months after initiation of therapy was done in 81 and 41 of these patients, respectively. Mean initial levels of CRP in tuberculosis group was 18.52μg/mL while in the control group it was 2.77μg/mL ($P < 0.001$). The elevated CRP levels fell significantly to 5.93μg/mL after 1 month of treatment ($P < 0.001$) and by 3–6 months of treatment had fallen to normal values. The fall in CRP levels correlated with clinical response. So they concluded that CRP is a sensitive indicator of activity of tuberculosis and the return to normal values of initially elevated CRP levels may indicate a good therapeutic response (Bajaj et al. 1989).

The antibiotic therapy of infants and children with pneumonia is based on the clinician's diagnostic evaluations of the most common infecting pathogens, the susceptibilities of the host toward the infecting pathogens, and the gravity of the illness. Majority of the lower respiratory tract infections in children have a viral etiology, but bacterial etiology of pneumonia could not be ruled out. The causative pathogens are *Streptococcus pneumoniae* and *Staphylococcus aureus*. Pneumonia with viral etiology was not managed by antibiotic therapy. In replacement, protein-conjugated vaccines employing inflammatory biomarker like CRP could be employed. In case of invasive bacterial infection or viral etiology, new ancillary tests like blood culture, CRP level, and white blood cell count should be employed (Bradley 2002).

CRP and while blood cell counts have good negative predictive value and specificity in bacterial pneumonia where polymerase chain reaction technique remains too expensive (Le Roux 1998).

To withhold antimicrobial therapy, or its early discontinuance for subjects suspected of having neonatal sepsis, bacteremia, or pneumonia, CRP could be employed regardless of immune status. CRP was used in differential diagnosis of bacterial vs. viral diseases, acute otitis media, and lower respiratory tract infection (Jaye and Waites 1997).

8.9 Management of Acute Respiratory Tract Infections

Children become the future of the world after decades. According to their age, they are categorized as neonates, infants, children, etc. Acute respiratory tract infection carries a higher percentage in suffering, attending physician chamber, hospital admission, and overall a huge burden for the family in terms of money, anxiety, leave in workplace, etc. The signs and symptoms of acute respiratory tract infections are different in different age group. Neonates and infants, toddler, etc. can be presented with crying, grunting, chest retraction, and little older age group can be presented with complain of earache, cough, throat pain, respiratory distress, or difficulty in sleeping or sitting in a particular posture (as explained by parents). According to anatomical position, we can divide the acute respiratory tract infections into two parts: upper respiratory tract infections which are above the vocal cord like sinusitis, pharyngitis, tonsillitis, epiglottitis, etc. and lower respiratory tract diseases which are below the vocal cord like pneumonia, bronchiolitis, tuberculosis, etc. and pneumonia is the major cause of under-five mortality in children.

We have to diagnose whether the infections are viral, bacterial, or allergic or other factors. The lines of management are different according to etiologies and stage of suffering. In medical science, to make a diagnosis, contribution of patients' history is about 60–70%, clinical examination about 20–30%, and nearly 15–20% from laboratory investigations. In Infants, toddlers, preschool children and little older children getting proper history is often difficult as they do not express themselves properly. Furthermore, to clinically examine a child is also difficult as they turn restless. To prepare a patients' history out of this and diagnose accurately is very challenging. The whole process of diagnosis should be made with great care. In case of infants, the total WBC count is usually high. CRP estimation has high importance in correct diagnosis as serial measurement of CRP level helps the clinicians to determine the children's responsiveness in treatment. Gradual decrease in CRP level reflects therapeutic responsiveness and an increase in CRP concentration showed the worsening health status of the child.

For effective management (Florman et al. 1987), it is recommended that

1. The suffering patient with acute respiratory infections who is to be treated as an outpatient may need routine laboratory (if any) investigations.
2. If a child is admitted to hospital, then serum CRP test, cold agglutinin test (according to appropriate age), urine bacterial antigen tests will help to identify (type/strain) the etiology of the infection (or allergic infection also). Apart from viral, bacterial, or allergic infections, parasitic infections are also considered.
3. If antibiotic therapy is to be given, blood culture should be done before starting to check the culture positivity and sensitivity of the strain (toward the antibiotic). Diagnostic (bacterial or viral cause) or prognostic (self-limiting or life-threatening infection) ambiguity makes it complicated for clinicians to know when to prescribe and when to hold back antibiotic treatment.
4. For the child admitted to the hospital with viral lower respiratory infection or possible chlamydial infection, specific therapy is needed. Nasopharyngeal secretions should be studied for antigens of respiratory viruses like RSV, H1N1, Influenza, Parainfluenza and also chlamydia and other common pathogens, which helps us to move toward targeted therapy. Serum IgM level of different pathogen may also helpful for acute infections. For chlamydia infections and other viral infections, it should be diagnosed and move toward targeted therapy. Serum for IgM level may also be helpful.

5. For the child who has been intubated for additional respiratory support, a specimen of bronchoalveolar lavage is to be collected for Gram stain, fungal stain, Ziehl-Neelsen staining (ZN stain) and culture sensitivity of the lavage, viral markers (antigen) determination for more accurate diagnosis that will help for specific management.

8.10 Clinical Study Design

Dr. Das ascertains a correlation between CRP value in children with upper respiratory tract infection and lower respiratory tract infection including tuberculosis. The aim of this study is to find the contributory role of CRP in URTI and LRTI patients in positive diagnosis. The level of CRP is correlated with the administration of antibiotic among patients in three hospitals, namely Daffodil Polyclinic, Rajarhat, Kolkata; Astha Polyclinic, Ghusuri, Howrah; and Nibedita Polyclinic, Bally, Howrah, West Bengal.

The study of population includes 25 patients suffering from URTI (Group I) and 20 patients suffering from LRTI (Group II) who went for medical check-up in OPD with subsequent follow-up. Some patients were admitted under Dr. Das in nursing homes of Kolkata over a period of 1 year from 2017 till the end of the year.

The age of the children is strictly to be taken below 14 years and neonatal and infant cases were excluded.

8.10.1 Materials and Methods

Pediatric population of more than 1–14 years were considered for the study over a period of 1 year from January 2017 to December 2017 taken into consideration with the standard protocols for diagnosis. For URTI, only fever with common cold that is rhinorrhea, sinusitis, acute pharyngitis, tonsillitis, adenoids, croup is considered, whereas in LRTI only radiography proved pneumonia and acute exacerbation of asthma is considered. Patients' blood samples were sent for laboratory investigations including CRP test. The levels of CRP of the patients were taken for this study (Table 8.1). Disease remission is taken into account by follow-up visits or telephonic conversation with the patients. Remission is taken based on the absence of the three symptoms, fever, cough, and breathing difficulty (if present), at the time of diagnosis. Generalized symptoms at the time of presentation include headache and chest pain.

The relationship between CRP values of two groups is calculated and relationship with remission is also calculated.

Study Design

Prospective observational study (cohort) taking only the values of CRP, age of patient, and antibiotic treatment.

Study Setting and Timeline

Pediatric population from more than 1 year to 14 years of age group come on OPD basis, they then may be admitted depending upon the disease severity by the clinician.

Place of Study

In three hospitals, namely Daffodil Polyclinic, Rajarhat, Kolkata; Astha Polyclinic, Ghusuri, Howrah; and Nibedita Polyclinic, Bally, Howrah.

Period of Study

Timeline is for 1 year from first January 2017 to 23 Dec 2017.

Study Population

Pediatric population from more than 1 year to 14 years of age group.

Sample Size

Sample size in this study includes the URTI group ($n = 25$) and LRTI group ($n = 20$). Both groups contain OPD patients; however, patients may be admitted after first OPD visit.

Inclusion Criteria

1. Pediatric population from more than 1 year to 14 years of age group.
2. All patients in this age group who visited the OPD of Dr. Das.
3. Patients having symptoms of respiratory tract infection were grouped based on the symptoms of URTI or LRTI. The most frequent medical symptom observed was acute cough. Wheezing, shortness of breath, and chest tightness were also recorded.

Exclusion Criteria

1. Subjects with severe form of disease needing ICU admission like acute epiglottitis were excluded.
2. Associated congenital diseases were excluded.
3. Any comorbidities like type 1 diabetes mellitus, Addison's disease, and hypothyroidism were excluded.
4. Patient already on antibiotics dose prior to first OPD visit. That could modify the study result.

Study Variables

Age, sex, days at presentation, CRP test given to each subject, antibiotics required or not was noted (Table 8.1), days of remission which is defined in this study as the absence of at least fever, cough, and respiratory difficulty (if present) at onset.

Table 8.1 Clinical profile of patients

Sl no	Sex	Age	disease	Days at presentation	CRP mg/dL	Antibiotics	Admission	Remission (days)
1	M	8	URTI	8	12	No	No	14
2	M	13	URTI	4	8	No	No	11
3	F	11	LRTI	3	86	Y	Y	16
4	M	7	LRTI	5	112	Y	Y	17
5	F	14	LRTI	2	45	Y	No	14
6	F	8	URTI	3	23	Y	No	9
7	M	14	LRTI	3	75	Y	Y	14
8	M	9	URTI	2	8	No	No	12
9	F	10	LRTI	3	86	Y	Y	16
10	F	14	LRTI	5	102	Y	Y	13
11	M	7	URTI	2	0.8	No	No	9
12	F	12	URTI	2	18	Y	No	10
13	M	11	LRTI	3	98	Y	Y	14
14	F	8	URTI	4	9	No	No	8
15	M	5	URTI	3	7	No	No	10
16	M	3	URTI	1	11	No	No	14
17	F	7	LRTI	2	7	Y	No	16
18	M	12	URTI	2	8	Y	No	8
19	F	11	LRTI	2	54	Y	No	13
20	F	9	URTI	2	3	No	No	8
21	F	5	LRTI	4	55	Y	No	14
22	M	9	LRTI	2	38	Y	No	11
23	F	10	LRTI	1	86	Y	Y	15
24	F	4	URTI	1	3	No	No	9
25	M	8	URTI	3	4	No	No	9

26	F	14	URTI	2	14	Y	No	12
27	F	11	LRTI	1	18	Y	No	9
28	F	9	URTI	4	9	No	No	7
29	M	11	URTI	4	7	No	No	14
30	M	13	URTI	2	8	No	No	11
31	F	12	LRTI	3	71	Y	No	15
32	M	8	LRTI	3	92	Y	Y	14
33	F	4	LRTI	2	73	Y	No	8
34	F	10	URTI	3	2	No	No	9
35	F	14	LRTI	1	15	Y	No	9
36	M	7	LRTI	2	62	Y	No	10
37	M	3	URTI	2	3	No	No	8
38	F	7	URTI	4	25	Y	No	14
39	M	14	LRTI	2	47	Y	No	11
40	M	12	URTI	1	8	Y	No	7
41	F	8	LRTI	2	52	Y	No	9
42	F	9	URTI	3	5	No	No	9
43	M	11	URTI	2	1	No	No	7
44	F	12	URTI	1	6	Y	No	6
45	F	7	URTI	2	13	Y	No	8

8.10.2 Results and Interpretation

Age and Sex

A. URTI cohort—there were 25 patients in this cohort, of them 13 boys and 12 girls, the age distribution is as follows:

Age	Number of patient
0–4	2
4–8	5
8–12	11
12–14	7

The mean age was 8.96 years with SD of 3.039

B. LRTI cohort—there were 20 patients in this cohort, of which 7 boys and 13 girls, the age distribution are as follows:

Age	Number of patients
0–4	0
4–8	5
8–12	9
12–14	6

The mean age was10.05 years with SD 3.119

CRP

A.URTI cohort—there were 25 patients in this cohort, the CRP distribution is as follows:

CRP value (mg/L)	Number of patients
0–7	11
8–15	11
16–23	2
24–30+	1

The mean CRP value was 8.632 and the Standard Deviation was σ: 6.115 the variance is 37.3941

B. LRTI cohort—there were 20 patients in this cohort, there were 7 boys and 13 girls, the age distribution is as follows:

CRP value (mg/L)	Number of patients
0–30	2
30–60	6
60–90	8
90–120+	4

The mean CRP value was 66.95, the Standard Deviation was σ: **25.935**, the variance was 672.6475

The difference of mean of CRP in two groups is = −58.268, with standard error 5.351, value of *t* statistics is −10.890 with *P* value <0.0001, which is significant

8.11 Discussion on the Clinical Study

As in authors study, the significance of CRP determination is high in LRTI compared to URTI patients. It also can be seen that patient with high CRP needs antibiotics as treatment measures.

ARTIs are among the most frequent reasons to consult in primary care for children and a common reason to prescribe antibiotics. Most of these prescriptions increase only the emerging problem of bacterial resistance. The use of so many antibiotics over the past few years has made almost all disease-causing bacteria resistant to antibiotics, which were commonly used to treat them. Studies showed that indiscriminate use of antibiotics is more pronounced in ARTIs (Little et al. 2013; Butler et al. 2009).

The blow of antibiotic resistance on human health are most commonly highest in countries with the lowest income because the spread of resistant bacterial strains is accompanied by poor hygienic condition, contaminated food, overcrowding, polluted water, and increased susceptibility to infection consequential from malnutrition or HIV. Diagnostic uncertainty makes it complicated for clinicians to know when to prescribe and when to withhold antibiotic treatment.

CRP is a diagnostic marker that helps to distinguish grave infections like pneumonia and sepsis from other self-limiting illnesses (Hopstaken et al. 2003; van der Meer et al. 2005). The effect of CRP in reducing antibiotic therapy for LRTI prescribing for lower respiratory tract infections has been studied in developed countries (Cals et al. 2010).

References

Aaronson AL, Kaplan MA, Goldin M, Libretti A, Goldman B (1955) C-reactive protein in bronchial asthmatic patients. II. Further evaluation. Ann Allergy 13(5):586–591

Acharya D, Prasanna KS, Nair S, Rao RS (2003) Acute respiratory infections in children: a community based longitudinal study in south India. Indian J Public Health 47:7–13

Albuquerque VVS, Kumar NP, Fukutani KF, Vasconcelos B, Arriaga MB, Silveira-Mattos PS, Babu S, Andrade BB (2019) Plasma levels of C-reactive protein, matrix metalloproteinase-7 and lipopolysaccharide-binding protein distinguish active pulmonary or extrapulmonary tuberculosis from uninfected controls in children. Cytokine 123:154773. https://doi.org/10.1016/j.cyto.2019.154773. Epub 2019 Jul 9

Andreeva E, Melbye H (2014) Usefulness of C-reactive protein testing in acute cough/respiratory tract infection: an open cluster-randomized clinical trial with C-reactive protein testing in the intervention group. BMC Fam Pract 15:80. https://doi.org/10.1186/1471-2296-15-80

Appenzeller C, Ammann RA, Duppenthaler A, Gorgievski-Hrisoho M, Aebi C (2002) Serum C-reactive protein in children with adenovirus infection. Swiss Med Wkly 132(25-26):345–350

Bajaj G, Rattan A, Ahmad P (1989) Prognostic value of 'C' reactive protein in tuberculosis. Indian Pediatr 26(10):1010–1013
Bradley JS (2002) Management of community-acquired pediatric pneumonia in an era of increasing antibiotic resistance and conjugate vaccines. Pediatr Infect Dis J 21(6):592–598. discussion 613-4
Brouard J, Vabret A, Nimal-Cuvillon D, Bach N, Bessière A, Arion A, Freymuth F (2007) Epidemiology of acute upper and lower respiratory tract infections in children. Rev Prat 57 (16):1759–1766
Bryce J, Boschi-Pinto C, Shibuya K, Black RE, WHO Child Health Epidemiology Reference Group (2005) WHO estimates of the causes of death in children. Lancet 365(9465):1147–1152
Butler CC, Hood K, Verheij T, Little P, Melbye H, Nuttall J, Kelly MJ, Mölstad S, Godycki-Cwirko M, Almirall J, Torres A, Gillespie D, Rautakorpi U, Goossens H (2009) Variation in antibiotic prescribing and its impact on recovery in patients with acute cough in primary care: prospective study in 13 countries, 338. BMJ. https://doi.org/10.1136/bmj.b2242. Article
Cals JW, Chappin FH, Hopstaken RM, van Leeuwen ME, Hood K, Butler CC, Dinant GJ (2010) C-reactive protein point-of-care testing for lower respiratory tract infections: a qualitative evaluation of experiences by GPs. Fam Pract 27(2):212–218. https://doi.org/10.1093/fampra/cmp088. Epub 2009 Dec 18
Calviño O, Llor C, Gómez F, González E, Sarvisé C, Hernández S (2014) Association between C-reactive protein rapid test and group A streptococcus infection in acute pharyngitis. J Am Board Fam Med 27(3):424–426. https://doi.org/10.3122/jabfm.2014.03.130315
Çetin AÇ, Olgun Y, Özses A, Erdağ TK (2017) A new trend in the management of pediatric deep neck abscess: achievement of the medical treatment alone. Turk Arch Otorhinolaryngol 55 (2):57–63. https://doi.org/10.5152/tao.2017.2181. Epub 2017 May 22
Chalmers JD, Singanayagam A, Hill AT (2008) C-reactive protein is an independent predictor of severity in community-acquired pneumonia. Am J Med 121(3):219–225. https://doi.org/10.1016/j.amjmed.2007.10.033
Cutrera R, Baraldi E, Indinnimeo L, Miraglia Del Giudice M, Piacentini G, Scaglione F, Ullmann N, Moschino L, Galdo F, Duse M (2017) Management of acute respiratory diseases in the pediatric population: the role of oral corticosteroids. Ital J Pediatr 43(1):31. https://doi.org/10.1186/s13052-017-0348-x
Daniel BD, Grace GA, Natarajan M (2019) Tuberculous meningitis in children: clinical management & outcome. Indian J Med Res 150(2):117–130
Florman AL, Cushing AH, Umland ET (1987) Rapid noninvasive techniques for determining etiology of bronchitis and pneumonia in infants and children. Clin Chest Med 8(4):669–679
Fujita M, Ueki S, Ito W, Chiba T, Takeda M, Saito N, Kayaba H, Chihara J (2007) C-reactive protein levels in the serum of asthmatic patients. Ann Allergy Asthma Immunol 99(1):48–53
Ge YL, Liu CH, Wang MH, Zhang JB, Chen H, Zhu XY, Li ZZ, Li HL, Cui ZY, Li LQ, Fu AS, Wang HY (2019) Anemia combined significantly increased high-sensitivity C reactive protein and lung lesions lead to the diagnosis of granulomatosis with polyangiitis proven by lung biopsy and anti-neutrophil cytoplasmic antibody test. Clin Lab 65(1). https://doi.org/10.7754/Clin.Lab.2018.180733
Gupta BH, Bharat V, Bandopadhyay D, Chamlial S (2013) Role of cerebrospinal fluid adenosine deaminase estimation in diagnosis of tuberculous meningitis. J Indian Med Assoc 111 (9):603–605, 608. PMID: 24968523
Guo W, Gao B, Li L, Gai W, Yang J, Zhang Y, Wang L (2019) A community-acquired lung abscess attributable to odontogenic flora. Infect Drug Resist 12:2467–2470. https://doi.org/10.2147/IDR.S218921. eCollection 2019
Hart CA, Cuevas LE (2007) Acute respiratory infection in children. Infant 7(1). Recife Jan/March. https://doi.org/10.1590/31519-3829007000100003
Hedlund J (1995) Community-acquired pneumonia requiring hospitalisation. Factors of importance for the short-and long term prognosis. Scand J Infect Dis Suppl 97:1–60

Heiskanen-Kosma T, Korppi M (2000) Serum C-reactive protein cannot differentiate bacterial and viral aetiology of community-acquired pneumonia in children in primary healthcare settings. Scand J Infect Dis 32(4):399–402

Hopstaken RM, Muris JW, Knottnerus JA, Kester AD, Rinkens PE, Dinant GJ (2003) Contributions of symptoms, signs, erythrocyte sedimentation rate, and C-reactive protein to a diagnosis of pneumonia in acute lower respiratory tract infection. Br J Gen Pract 53 (490):358–364

Jain N, Lodha R, Kabra SK (2001) Upper respiratory tract infections. Indian J Pediatr 68:1135–1138

Jaye DL, Waites KB (1997) Clinical applications of C-reactive protein in pediatrics. Pediatr Infect Dis J 16(8):735–746. quiz 746-7

Jullien S, Pradhan D, Bassat Q (2020) Pneumonia in Bhutanese children: what we know, and what we need to know. Pneumonia (Nathan) 12.:1. https://doi.org/10.1186/s41479-019-0065-x. eCollection 2020

Jun JB (2018) Klebsiella pneumoniae liver abscess. Infect Chemother 50(3):210–218. https://doi.org/10.3947/ic.2018.50.3.210

Kaplan MA, Aaronson AL, Henderson J, Goldin M (1955) C-reactive protein levels and anti-streptolysin O titres in bronchial asthma: a preliminary report. Ann Allergy 13(1):29–34

Kliegman R, Stanton B, St Geme JW III, Schor NF, Behrman RE, Nelson WE (2020) Nelson textbook of pediatrics, 21st edn. Elsevier, ©2020, Philadelphia, Pennsylvania

Kumar NP, Anuradha R, Andrade BB, Suresh N, Ganesh R, Shankar J, Kumaraswami V, Nutman TB, Babu S (2013) Circulating biomarkers of pulmonary and extrapulmonary tuberculosis in children. Clin Vaccine Immunol 20(5):704–711. https://doi.org/10.1128/CVI.00038-13. Epub 2013 Mar 13

Kusel MM, de Klerk NH, Kebadze T, Vohma V, Holt PG, Johnston SL, Sly PD (2007) Early-life respiratory viral infections, atopic sensitization, and risk of subsequent development of persistent asthma. J Allergy Clin Immunol 119(5):1105–1110. Epub 2007 Mar 13

Kwas H, Guermazi E, Zendah I, Jemia EB, Khattab A, Khouaja I, Ghedira H (2015) C-reactive protein and pulmonary tuberculosis: what correlation with disease severity. Eur Respir J 46: PA2751. https://doi.org/10.1183/13993003.congress-2015.PA2751

Lanata CF, Rudan I, Boschi-Pinto C, Tomaskovic L, Cherian T, Weber M, Campbel H (2004) Methodological and quality issues in epidemiological studies of acute lower respiratory infection in children in developing countries. Int J Epidemiol 33:1362–1372

Le Roux P (1998) Investigations for etiologic causes in community-acquired pneumonia in children (besides imagery). Arch Pediatr 5(Suppl. 1):28s–32s

Lin MS, Chong IW, Hwang JJ, Wang TH, Huang MS, Tsai MS, Chiang PC (1990) Quantitative C-reactive protein in pulmonary infections. Gaoxiong Yi XueKeXue Za Zhi 6(8):440–448

Little P, Stuart B, Moore M, Coenen S, Butler CC, Godycki-Cwirko M, Mierzecki A, Chlabicz S, Torres A, Almirall J, Davies M, Schaberg T, Mölstad S, Blasi F, De Sutter A, Kersnik J, Hupkova H, Touboul P, Hood K, Mullee M, O'Reilly G, Brugman C, Goossens H, Verheij T, GRACE consortium (2013) Amoxicillin for acute lower-respiratory-tract infection in primary care when pneumonia is not suspected: a 12-country, randomised, placebo-controlled trial. Lancet Infect Dis 13(2):123–129. https://doi.org/10.1016/S1473-3099(12)70300-6. Epub 2012 Dec 19

Liu L, Oza S, Hogan D, Chu Y, Perin J, Zhu J, Lawn JE, Cousens S, Mathers C, Black RE (2016) Global, regional, and national causes of under-5 mortality in 2000–15: an updated systematic analysis with implications for the Sustainable Development Goals. Lancet 388 (10063):3027–3035. https://doi.org/10.1016/S0140-6736(16)31593-8. Epub 2016 Nov 11

Määttä AM, Kotaniemi-Syrjänen A, Malmström K, Malmberg LP, Sundvall J, Pelkonen AS, Mäkelä MJ (2017) Vitamin D, high-sensitivity C-reactive protein, and airway hyperresponsiveness in infants with recurrent respiratory symptoms. Ann Allergy Asthma Immunol 119(3):227–231. https://doi.org/10.1016/j.anai.2017.06.014. Epub 2017 Jul 28

Matthew J, Pinto Pereira LM, Pappas TE, Swenson CA, Grindle KA, Roberg KA, Lemanske RF Jr, Lee W-M, Gern JE (2009) Distribution and seasonality of rhinovirus and other respiratory viruses in a cross-section of asthmatic children in Trinidad, West Indies. Italian J Pediatr 35 (16):1–10

Melbye H, Hvidsten D, Holm A, Nordbø SA, Brox J (2004) The course of C-reactive protein response in untreated upper respiratory tract infection. Br J Gen Pract 54(506):653–658

Monadi M, Firouzjahi A, Hosseini A, Javadian Y, Sharbatdaran M, Heidari B (2016) Serum C-reactive protein in asthma and its ability in predicting asthma control, a case-control study. Caspian J Intern Med 7(1):37–42

Nasser HA, Ezz NZA, Abdel-Mageed HM, Radwan RA (2019) Body mass index and C-reactive protein are potential predictors of asthma development in Egyptian polycystic ovary syndrome patients. J Med Biochem 38(4):427–436. https://doi.org/10.2478/jomb-2019-0012. eCollection 2019 Oct

Oymar K, Skejerren HO, Mikalsen IB (2014) Acute bronchiolitis in infants, a review. Scand J Trauma Resusc Emerg Med 22:23. https://doi.org/10.1186/1757-7241-22-23

Parikh K, Hall M, Blaschke AJ, Grijalva CG, Brogan TV, Neuman MI, Williams DJ, Gerber JS, Hersh AL, Shah SS (2016) Aggregate and hospital-level impact of national guidelines on diagnostic resource utilization for children with pneumonia at children's hospitals. J Hosp Med 11(5):317–323. https://doi.org/10.1002/jhm.2534. Epub 2016 Jan 13

Prasad PN, Venkitasubramanian TA, Viswanathan R (1965) Studies on C-reactive protein in bronchial asthma. Indian J Chest Dis 7:1–7

Razi E, Ehteram H, Akbari H, Chavoshi V, Razi A (2012) Evaluation of high-sensitivity C-reactive protein in acute asthma. Tanaffos 11(1):32–37

Reuter S, Moser C, Baack M (2014) Respiratory distress in the newborn. Pediatr Rev 35 (10):417–28; quiz 429. https://doi.org/10.1542/pir.35-10-417

Ritland N, Melbye H (1991) C-reactive protein, SR and white blood cell count in acute lower respiratory tract diseases. The usefulness of blood tests in diagnosis of pneumonia. Tidsskr Nor Laegeforen 111(18):2249–2252

Rubner FJ, Jackson DJ, Evans MD, Gangnon RE, Tisler CJ, Pappas TE, Gern JE, Lemanske RF Jr (2017) Early life rhinovirus wheezing, allergic sensitization, and asthma risk at adolescence. J Allergy Clin Immunol 139(2):501–507. https://doi.org/10.1016/j.jaci.2016.03.049. Epub 2016 May 10

Rudan I, O'Brien KL, Nair H, Liu L, Theodoratou E, Qazi S, Lukšić I, Fischer Walker CL, Black RE, Campbell H, Child Health Epidemiology Reference Group (CHERG) (2013) Epidemiology and etiology of childhood pneumonia in 2010: estimates of incidence, severe morbidity, mortality, underlying risk factors and causative pathogens for 192 countries. J Glob Health 3 (1):010401. https://doi.org/10.7189/jogh.03.010401

Ruiz-González A, Utrillo L, Bielsa S, Falguera M, Porcel JM (2016) The diagnostic value of serum C-reactive protein for identifying pneumonia in hospitalized patients with acute respiratory symptoms. J Biomark 2016:2198745. https://doi.org/10.1155/2016/2198745. Epub 2016 Aug 16

Savitha MR, Nandeeshwara SB, Kumar MJ, ul-Haque F, Raju CK (2007) Modifiable risk factors for acute lower respiratory tract infections. Indian J Pediatr 74:477–482

Schalek P, Hornáčková Z, Hahn A (2015) The relationship of C-reactive protein levels and positive culture with quality of life in acute rhinosinusitis. Patient Prefer Adher 9:185–189. Published 2015 Jan 27. https://doi.org/10.2147/PPA.S76915

Selvaraj K, Chinnakali P, Majumdar A, Krishnan IS (2014) Acute respiratory infections among under-5 children in India: a situational analysis. J Nat Sci Biol Med 5(1):15–20. https://doi.org/10.4103/0976-9668.127275

Seppä Y, Bloigu A, Honkanen PO, Miettinen L, Syrjälä H (2001) Severity assessment of lower respiratory tract infection in elderly patients in primary care. Arch Intern Med 161 (22):2709–2713

Shann F, Hart K, Thomas D (1984) Acute lower respiratory tract infections in children: possible criteria for selection of patients for antibiotic therapy and hospital admission. Bull WHO 62:749–753

Sharma D, Kuppusamy K, Bhoorasamy A (2013) Prevalence of acute respiratory infections (ari) and their determinants in under five children in urban and rural areas of Kancheepuram district, South India. Ann Public Med Trop Health 6(5):513–518

Shimoda T, Obase Y, Kishikawa R, Iwanaga T (2015) Serum high-sensitivity C-reactive protein can be an airway inflammation predictor in bronchial asthma. Allergy Asthma Proc 36(2):e23–e28. https://doi.org/10.2500/aap.2015.36.3816

Temani K, Amrita M, Bairwa AL (2016) Assessment of prevalence of acute respiratory tract infection and risk factors in under five children in anganwadi of Kota city. Indian J Child Health 3(3):234–237

Theodoratou E, Zhang JS, Kolcic I, Davis AM, Bhopal S, Nair H, Chan KY, Liu L, Johnson H, Rudan I, Campbell H (2011) Estimating pneumonia deaths of post-neonatal children in countries of low or no death certification in 2008. PLoS One 6(9):e25095. https://doi.org/10.1371/journal.pone.0025095. Epub 2011 Sep 22

Troeger C, Blacker B, Khalil IA, Rao PC, Cao J, Zimsen SRM et al (2018) Estimates of the global, regional, and national morbidity, mortality, and aetiologies of lower respiratory infections in 195 countries, 1990–2016: a systematic analysis for the global burden of disease study 2016. Lancet Infect Dis 18(11):1191–1210. https://doi.org/10.1016/S1473-3099(18)30310-4

Tuft HS, Scherr MS (1956) The significance of C-reactive protein in asthmatic patients. J Allergy 27 (4):344–351

Ulrich A, Pillat MM, Tarnok A (2020) Dengue fever and COVID 19 (SARS COV 2) and antibody dependent enhancement (ADE) a perspective. Cytometry A. https://doi.org/10.1002/cyto.a.24047

UNICEF (2008) The State of Asia-Pacific's Children 2008: Child Survival. Available from: http://www.unicef.org/sapc08/report/report.php. (Last accessed on 2013)

UNICEF (2018) Estimates of child cause of death, acute respiratory infection

van der Meer V, Neven AK, van den Broek PJ, Assendelft WJ (2005) Diagnostic value of C reactive protein in infections of the lower respiratory tract: systematic review. BMJ 331 (7507):26. Epub 2005 Jun 24

Vashishtha VM (2010) Current status of tuberculosis and acute respiratory infections in India: Much more needs to be done! Indian Pediatr 47:88–89

Vázquez EG, Martínez JA, Mensa J, Sánchez F, Marcos MA, de Roux A, Torres A (2003) C-reactive protein levels in community-acquired pneumonia. Eur Respir J 21(4):702–705

Wagenmann M, Naclerio R (1992) Complication of sinusitis. J Allergy Clin Immunol 90(3) Pt 2:552–554. https://doi.org/10.1016/0091-6749(92)90184-4

Walker CLF, Rudan I, Liu L, Nair H, Theodoratou E, Bhutta ZA et al (2013) Global burden of childhood pneumonia and diarrhoea. Lancet 381(9875):1405–1416. https://doi.org/10.1016/S0140-6736(13)60222-6

Wang JL, Chen KY, Fang CT, Hsueh PR, Yang PC, Chang SC (2005) Changing bacteriology of adult community-acquired lung abscess in Taiwan: *Klebsiella pneumoniae* versus anaerobes. Clin Infect Dis 40:915–922. https://doi.org/10.1086/428574

Wardlaw TM, Johansson EW, Hodge MJ. UNICEF (2006) World Health Organization. Pneumonia: The Forgotten Killer of Children. Available from: http://www.who.int/maternal_child_adolescent/documents/9280640489/en. (Last accessed on 2013 Mar 13)

Williams BG, Gouws E, Boschi-Pinto C, Bryce J, Dye C (2002) Estimates of world-wide distribution of child deaths from acute respiratory infections. Lancet Infect Dis 2(1):25–32

Wilson D, Badri M, Maartens G (2011) Performance of serum C-reactive protein as a screening test for smear-negative tuberculosis in an ambulatory high HIV prevalence population. PLoS One 6 (1):e15248. https://doi.org/10.1371/journal.pone.0015248

Yang BY, Markevych I, Harris C, Standl M, Schikowski T, Koletzko S, Herberth G, Bauer CP, von Berg A, Berdel D, Dong GH, Heinrich J (2019) High-sensitivity C-reactive protein and allergic endpoints in German adolescents. Arch Allergy Immunol 179(2):152–157. https://doi.org/10.1159/000497320. Epub 2019 Apr 3

Zhang JJ, Dong X, Cao YY, Yuan YD, Yang YB, Yan YQ, Akdis CA, Gao YD (2020) Clinical characteristics of 140 patients infected with SARS-CoV-2 in Wuhan, China. Allergy. https://doi.org/10.1111/all.14238

Role of C-Reactive Protein (CRP) in Sepsis: Severity and Outcome

9

Sheikh Hasan Habib and Waliza Ansar

Contents

9.1 Introduction 252
9.2 History of Sepsis 254
9.3 To Understand and Define Sepsis 255
9.4 Epidemiology of Sepsis 262
9.5 Pathophysiology of Sepsis 263
9.6 Liver-Mediated Pro-Inflammatory Response, APPs, and Sepsis 267
9.7 Biomarkers of Sepsis 271
9.8 Role of CRP in Sepsis 275
9.9 Management of Sepsis 279
9.10 Discussion 281
References 283

Abstract

Different types of infections in a community depend on the defense mechanism of the patient and surroundings of the patient. Early intervention can prevent hospital admission of the patient. Though the response to the treatment needs overall physiological condition of the patient. Healthcare-associated infections are common cause of increased cost of total care of patient in a hospital, morbidity, and mortality. Rapid identification of the source of infection and involvement in the body are helpful for better outcome. The infection may be viral, bacterial, parasitic, and fungal. The infection may be primary or secondary

S. H. Habib
K. P. C. Medical College and Hospital, Kolkata, West Bengal, India

Intensive Therapeutic Unit, Woodlands Multispecialty Hospital Limited, Kolkata, West Bengal, India

W. Ansar (✉)
Department of Zoology, Behala College, Kolkata, West Bengal, India

W. Ansar, S. Ghosh (eds.), *Clinical Significance of C-reactive Protein*,
https://doi.org/10.1007/978-981-15-6787-2_9

or others. Managing infection in a patient and response to the treatment are multifactorial like treating infection in an immune-compromised patient is a big challenge. The challenge also is more to the physician during treating a geriatric patient. Collection of history from the patient/relatives of patient is an initial tool for probable diagnosis. The methodical approaches of investigation, proper interpretation of the result, early initiation of treatment, and response to the treatment are essential for patient's final outcome. The final outcome is also warranted by appropriate microbial and other supportive care of the hospital. Complexity of the situation needs close liaison between microbiologist, pathologist, infectious disease consultant, hematologist with quality of laboratory, and promptness of interrelationship within the hospital unit. While treating a subject, sometimes protective isolation for neutropenic and immune-compromised patients and source isolation of colonized and infected patients are essential. A new onset of fever in a patient admitted to hospital should be investigated, treated, and attended very sincerely. The reason could be manifold much as non-infectious causes, immunological cause, mild infection, or initial presence of severe infection. With increasing incidence of geriatric patient population, comorbidity, prolonged ICU stay, invasive therapies, use of broad-spectrum antibiotics, in-dwelling catheters, there is a rising incidence of fungal infection also. Candidemia, the common invasive fungal infection, is the fourth most common cause of bloodstream infection. Here, the sepsis markers like CRP have immense role since presentation of the patient to the healthcare facility till the outcome of the patient according to rising or decreasing CRP concentration of the patient.

Keywords

Sepsis · Candidemia · Epidemiology · ICU · Broad-spectrum antibiotics · Biomarker · C-reactive protein · Procalcitonin · IL-6

Abbreviations

ABG	Arterial blood gas
ALL	Acute lymphoblastic leukemia
APACHE	Acute physiology and chronic health evaluation
APPs	Acute phase proteins
APTT	Activated partial thromboplastin time
BNP	Natriuretic peptide
CBG	Capillary blood glucose
cfDNA	Cell-free DNA
CRP	C-reactive protein
CRP/ALB	C-reactive protein/albumin
CSF	Cerebrospinal fluid
CT	Computed tomography
CVP	Central venous pressure

DAMPs	Danger-associated molecule patterns
DP	Death probability
ESR	Erythrocyte sedimentation rate
FIO_2	Fraction of inspired oxygen
GAS	Group A streptococcus
GBD	Global Burden of Diseases, Injuries, and Risk Factors Study
HCT	Hematocrit
hs-CRP	High-sensitivity CRP
ICU	Intensive care unit
IFN-γ	Interferon gamma
iGAS	invasive GAS
IL-1	Interleukin 1
IL10	Interleukin 10
IL-1β	Interleukin 1 beta
IL-6	Interleukin 6
LBP	Lipopolysaccharide-binding protein
LDLR	Low-density lipoprotein receptor
lncRNA	Long noncoding RNAs
LOS	Length of stay
LPS	Lipopolysaccharides
MALAT1	Metastasis-associated lung adenocarcinoma transcript 1
MDSCs	Myeloid-derived suppressor cells
MPO	Myeloperoxidase
NE	Neutrophil elastase
NET	Neutrophil extracellular traps
NEU count	Neutrophil count
NHIRD	National Health Insurance Research Database
NTproBNP	N-terminal pro b-type natriuretic peptide
PAI-1	Plasminogen activator inhibitor-1
PCO_2	Partial pressure of carbon dioxide
PCR	Polymerase chain reaction
PCSK9	Proprotein convertase subtilisin/kexin type 9
PCT	Procalcitonin
PIRO	Predisposition, infection, response, organ dysfunction
PLA2-II	Group II phospholipase A2
PO_2	Partial pressure of oxygen
POD	Postoperative period
PTT	Partial thromboplastin time
qSOFA	quick SOFA
RNS	Reactive nitrogen species
ROS	Reactive oxygen species
SAA	Serum amyloid A protein
SAPS 3	Simplified Acute Physiology Score
SCCM-ACCP	American College of Chest Physicians/Society of Critical Care Medicine Consensus Conference

SIRS	Systemic inflammatory response syndrome
SOFA	Sequential Organ Failure Assessment
sRAGE	Soluble form of the receptor for advanced glycation end products
sTREM-1	Soluble triggering receptor expressed on myeloid cells-1
suPAR	Soluble urokinase plasminogen activator receptor
TNF-α	Tumor necrosis factor-α
WBC	White blood cell
WHO	World Health Organization

Highlights

1. Sepsis is a multifactorial clinical condition that occurs when an infection exceeds the boundaries of the local environment and a generalized infective response, with the release of varied pro-inflammatory mediators. The body's response to an infection injures its own tissues and organs. It is associated with both significant morbidity and potential mortality. It may be initiated in response to a non-infectious condition (like pancreatitis, trauma, etc.), then the process is referred to as systemic inflammatory response syndrome (SIRS).
2. A characteristics septic response develops with the invasion of the infection by microorganisms, or their toxic products, beyond the local environment. In addition, some individuals may be genetically susceptible to developing sepsis.
3. It is difficult to ascertain the global epidemiological burden of sepsis. It is estimated that approximately 3 million newborns and 1.2 million children suffer from sepsis worldwide each year. Resistant pathogens cause three out of every ten deaths in neonatal sepsis. Sepsis can be the clinical manifestation of infections acquired both in the community setting or in healthcare facilities (Fleischmann-Struzek et al. 2018; https://www.who.int/news-room/fact-sheets/detail/sepsis). These infections are often resistant to antibiotics and can rapidly lead to deteriorating clinical conditions.
4. CRP is an important biomarker of sepsis. It is used in sepsis diagnosis, but not always routinely recommended in some current clinical practices. CRP can differentiate between routine infection and sepsis. Elevations of CRP were added to the updated definition of sepsis in 2003.

9.1 Introduction

Sepsis is a medical emergency and is a major cause of mortality worldwide. It is primarily described as the body's systemic immunological response to an infectious process that can culminate to end-stage organ dysfunction and final death. In spite of significant advancements in the perceptive of the pathophysiology of this clinical condition, advancements in hemodynamic monitoring tools, measurement of biomarkers and resuscitation measures, sepsis remains one of the major causes of morbidity and mortality in critically ill hospitalized patients (Kaukonen et al. 2014).

Sepsis can be acquired both in the community setting or in healthcare facilities. It is especially common in the elderly patients and is likely to increase considerably as the population ages. The burden of sepsis is in all likelihood is highest in low- and middle-income countries (Say et al. 2014; https://www.who.int/news-room/fact-sheets/detail/sepsis). Many patients with sepsis may remain undiagnosed as occurrence of infection and its relation to organ dysfunction is poorly acknowledged outside of the intensive care unit (ICU).

Sepsis is defined as the combined interflow of invasion of pathogens into the blood flow together with the diverse host response to this invasion. Thus, sepsis may be summed as the systemic inflammatory response syndrome (SIRS) caused by infection. The clinical variables prediction of SIRS with nonspecific markers, such as white blood cell count (WBC), in the presence of a known or even suspected infection. Sepsis may be complicated by arterial hypotension (septic shock) or remote organ dysfunction (severe sepsis), which considerably worsens outcome of patients with infection. To optimize the outcome, the current guidelines recommend antimicrobial therapy (which is anti-infectious therapy) and the surgical source control should be initiated as soon as possible. Indeed, fulfillment to sepsis guidelines improves outcome and time to antimicrobial therapy is an independent interpreter for death. Conventional early diagnosis relies on the recognition of SIRS caused by different infections and the new inception of organ dysfunction. As SIRS lacks the specificity to be clinically significant, this concept has been widely criticized. The main problem in the clinical setup lies in discriminating SIRS with and without infection. Thus, confirming the infecting entity as the main cause of a severe inflammatory response is the main challenge in the diagnosis of sepsis.

Thus, the original sepsis guidelines were modified by group of experts and developing the concept of **PIRO** (predisposition, infection, response, organ dysfunction) for an improved diagnosis of stages and characterization of sepsis in patients with. The clinical **predisposition (P)** includes age, alcohol abuse, steroid or immunosuppressive therapy of the patients as assessed by immunologic monitoring and genetic factors. Site-specific **infection (I)** includes pneumonia, peritonitis, etc. as diagnosed by X-rays, CT scan, and bacteriology tests. The **response (R)** includes malaise, temperature, heart rate, and respiratory rate as monitored by WBC count, C-reactive protein (CRP), and procalcitonin (PCT) levels and modified APTT tests. The **organ dysfunction (O)** clinically includes arterial pressure, urine output, and Glasgow Coma Score measured by PaO_2/FIO_2, creatinine, bilirubin, platelets count test (Vincent and Abraham 2006).

The future tools of PIRO system include the detection of biomarkers and microbial nucleic acids by polymerase chain reaction (PCR). In current diagnosis strategy, measurement of biomarkers as well as PCR-based pathogen detection is allowed for a more rapid diagnosis of sepsis available within one working day time span. The severity of sepsis is graded according to the associated perfusion abnormalities and organ dysfunction (Bone et al. 1992; Levy et al. 2003; Dellinger et al. 2008).

The application of different diagnostic criteria of sepsis, severe sepsis, or septic shock is quite easy in patients just admitted to the ICU with a very serious disease. But, it is always more difficult or even impossible in patients who are staying in the ICU for weeks, as their clinical outcome is marked by progress without definite recovery and may be followed by numerous relapses. Parameters like body temperature, pulse rate, respiratory rate, and WBC count, as well as organ function and tissue perfusion, are frequently abnormal in sepsis. Different laboratory cultures confirm the type of infection in only a few number of cases when bacteria are found in the cerebrospinal fluid or in the blood.

In most ICU patients with slowly developing disease, bacterial colonization in multiple sites is the rule, and some degree of inflammation is nearly inevitable. The well-known dilemma of the clinicians revolves around the daily decision of persisting inflammation versus new infection and whether to wait and observe the patient versus to start a new course of antimicrobial therapy. Appropriate dosage and timely treatment with antimicrobials could save lives of ICU patients, whereas prolonged and excessive dosage favors the emergence of multi-resistant bacterial strains. The available diagnostic tools are of modest help in sepsis.

9.2 History of Sepsis

The last 100 years had evidenced wide advances in our understanding of sepsis. Some of the first descriptions of bacteria include Leeuwenhoek's "animalcules," but it was another 200 years of rolling before the linkage between bacteria and infection finally began to be modeled by eminent founders of modern microbiology like Koch, Semmelweis, Pasteur, and Lister. Our modern understanding of the term "sepsis" finally, realized in 1914, when Schottmueller noted that the release of pathogenic microbes into the bloodstream was responsible for systemic signs and symptoms (Schottmueller 1914; Vincent and Abraham 2006; Reinhart et al. 2012).

The world famous infectious diseases (pulmonary, intestinal, and urino-genital) which were categorized as nosological entity were researched by medical historians to form a history of modern medicine. Sepsis was never grouped as this kind of disease. So we may start from a very simple description of sepsis in ancient Egypt to its running in premodern times. From Egyptian period to modern times, it had bodied clinical entity with infection/wound, fever, local signs of inflammation, and final secondary systemic responses.

The word "sepsis" is derived from the ancient Greek (σηψις), which means the bacterial decomposition of plant- or animal-based organic materials. In Homeric poems, the word "sepsis" was used as "sepo" (σηπω), which means "I rotted." The word is used in the poem in the 24th song of the Iliad in a sad description where Priam was led by Hermes to beg from Achilles the body of Hector from the Greek camp. Hippocrates (in Hippocratic corpus) also mentioned the term sepsis between 460 and 370 BC with a representative word "sepidon," which means "distortion, dissolution of a web structure." The term was subsequently used by academicians like Aristotle, Plutarch, and Galen with same relevant meaning and it continues to hold its meaning with virtually no change in meaning for over a period of 2700 years (Gül et al. 2017). This word is little used in medieval medical work. During the Middle Ages and Arabic period, sepsis became the silent partner with barber surgeons. Matthaeus Silvaticus, a famous physician mentioned sepsis in his medical encyclopedias in late Middle Ages. In 1858, in European language, the word sepsis was found in Oxford English Dictionary (Botero and Pérez 2012).

Over the time course of the twentieth century, numerous experimental works and clinical trials were able to congregate the relevance of the immune host response toward infection as the manifestations of sepsis. However, due to heterogeneity of the disease molecular pathways, it posed serious problem in diagnosing, treating, and studying sepsis. Finally, in 1991, in an SCCM-ACCP conference, Bone et al. 1992 laid the groundwork for the first consensus definition of sepsis. After that, there have been significant advances in the pathophysiology of sepsis in the last two decades. We have an improved perceptive of morphology, biochemistry, immunology, and cell biology as well as alterations in circulation and organ function. This considerate thought has led to the changes in the definition of sepsis. This has also demonstrated to improved management of sepsis leading to changes in the spectrum of epidemiology of the sepsis. The inflammatory, hemodynamic, organ dysfunction and tissue perfusion parameters of 2001 were mentioned by (Bone et al.1992; Levy et al. 2003).

9.3 To Understand and Define Sepsis

A modern view of sepsis was proposed by Semmelweis, Pasteur, and Lister in the nineteenth century (Marshall 2004). Ever since then, diverse spectrum of researches on sepsis has been conducted till date. Studies include the mechanism of sepsis through identification of pathogens, coagulation cascades involved, varied immune responses, and organ damage (Hotchkiss et al. 2013; Crouser et al. 2008; Dhainaut et al. 2001; Yan et al. 2014). The upcoming researches have enlightened the knowledge of the pathogenesis of sepsis. But still, sepsis morbidity and mortality rates remain high. The incidence of sepsis is expected to increase to 1.5% annually because of the increasing number of old patients, who are more susceptible to sepsis. Studies have also predicted that by the upcoming year 2029, the annual incidence of sepsis cases may increase from the incidence in the year 2001 by as much as 50% (Gaieski et al. 2013; Yan et al. 2014).

Recent research consensus defined sepsis as the systemic host response of the body to infection caused by invasion of pathogens into the bloodstream. Thus, sepsis consists of the SIRS caused by infection (Fig. 9.1). By simply defining sepsis as "the host response to an infection" does not necessarily cover its spectrum of negative connotation, as it mainly involves host low/poor responses to infection, where the body is not well adapted to the infection.

Sepsis is a dysregulated systemic inflammatory response(s) triggered by acute infection, toxin, or trauma. In infected patients, the response(s) progress from inflammation to sepsis, subsequent to severe sepsis (remote organ dysfunction), septic shock (arterial hypotension), and lastly to multiple organ dysfunction syndrome (Fig. 9.2). All these significantly worsens outcome of patients with infection. Subjects with a weakened defense system like infants, children, and the orderly (over 65 years)—are mostly susceptible to sepsis and sepsis-related death (Iwashyna et al. 2012; Yan et al. 2014). The modern notion of sepsis diversify across bacterial, fungal, viral, and parasitic intruders, focuses mainly on the host response as the

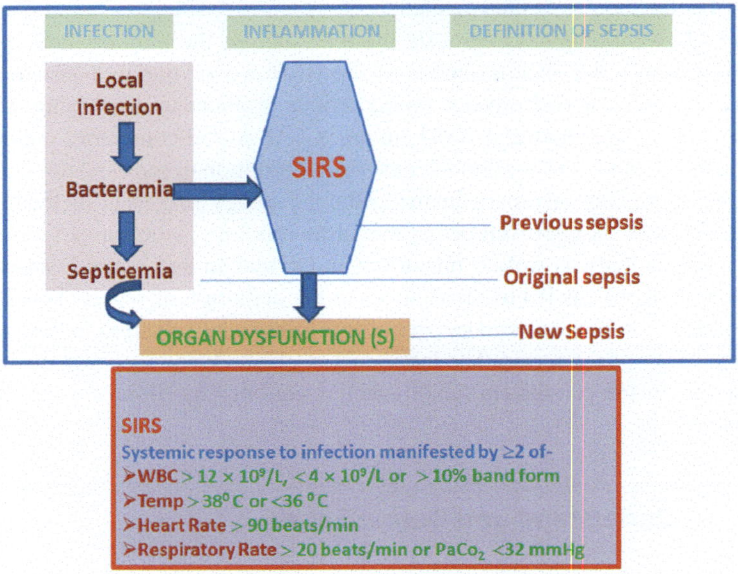

Fig. 9.1 Systemic inflammatory response syndrome (SIRS). The criteria for SIRS were noted and were taken from ACCP/SCCM Conference Term Criteria 1992 (Bone et al. 1992; Singer et al. 2016; Huerta and Rice 2019)

prime source of morbidity and mortality and requires only that suspected infection be proved, but recognition and treatment in many cases do not have such confirmation.

Current treatment guidelines recommend the fast usage of anti-infectious therapy like antimicrobial therapy and surgical source control to optimize best outcome. Indeed, compliance to sepsis guidelines improves outcome (American College of Chest Physicians/Society of Critical Care Medicine Consensus Conference 1992; ACCP-SCCM). The time course of delivery of antimicrobial therapy is an independent predictor for death. Traditional diagnosis relies on the diagnosis of SIRS (Fig. 9.1) caused by infection and the new inception of organ dysfunction. But a clear-cut diagnosis of SIRS with specificity needs a lot of meaningful clinics. The bigotry of SIRS with and without infection remains the main dilemma in the clinical setup.

Thus, the main challenge in the diagnosis of sepsis is to confirm the source of infection as cause of a severe inflammatory response. For improved characterization and staging of patients with sepsis, experts reformulated the original sepsis guidelines and build up the PIRO (predisposition, infection, response, organ dysfunction) concept. Within the PIRO system, naming the biomarkers and detection of pathogenic nucleic acids by PCR remain as future tools to describe the infection process and response.

These detection methodologies (PCR and biomarker identification) provide a more rapid diagnosis of sepsis as they are accessible within one working day

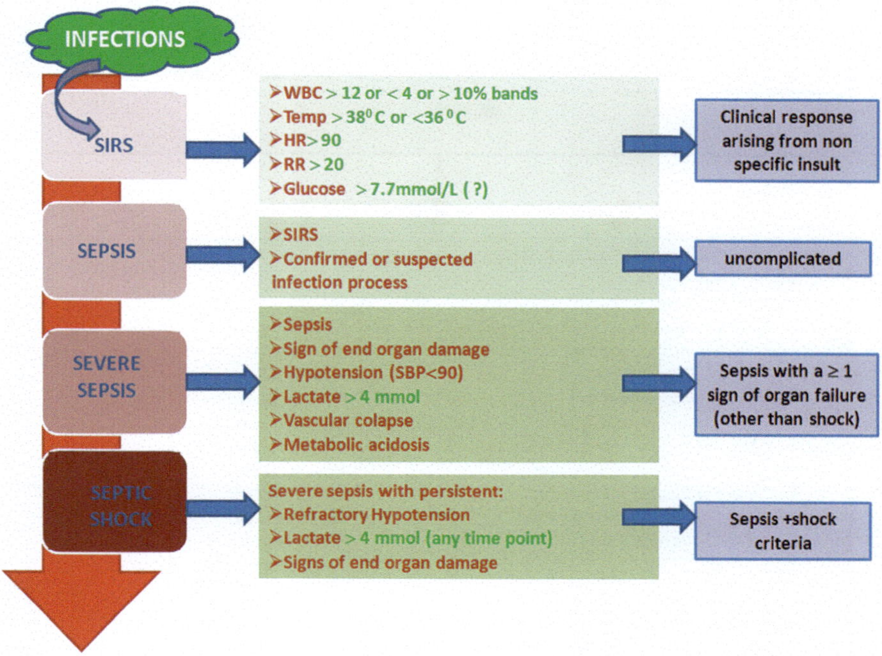

Fig. 9.2 Sepsis is a continous process: A definition. The continuous process of sepsis starts with SIRS, sepsis, severe sepsis and Septic shock. The criterion for each definition is mentioned (http://www.google.com/url?sa=i&url=http%3A%2F%2Feminduction)

(Ferrer et al. 2009; MacArthur et al. 2004; Levy et al. 2003; Bloos and Reinhart 2014). The laboratory criteria for sepsis including different parameters were also recorded (Fig. 9.3).

Any bloodstream infection does not lead to sepsis. A better understanding of the pathogenesis of sepsis is needed for its high mortality and morbidity rate. Sepsis was defined (in 1991) as the presence of an inflammatory response to infection with the presence of no less than 2 (fever or hypothermia, leukocytosis or leucopenia, tachycardia, and tachypnea or hyperventilation) out of the 4 SIRS criteria realized during infection (ACCP/SCCM Consensus Conference 1992). In the early 2000s, markers of organ dysfunction were included in the definition of sepsis. Though the definition was an extended one, the SIRS criteria, as mentioned in the formal sepsis definition, still remained a part of the recent definition (Levy et al. 2003; Huerta and Rice 2019) (Fig. 9.3).

In 2016, the sepsis definition was significantly remodeled to encompass a more modern understanding of the disease. Further, it had become apparent that any normal response to infection may lead to inflammatory response and SIRS criteria had been found to be a bad marker of sepsis as it lacks both specificity and sensitivity. Interestingly, many hospitalized patients with reasons unrelated to infection, at presentation, had at least $\geq$2 SIRS criteria (Churpek et al. 2015; Huerta and

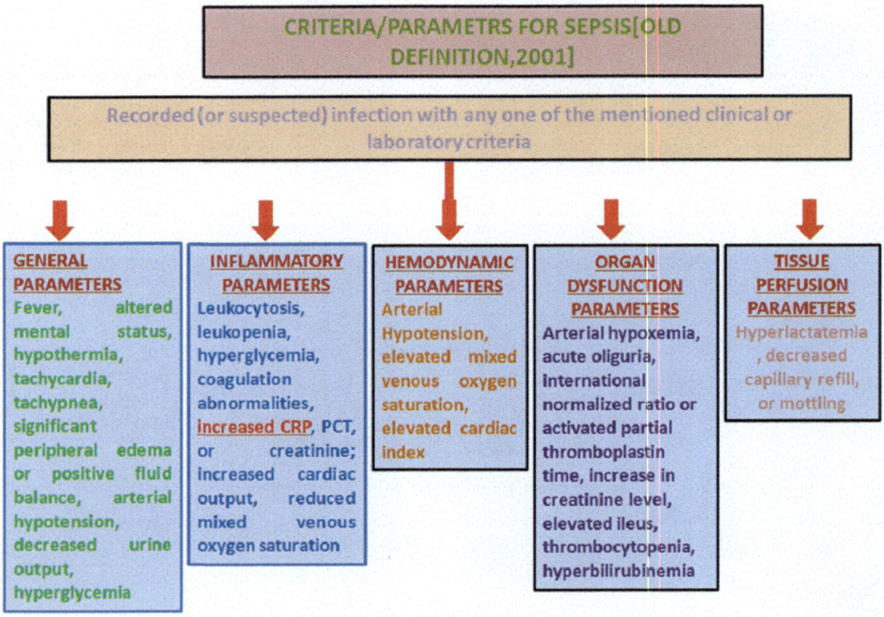

Fig. 9.3 Criteria/parameter of sepsis according to old definition, 2001. The parameters recorded for old definition of sepsis were general, inflammatory, hemodynamic, organ dysfunction and tissue perfusion (Bone et al. 1992). Increased CRP was recorded in inflammatory parameters. *PCT* procalcitonin

Rice 2019). In 2015, another report showed that sepsis were not detected in ICU using the SIRS criteria (Kaukonen et al. 2015; Huerta and Rice 2019) (Fig. 9.4).

So, in 2016, sepsis was redefined as a "life-threatening organ dysfunction caused by a dysregulated host response to infection" (Singer et al. 2016; Huerta and Rice 2019) (Fig. 9.5), with the basis of identified organ dysfunction and an increase of at least 2 points in the Sequential Organ Failure Assessment (SOFA) Score. SOFA Score is a marker of severity which includes various clinical and laboratory tests (total bilirubin, platelet count, creatinine) and the partial pressure of oxygen in arterial blood (Vincent et al. 1996; Huerta and Rice 2019).

As a part of the new Sepsis-3 definitions employing SOFA as a grading score for defining acute organ dysfunction (SOFA Score) (Vincent et al. 1996). This score distributes points according to pathological alterations in six different organ systems; a rise in the total SOFA Score by at least 2 points (in negative patient history, the score of 0 is assigned) pointing acute organ dysfunction, and the diagnosis of sepsis is fulfilled if an undercurrent an infection is present. It is a case of "septic shock" when hypotension (in spite of adequate fluid management, mean blood pressure of 65 mmHg can only be achieved by using vasopressors) and serum lactate levels (more than 2 mmol/L) present. Sepsis-3 uses qSOFA ("quick SOFA") (Singer et al. 2016;) screening tool for early recognition of organ dysfunction due to infection. It uses the criteria of altered mental status, respiratory rate (more than 22 breaths/

Fig. 9.4 Definition of sepsis (old and New). The older definition of sepsis (1991) (Sepsis I) (Bone et al. 1992) was given based on some parameters. The stages of response towards infection are SIRS (systemic inflammatory Response Syndrome), Sepsis, Severe Sepsis, Severe sepsis. The modern sepsis 3 definition

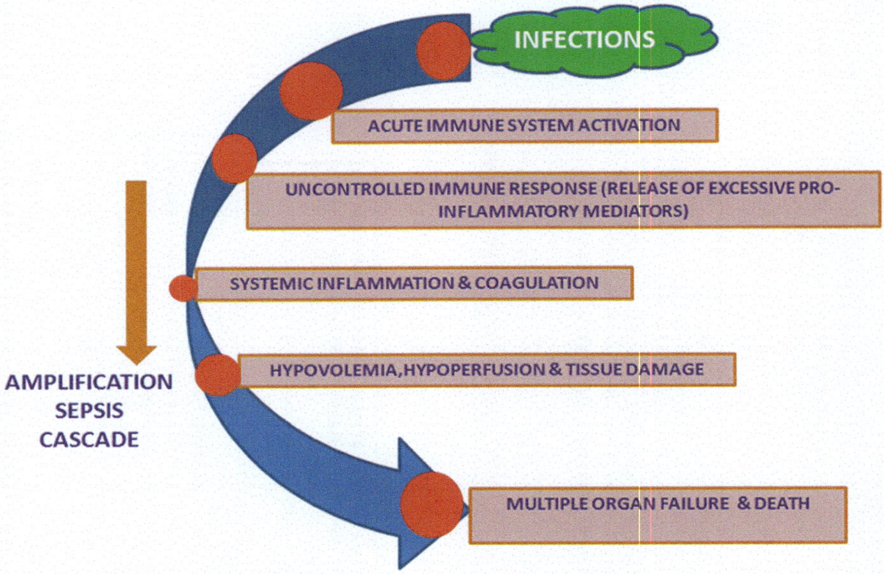

Fig. 9.5 Sepsis cascade. After the infection, the immune system gets activated and it turned out to be an uncontrolled immune activation which finally cascades to multi organ failure and death

minute), and systolic blood pressure (less than 100 mmHg). It is not used in ICU but used in rest homes, peripheral wards, and emergency departments. If two of these three qSOFA criteria are fulfilled, then it is a suspected case of organ dysfunction which should be confirmed by the physicians/intensive care specialists using the classic SOFA Score (Table 9.1).

The modern definition of sepsis (inclusive of organ dysfunction) is much related to the previous concept of severe sepsis, denoted by a subset of septic patients with evidence of organ dysfunction. As a consequence, the distinction between sepsis and severe sepsis was abolished from the new sepsis definitions (in 2016), with severe sepsis patients were now categorized as sepsis.

Now, septic shock, a more rigorous clinical manifestation of infection, is considered to be "a subset of sepsis in which particularly profound circulatory, cellular, and metabolic abnormalities are associated with a greater risk of mortality than with sepsis alone" (Singer et al. 2016; Huerta and Rice 2019).

Fig. 9.4 (continued) (2016) was given based on qSOFA screening. The general parameters, inflammatory parameters, Hemodynamic parameters, organ dysfunction parameters and tissue perfusion parameters were noted (Fig. 9.3). The qSOFA parameters for sepsis and septic shock was mentioned (JAMA. 2016; 315(8):801–810. https://doi.org/10.1001/jama.2016.0287; Gul et al. 2017). *qSOFA* quick sequential organ failure assessment, *HR* heart rate, *RR* respiratory rate, *MAP* mean arterial pressure, *GCS* Glasgow coma scale, *$PaCO_2$* partial pressure of carbon dioxide, *PaO_2* partial pressure of oxygen, *SOFA* sequential organ failure assessment

Table 9.1 Sofa screening score

SOFA score	1	2	3	4
Respiration parameter				
PaO_2/FiO_2 (rnm Hg)	<400	<300	<200	<100
Coagulation parameter				
Platelets $\times 10^3/mm^3$	<150	<100	<50	<20
Liver parameter				
Bilirubin (mg/dL)	1.2–1.9	2.0–5.9	6.0–11.9	>12.0
Renal parameters				
Creatinine (mg/dL) or	1.2–1.9	2.0–3.4	3.5–4.9	>5.0
Urine output (mL)			or <500	or <200
Cardiovascular parameter				
Hypotension	MAP < 70	Dopamine 55 or dobutamine (any)	Dopamine >5 or norepinephrine <0.1	Dopamine >15 or norepinephrine >0.1
Central nervous system parameter				
Glasgow coma score	13–14	10–12	6–9	<6

The SOFA is a scoring system used to determine the speed and pervasiveness of sepsis-related organ dysfunction for bedside clinical variables. It calculates morbidity rather than mortality. It computes the level of dysfunction in six systems (respiratory, cardiovascular, renal, gastrointestinal, coagulation and neurologic). SOFA is calculated at admission and once every 24 h. The computation, based on the average and worst scores during ICU stay with a prediction of clinical expectations. This scoring system should not be used to determine the success of treatment or the clinical method. If a person is normal, the SOFA score is 0, if the score is 4, it indicates a severe abnormality. It was calculated to provide the easiest daily description of organ dysfunction for use in clinical trials. *FIO_2* fraction of inspired oxygen, *MAP* mean arterial pressure, *PaO_2* partial pressure of oxygen

It should be noted that bloodstream bacterial infection is called bacteremia. Any bloodstream infections, with dysregulated immune response, may not lead to sepsis; also not all cases of sepsis are due to bloodstream infections (Bates et al. 1997; Huerta and Rice 2019).

Thus, in 1991, a consensus conference (ACCP-SCCM) was developed where the initial definitions of SIRS to infection was denoted as sepsis. In 2001, with the incorporation of the threshold values for organ damage, the definitions of sepsis and septic shock were revised. In 2016, the modern definitions of sepsis and septic shock have transformed dramatically. Furthermore, sepsis is now defined as life-threatening organ dysfunction caused by a dysregulated host response to infection (Fig. 9.5). The SOFA Score was also documented in the consensus describing organ dysfunction as an acute rise in SOFA Score by two points after suspected infection. These modern definitions eliminate the application of SIRS. The Sepsis-3

Task Force also launched qSOFA (a new bedside index) Score for physicians employed ICU to diagnose suspected infected patients from patients who would develop sepsis subsequently.

9.4 Epidemiology of Sepsis

The preceding 100 years have seen extensive changes in the epidemiology of sepsis. However, as definitions have modified during this time course and laboratory and imaging technologies have improved, it is only possible to draw fairly good diagnostic conclusions. The primary causative infectious site of sepsis has changed in due course, from the abdomen as main source before 1990 to the lungs in these recent years. Pneumonia is regarded as the one of the most common infections associated with sepsis, followed by other infections. The microbiology of severe sepsis and septic shock now includes both Gram-negative and Gram-positive organisms. Other pathogenic organisms include fungal or parasitic infection and in some cases some unidentified source also.

Many patients with sepsis may remain undiagnosed as occurrence of infection and related organ dysfunction is weakly recorded outside the ICU. Sepsis is especially common in the elderly group and is likely to precede substantially as the population ages (Rohde et al. 2013; Angus et al. 2001).

Globally, there were nearly 60.2 million incident cases of sepsis in the year 1990 (Rudd et al. 2020). In the United States, the incidence of severe sepsis cases in 2001 is nearly 300 cases per 100,000 population (Angus et al. 2001).

In Taiwan, Shen et al. showed the epidemiology of sepsis in their studies from 1997 to 2006 by using the longitudinal National Health Insurance Research Database (NHIRD) of Taiwan. They found that the incidence of sepsis increased from 135 to 217 per 100,000 people. Quite recently, another group in 2013 also used the NHIRD in Taiwan to investigate the epidemiology of sepsis and showed that the hospital mortality rate was 19.3% (Henriksen et al. 2015).

In Spain from 2006 to 2011, out of the 22,070,672 hospital discharges that occurred, and after several thorough processes of verification and depuration of the obtained database, researchers identified 240,939 overnight hospitalizations with severe sepsis. The overall occurrence of severe sepsis was 86.97 cases per 100,000 populations. In women, it was of 71.90 cases per 100,000 populations whereas in men, the crude prevalence rate was 102.47 cases per 100,000 populations (Bouza et al. 2014).

From 2002 to 2012, in Taiwan, 1,259,578 patients with sepsis were identified, of which nearly 33.1% were adults, 51.8% were elderly, and 15.1% were oldest old. On 2012, the incidence of sepsis in the oldest old was 9414 cases per 100,000 populations. However, the mortality decreased by 34% for adults, 22% for oldest old, and 24% for elderly, despite the increasing trend in incidence (Lee et al. 2018).

Unlike the huge prevalence, however, case fatality of sepsis was found to decrease annually (Dombrovskiy et al. 2007; Lagu et al. 2012). Unlike the huge prevalence, however, case fatality of sepsis was found to decrease annually (Dombrovskiy et al. 2007; Lagu et al. 2012).

The annual number of sepsis-related deaths in the United States reached 258,000 in 2009, according to Sepsis Alliance (an organization whose goal is to raise awareness of this condition to facilitate its early detection and treatment for successful outcomes). In the United States, in the year 2000, nearly 8.7% yearly increase was reported by Martin et al. for the occurrence of sepsis. The incidence of sepsis is 240.4 cases per 100,000 inhabitants (Martin et al. 2003; Bloos and Reinhart 2014).

The incidence of different severity levels of sepsis rise from 623 per 100,000 people in 2010 to 647 per 100,000 in 2014, but the mortality increased from 264 to 297 per 100,000 people during the same 5-year period (Chen et al. 2019).

Sepsis is the most common cause of patient mortality in ICU, with a global incidence of approximately 18 million cases per year and a mortality rate of 28–40% (Hotchkiss and Nicholson 2006; Yan et al. 2014).

Group A streptococcus (GAS) infection is responsible for mild to very severe disease in patients. In Ireland, the epidemiology of an upsurge in invasive GAS (iGAS) infections from 2012 to 2015 was investigated. Annual incidences of iGAS cases ($n = 561$) in the study group were 2.33–3.66 per 100,000 population (Meehan et al. 2018).

Sepsis remains a huge disease burden for many countries. There were up to 5.3 million deaths, 19.4 million severe sepsis cases, and 31.5 million sepsis cases in the world annually (Fleischmann et al. 2016). The epidemiology of sepsis varies over time and varies among countries (World health Organisation 2011).

During 2003 and 2004 in Denmark, active surveillance of invasive group A streptococcal (GAS) infections was conducted as a part of the Strep-EURO (program initiated in 2002, for group A Streptococcal disease Eurosurveillance) initiative. During the study period, 278 cases were reported, equivalent to a mean annual incidence of 2.6 cases per 100,000 inhabitants (Luca-Harari et al. 2008).

In 2017, nearly 48.9 million incident cases of sepsis were noted worldwide and 11.0 million sepsis-related deaths were accounted, signifying 19.7% of all global deaths. The mortality rate decreased by 52.8% from 1990 to 2017. Sepsis-related mortality and incidence vary significantly across different regions, with the highest burden centered in Oceania, sub-Saharan Africa, East Asia, south Asia, and south East Asia. This data was interpreted from Global Burden of Diseases, Injuries, and Risk Factors Study (GBD) 2017 to estimate the global, regional, and national occurrence/incidence of sepsis and sepsis-related mortality across 195 countries and territories, 282 different underlying causes, in both sexes, and 23 different age groups, for the years 1990 to 2017 were recorded (Rudd et al. 2020).

9.5 Pathophysiology of Sepsis

Despite the prevalent use of antibiotics, bacterial infections remain a prime cause of septic shock, which may develop a few hours after infection by certain bacteria, including *Staphylococcus aureus*, *E. coli*, *Pseudomonas aeruginosa*, *Klebsiella*

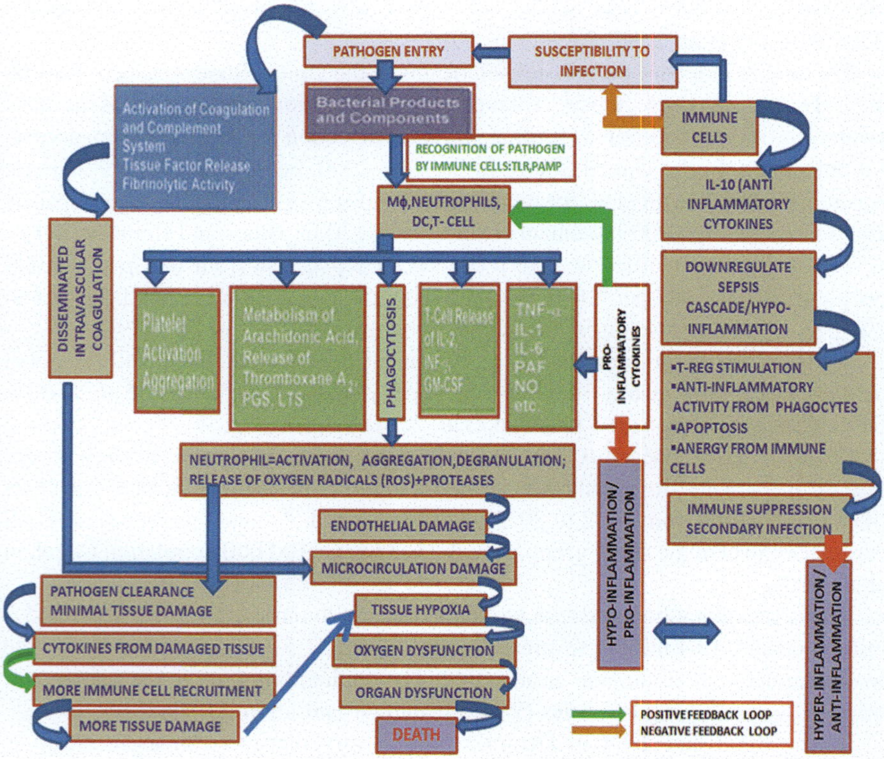

Fig. 9.6 Pathophysiology of sepsis. The pathophysiology of sepsis cascade involves inflammatory, coagulation, apoptosis and organ dysfunction. The pro- and anti-inflammatory pathways were shown

pneumoniae, *Enterobacter aerogenes*, and *Neisseria meningitidis* (Owen et al. 2013). Sepsis results from septicemia, the infections of the blood, in particular those involve Gram-negative bacteria like *Salmonella*. Other pathogens can also cause sepsis. The systemic response in sepsis due to infection causes fever, elevated heartbeat and breathing rate, low blood pressure, and compromised organ function due to circulatory dysfunction. Sepsis is not only caused by bacterial infection but also by injury, trauma, ischemia (reduce blood supply to an organ or a tissue), and by certain cancers. Sepsis is the most common cause of death worldwide in hospital ICUs (Owen et al. 2013). With the entry of the pathogen, the host response to infection begins. Toll-like receptors and other pathogen-associated molecular patterns help in pathogen recognition by multiple innate immune cells such as dendritic cells, macrophages, and some epithelial cells (Fig. 9.6). These receptors are highly conserved in pathogen species (Schulte et al. 2013; Huerta and Rice 2019). Activation of innate immune cell is crucial for host defense against invading microbes and for the consequent generation of an adaptive immune response (Medzhitov 2007). In the pathogenesis of severe sepsis and multi-organ failure, pro-inflammatory mediators produced by innate immune cells are regarded as the key players (Rittirsch et al. 2008). For immune homeostasis and host survival, tight regulation of pro-inflammatory pathways is therefore important (Fig. 9.6). A complex web of regulatory and activating pathways controls the innate immune responses. After the recognition of the pathogens, pro-inflammatory cytokines (IL-6, IL-1, and TNF-α) are released from immune cells. These immune cells eliminate pathogens through the process of phagocytosis and releases reactive oxygen species (ROS, cytotoxic substances), resulting in removal of the infection with minimal tissue damage. Large amounts of ROS are released from immune cells and damaged neighboring tissues. Different host cells that are damaged by the body's host response start releasing their own pro-inflammatory cytokines in reaction to the damage. Thus, a positive feedback loop of progressive inflammation continues with the recruitment of immune cells and further tissue damage (Freeman and Natanson 2000).

Apart from pro-inflammatory pathways, the anti-inflammatory response is also evidently involved in sepsis. IL-10, the anti-inflammatory cytokines, are released in response to infection to downregulate the levels of pro-inflammatory cytokines (Schulte et al. 2013) (Fig. 9.6).These anti-inflammatory cytokines also increase regulatory T cells and reduce phagocytic activity (Schulte et al. 2013). The anti-inflammatory response reduces host inflammation and also suppresses damage to host tissues, before recovery from the initial infection. The resultant immune-suppression leads to vulnerability to secondary infections. The sepsis cascade through different molecular pathways continues with such feedback loops (Huerta and Rice 2019).

It is known that sepsis is characterized by a dysregulated inflammatory response to infection (Hotchkiss and Karl 2003). Therefore, inflammation accounts for serious complications in sepsis. Though strategic targeting of key pro-inflammatory mediators is needed (Fisher et al. 1996; Riedemann et al. 2003), little success is obtained as sepsis demonstrates a complex pathogenesis controlled by a variety of pathogen and host factors. Notably, patients with chronic liver diseases have appreciably increased risk of acquiring sepsis and its associated complications. Higher sepsis-related mortality is common in cirrhotic patients (Foreman et al. 2003). So, hepatic acute phase response plays a major role in the regulation of sepsis.

IL-6, the major pro-inflammatory mediators in sepsis, is also known to be a strong activator of APP production in the liver. The resultant acute phase response, induced by IL-6, also acts as a counter-regulator of the initial inflammatory responses. The central and peripheral immune cell composition undergoes spectacular changes during infections and endotoxemia (Sander et al. 2010). Although the application of antibiotics dramatically decreased the mortality of bacterial infections in the last century, sepsis-related mortality has still remained soaring, even with appropriate antibiotic therapy, organ function support, and fluid replacement (Angus et al. 2001; Hotchkiss and Karl 2003). Thus, there is an imperative requirement for targeted sepsis therapies. Various researches/therapies targeted to inhibit key inflammatory mediators like IL-1, TNF, IL-1β, leukotrienes, or the endotoxin component lipid A have been vitalized in clinical trials (Fisher et al. 1996; Parrish et al. 2008; Owen et al. 2013). In contrast, a more promising strategy includes the regulation of endogenous regulatory mechanisms (Parrish et al. 2008). Cytokines critical in the later stages of sepsis are IL-6, MIF, and CCL-8 (Owen et al. 2013).

The underlying cause of sepsis is a surge production of pro-inflammatory and fever-inducing cytokines like TNF-α and IL-1β. The cytokine imbalance triggers abnormal high body temperature, fever, changes in the respiratory rate, and high WBC counts, subsequent with capillary leakage, tissue injury, extensive blood clotting, and final lethal organ failure (Owen et al. 2013). The major source of sepsis from Gram-negative bacteria is the cell wall constituent LPS (lipopolysaccharide, also known as endotoxin). LPS is a highly potent stimulator of innate immune mediators, such as the TNF-α, IL-1, and IL-6, chemokines, and some antimicrobial components (Owen et al. 2013). Systemic infections activate different blood cells such as neutrophils, monocytes, vascular endothelial cells, resident macrophages, and other cells in the spleen, liver, etc. (Owen et al. 2013). Disseminated infections activate blood cells including monocytes and neutrophils, as well as resident macrophages and other cells in the spleen, liver, and other tissues also. The release of pro-inflammatory mediators systemically activate vascular endothelial cells, triggering their expression of cytokines, chemokines, adhesion molecules, and various clotting factors that amplify the inflammatory response (s). Different enzymes, ROS, and reactive nitrogen species (RNS) released by stimulated neutrophils and other cells damage the tissue vasculature. This damage, together with TNF-α-activated vasodilation and amplified vascular permeability, results in excessive fluid loss into the tissues. The final outcome of this is the lowering of blood pressure. The release of clotting factors by vascular endothelial cells is stimulated by TNF-α, resulting in intravascular blood clotting in capillaries. These adverse effects on the blood vessels finally damage the lungs and kidneys in particular, as these organs are highly vascularized. High circulating TNF-α and IL-1 levels also unfavorably affect the organ the heart (Owen et al. 2013).

Activated protein C, a plasma protein, produced in the liver with anticoagulant and anti-inflammatory functions is also used in sepsis treatment (Bernard et al. 2001). Myeloid-derived suppressor cells (MDSCs) have potent host-protective anti-inflammatory functions during polymicrobial infection (Sander et al. 2010).

Coagulation pathways (non-inflammatory) also donate to sepsis (Fig. 9.6). Increased tissue factor activity and resultant increase in fibrin deposition are features of sepsis. Furthermore, an increase in plasminogen activator inhibitor-1 (PAI-1) decreases the fibrinolysis process. The increased fibrin deposition and decreased fibrin breakdown causes a prothrombotic state. Normally, activated coagulation system during infection causes bacterial trapping and hypoxia to the affected area, restraining the pathogens from multiplication and spreading (Fourrier 2012).

In sepsis, dysregulated coagulation system leads to areas of thrombosis leading to organ damage. The resultant sepsis-associated coagulopathy causes abnormalities in coagulation factors (raised prothrombin or partial thromboplastin time), thrombocytopenia (low platelet count), or increased fibrin split products. In severe cases, sepsis can lead to disseminated intravascular coagulation, thrombosis, and hemorrhage throughout the body. Apoptosis and mitochondrial dysregulation are important characteristics in septic patients (Huerta and Rice 2019) (Fig. 9.6).

9.6 Liver-Mediated Pro-Inflammatory Response, APPs, and Sepsis

Sepsis is characterized by a dysregulated inflammatory response toward infection (Hotchkiss and Karl 2003). Excessive inflammation develops serious complications, but strategies modulation of key pro-inflammatory mediators had limited success. Sepsis involves a complex pathogenesis of the intertwining of a variety of pathogen (s) and host molecules. Interestingly, patients with chronic liver diseases have appreciably increased risk of having sepsis and its related complications, thus have comparatively higher sepsis-related mortality. A vast population of cirrhotic patients died of sepsis (Foreman et al. 2003). The underlying causes for this clinical observation are not well defined. The liver hepatocyte cells secrete the acute phase proteins (APPs), which serum concentration may change by >25% during inflammation (Gabay and Kushner 1999). APP are key players in innate immune response to infection (Medzhitov 2007). APPs have anti-inflammatory properties (Zouki et al. 1997), act as potent opsonins (Shah et al. 2006), and activate innate immune cells like neutrophils (Cheng et al. 2008). The diversified role of APPs as pro- and anti-inflammatory molecules (Gabay and Kushner 1999) may them quite dusky in their role during infections. APP production from hepatocytes is induced by IL-6 (Ritchie and Fuller 1983). But, in IL-6-deficient mice, APP production is only partly impaired depending on the varied stimulus (Kopf et al. 1994). The intruded pathogens express some pattern molecules (Pathogen-Associated Molecular Patterns) which are identified by immune cells by pattern-associated receptors. In case of trauma/surgery, DAMP molecules are expressed on damaged tissue of host cells. Depending on intensity or trauma or level of evading pathogens, an amplified inflammatory cascade forms (Fig. 9.7).

Activation of innate immune cell is crucial for host defense against invading pathogens and for the consequent generation of an adaptive immune response (Medzhitov 2007). In case of severe sepsis and multi-organ failure, the key elements in the pathogenesis are the pro-inflammatory mediators produced by various innate immune cells (Rittirsch et al. 2008). Regulation of pro-inflammatory pathways is therefore necessary for host survival and immune homeostasis.

Innate immune responses are controlled by a network of regulatory and complex activating pathways. Thus, the role of hepatic acute phase response in sepsis is highly related. IL-6, the major pro-inflammatory mediators in sepsis is strongly counter-regulated by the release of APPs with resultant acute phase response to succumb the initial inflammatory reaction. During infections and endotoxemia, hepatic APPs affect the peripheral and central immune cell composition which undergoes dramatic changes in function. In sepsis or in other infections of blood, liver plays a pivotal role in balancing the pro- and anti-inflammatory responses (Fig. 9.8).

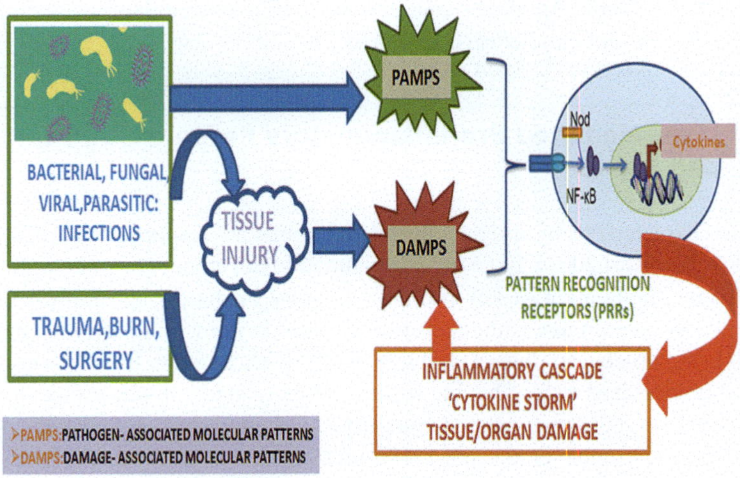

Fig. 9.7 Inflammation mechanism in sepsis. The first step after the initiation of the host response to the pathogen is the triggering of innate immune cells, constituted mainly by monocytes, macrophages, neutrophils, and natural killer cells (NK Cells). This pathway mediates through binding of pathogen-associated molecular patterns (PAMPs), like fungal β-glucans and bacterial endotoxins to specific pattern recognition receptors, on these immune cells. These receptors include damage-associated molecular patterns (DAMPs; molecules released from dead or damaged host cells or other intracellular material) (Gyawali et al. 2019; de Jong et al. 2010)

Interestingly, a growing body of evidence also highlights the capacity of MDSCs, a heterogeneous, immature population of myeloid (precursor) cells characterized by the cell surface expression of the myeloid lineage marker CD11b and Gr1 to restrain T cell responses (Movahedi et al. 2008; Gabrilovich and Nagaraj 2009). MDSCs also accumulate in the spleens of mice during polymicrobial sepsis and repress T cell functions (Delano et al. 2007). However, to which point MDSCs modify the path of disease or if they crucially contribute to immune regulation during sepsis remains undecided.

Of the many projected markers for sepsis, APPs have possibly been most vividly assessed. PCT has been employed particularly extensively in current years though the sensitivity and specificity of PCT for the diagnosis of sepsis is comparatively low (typically within 85–90%) (Shiferaw et al. 2016).

PCT level as an APP had been raised in other inflammatory conditions like trauma (Mimoz et al. 1998) or major surgery (Pierrakos and Vincent 2010; Meisner et al. 1998). In literature review, sometimes CRP is often reported as low-grade compared with PCT in terms of sensitivity, specificity, diagnosis of sepsis, and in monitoring or evaluation. But, still it has wide applicability in ICU patients, and it is commonly used in clinical practice for its greater availability. Increased concentrations of serum CRP are correlated with an elevated risk of multi-organ failure and related mortality (Lobo et al. 2003). CRP responses in septic patients may be helpful to assess and study the time course of therapeutic interventions (Schmit and Vincent 2008). Pathophysiology of sepsis is quite rapidly produced in high concentration after the onset of sepsis. Clinical studies also revealed elevated concentrations of TNF and IL-10 within the first 24 h after admission in ICU of the patient (Pinsky et al. 1993).

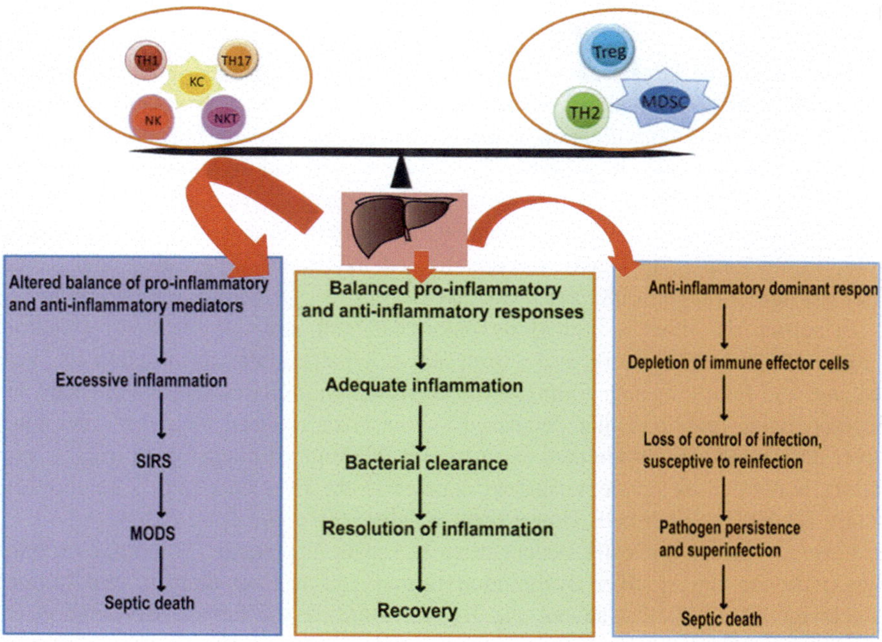

Fig. 9.8 Pro-inflammatory and anti-inflammatory liver response. The Pro-inflammatory and anti-inflammatory (immunosupression) response in the liver were mainly to regulate the outcome of sepsis (Yan et al. 2014). The role of the liver (as a lymphoid organ) in sepsis is widespread. Pre-existing liver dysfunction is a risk factor in the progression of infection to sepsis. Liver dysfunction after sepsis is also an independent risk factor for multiple organ dysfunction and sepsis-induced death. In sepsis, the liver-mediated immune response is accountable for removing bacteria and toxins. On the other hand, it also causes inflammation, immunosuppression and organ damage. The balance of pro-inflammatory and anti-inflammatory (immunosupression) liver response is vital for regulating the systemic immune responses and final sepsis outcome. *MDSC* myeloid derived stem cell, *Treg* T regulatory cells, *NK* natural killer cells, *KC* Kuppfer cells, *NKT* hepatic natural killer T cell

Another group of potential biomarkers that had wide applicability includes the inflammatory (pro- and anti-inflammatory) cytokines. They play a vital role in the release of cytokines in the bloodstream, the blood cytokine levels are rather unpredictable, and their time course is not evidently in pace with the spread of sepsis (Pinsky et al. 1993), making the biological interpretation difficult.

The diagnosis of sepsis is a great challenge. Clinical and standard biochemical laboratory tests are not always very helpful because most critically ill patients in ICU ward develop some kind of inflammatory response (s), whether or not they have sepsis.

Even clinico-microbiological assessment or culture test (positive/negative) for etiology is unreliable because many culture samples do not reveal microbes/causative pathogen in these ill patients. Thus, role of biomarkers in the diagnosis of sepsis is always dicey although relatively few biomarkers have been recorded as diagnostic markers.

The liver is the largest gland in the human body. Liver plays a central role in immunological and metabolic homeostasis. This organ is accountable for over 200 functions like detoxification, nutrient conversion, storage, hormonal balance, energy production, and coagulation. Thus, liver is a critical organ and plays important physiological functions in host survival following severe injury like sepsis (Fig. 9.8). Researches have shown that liver dysfunction and failure, particularly grave complications in sepsis, directly related to disease progression and death (Canabal and Kramer 2008; Yan et al. 2014).

In sepsis, the liver is injured by pathogens, inflammatory mediators, and/or toxins. The stages of injury move from active hepatocellular dysfunction to liver damage and finally to liver failure. Liver dysfunction includes subtle alterations in hepatocellular functions like decreased synthesis or reduced clearance function. Liver damage is actually defined as an irreversible injury to hepatocyte cells. Liver failure is defined as severe, sustained damage to the liver and loss of function in nearly 80–90% of liver cells (Jarrar and Chaudry 2001; Yan et al. 2014).

Liver dysfunction mostly occurs in early sepsis in patients. Inflammation and hypoperfusion during liver dysfunction further causes liver damage and failure (Recknagel et al. 2012; Canabal and Kramer 2008). Early hepatic dysfunction in sepsis patients is a specific and independent risk factor for poor outcome (Kramer et al. 2007; Yan et al. 2014) (Fig. 9.8).

In sepsis patients, mortality, the severity of immune response, with liver injury is associated with bacterial and/or toxin levels. Bacterial clearance is the most important course for the survival of sepsis patients. Liver plays a crucial role in bacterial and/or toxin clearance in sepsis as it is the organ accountable for sterilizing and detoxifying the blood-borne pathogens, toxins, and debris (Protzer et al. 2012). Nearly 60% of total bacteria injected intravenously in an animal model can be removed from the bloodstream and through circulation are trapped in the liver after 10 min of injection; over 80% of the circulating bacteria can be trapped in the liver within 6 h (Gonnert et al. 2012). The key inducer of inflammation in bacterial infection, LPS, is also cleared mainly by the liver (Deng et al. 2013).

Liver plays major multi-tasking like metabolic, digastric, hormonal, and host–defense functions to maintain the normal systemic homeostasis of the body. In sepsis, the liver plays varied functions ranging from clearance of bacteria, in mediating inflammatory responses, the release of APPs, in coagulation pathways, regulating renal failure, acute respiratory distress syndrome, acute lung injury, coagulopathy, and related hepatic encephalopathy. The liver-interceded immune response to sepsis acts both ways as it clears bacteria and toxins but also causes immune-suppression, inflammation, and organ damage (Fig. 9.8).

In early sepsis, Kupffer cells are accountable for producing inflammatory cytokines such as TNF-α, IL-1, IL-6, IFN-γ, and IL-8 along with the release of several early pro-inflammatory mediators. Kupffer cells also mediate sepsis-induced liver injury. After an invasion of harmful bacteria and/or endotoxins, Kupffer cells increase the release of pro-inflammatory cytokines along with secretion of monocyte chemoattractant protein1, different secondary mediators of tissue injury like nitric oxide (NO) and ROS (Yan et al. 2014; Koo et al. 1999) (Fig. 9.8).

9.7 Biomarkers of Sepsis

Diagnosing and treating sepsis and bacteremia includes attempts to attenuate clinical decision-making with varied diagnostic tests for better specificity and sensitivity. Traditionally, fever and leukocytosis were employed to define sepsis nonspecifically (Bone et al. 1992). The diagnosis of sepsis and evaluation of the clinical severity are convoluted by the highly diverse and nonspecific signs and symptoms of sepsis. However, the early diagnosis/evaluation of sepsis with its stratification with increasing severity of sepsis is quite crucial, as it will evolve to starting timely and specific treatment. Thus, biomarkers can have an important role in this complex process because they can designate the presence/absence or severity of clinical status of sepsis. Biomarkers can also differentiate between different etiology of infection (bacterial, viral, or fungal infection), differentiating Gram-negative from Gram-positive bacterial source and clarify local infection to systemic sepsis responses. Control of sepsis must be at the local and systemic level also. Application of CRP as a biomarker of sepsis comes under systemic control (Fig. 9.9).

Other potential applications of biomarkers include specific prognostication, contributing antibiotic therapy, predicting sepsis-associated complications, evaluating therapeutic responses and recovery from sepsis, development of organ dysfunction (if any, of heart, kidneys, liver, or multiple organ dysfunction).

Subsequently, the domain was expanded with the inclusion of erythrocyte sedimentation rate (ESR) and CRP to help in the diagnostic algorithm. Biomarkers can be employed for diagnosing or ruling out sepsis, so that patients may benefit from

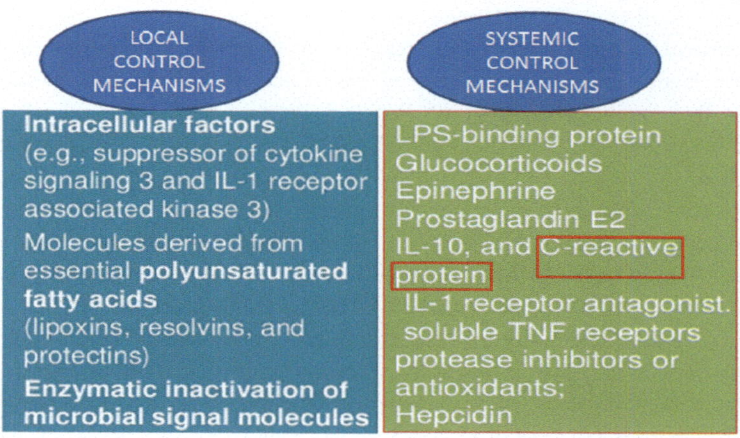

Fig. 9.9 Control mechanism in sepsis. The control mechanism of sepsis includes local mechanisms involving suppression of cytokine stimulation, enzyme inactivations etc. and the systemic control involves CRP, IL-1 receptor agonist etc

continuous monitoring, specific modal therapies, or assessing the response to therapies. Recently, early 180 biomarkers had been reviewed and evaluated in detail including PCT, lactate, soluble triggering receptor expressed on myeloid cells-1 (sTREM-1), cytokines like IL-6 and IL-8, and many more (Wu et al.2012; Pierrakos 2010). No single biomarker can be taken in the diagnosis/prognosis of sepsis. A combination of two or three biomarkers can be used in clinical evaluation. The relevance of CRP, PCT, and lactic acid were mentioned in Table 9.2.

As new knowledge/concept of sepsis evolved gradually, it became clear that not only direct causing pathogen effects but also a high-spirited inflammatory host response was responsible for the detrimental clinical and laboratory anomalies. As sepsis is an SIRS encompassing all organs and systems, it involves various interactions between a number of pathogen and host responses. The credential biomarkers of sepsis are wide–ranged including cytokines, receptor biomarkers, cell markers, coagulation factors, APPs, factors of organ dysfunction, vasodilation regulator/factors, vascular endothelial damage, and many others.

An array of laboratory tests can detect individual organ dysfunction, but none can yet precisely predict the cause of the organ failure. The use of biomarkers comes here in sepsis. The most prominent and useful biomarkers evaluated to date are PCT, CRP, and lactic acid (Table 9.2).

CRP directly binds to pathogens, activates the complement pathways, phagocytosis, and neutrophils. Its sensitivity in sepsis diagnosis is 0.89. By any major guidelines of sepsis treatment, CRP for sepsis diagnosis is not recommended (Póvoa 2002; Póvoa et al. 2006; Reny et al. 2002; Landry et al. 2017; Huerta and Rice 2019).

PCT is a precursor of calcitonin, which released as part of the inflammatory response. Its sensitivity in sepsis diagnosis is 0.77. Although Surviving Sepsis Guidelines support the use of PCT as an aid in diagnosing sepsis, it is not recommended by other guidelines (Wacker et al. 2013; Becker et al. 2008; Rhodes et al. 2017; Kalil et al. 2018; Huerta and Rice 2019).

Lactic acid is a metabolite produced during anaerobic metabolism. In sepsis, increased lactic acid concentrations is associated with increased mortality of patients. It is included as part of Sepsis-3 definition of septic shock (2016) (Garcia-Alvarez et al. 2014; James et al. 1999; Casserly et al. 2015; Huerta and Rice 2019).

CRP some prognostic value in sepsis, particularly when considering measurements in serial. PCT may be more reliable, but it may increase in non-septic conditions also like pancreatitis or cardiopulmonary bypass. To note that, the time duration may be more helpful than a single calculated value.

CRP has a high sensitivity for detection of early-onset sepsis (Hofer et al. 2012; Peters van Ton et al. 2018). Since it has a low specificity, it is not always used as a major biomarker of sepsis. PCT is known to be elevated in patients with invasive bacterial infections (Assicot et al. 1993; Peters van Ton et al. 2018). PCT cannot accurately distinguish patients with immunoparalysis from patients in a hyperinflammatory state. So whether it can be used in the detection of bacteremia is a major question asked by many researchers. The soluble form of the receptor for

Table 9.2 Sepsis biomarkers of our relevance[a]

Sepsis biomarkers	Normal level in blood	Functions	Sensitivity	Specificity	Role in sepsis	User guidelines	Limitations	References
CRP	0.8–3.0 mg/L (~10 mg/L in some individuals)	Directly binds to pathogens by binding with pathogen-surface molecule, opsonization, phagocytosis, activating classical complement pathway, activate neutrophils	0.89	0.59. The specificity increased to 100% if CRP was combined with a temperature >38.2 °C (has limited clinical transposition)	Wide availability, good reproducibility, and low cost biomarker. Used for sepsis diagnosis in critically ill patients. Helpful to evaluate the response to therapy in septic patients	Surviving Sepsis Guidelines does not currently recommend it for sepsis diagnosis	Slow kinetic, independent of infection severity, increased in patients with noninfectious causes of inflammatory conditions like rheumatologic disease, pancreatitis, and malignancy. Time course of CRP concentrations is more important than a single admission value.	Huerta and Rice (2019), Póvoa (2002), Póvoa et al. (2006), Reny et al. (2002), Landry et al. (2017), Pierrakos and Vincent (2010), Bloos and Reinhart (2014), Lelubre et al. (2013), Henriquez-Camacho and Losa (2014)
Procalcitonin	0.01 g/L (below the limit of detection)	Prohormone of calcitonin, a calcium regulator; released as part of the	0.77	0.79	Used for sepsis diagnosis in patients with SIRS response (AUC = 0.85)	Surviving Sepsis Guidelines support its use in sepsis diagnosis. Not	PCT level increases in non-infectious inflammatory conditions, like trauma,	Wacker et al. (2013), Becker et al. (2008), Rhodes et al. (2017), Kalil

(continued)

Table 9.2 (continued)

Sepsis biomarkers	Normal level in blood	Functions	Sensitivity	Specificity	Role in sepsis	User guidelines	Limitations	References
		inflammatory response, elevated in patients with bacterial infections				recommended in other guidelines	burns, and pancreatitis	et al. (2018), Huerta and Rice (2019), Pierrakos and Vincent (2010), Bloos and Reinhart (2014), Lelubre et al. (2013), Henriquez-Camacho and Losa (2014)
Lactic acid	Lactate differs substantially between arterial and venous levels (arterial level is more representative)	Metabolite produced during anaerobic glycolysis, increased in patients with tissue hypoxia	63%	80%	Elevated lactic acid level associated with increased mortality in septic patients	Its concentration higher than 2 mmol/L is part of Sepsis-3 definition of septic shock.	Increased in many non-septic causes of tissue hypoxia, seizures, ischemic limb and bowel. Has adverse events from certain medications	Garcia-Alvarez et al. (2014), James et al. (1999), Casserly et al. (2015), Weiss et al. (2019), Huerta and Rice (2019), Pierrakos and Vincent (2010), Bloos and Reinhart (2014), Lelubre et al. (2013); Henriquez-Camacho and Losa (2014)

[a]Normal concentration of CRP, procalcitonin and lactic acid are represented as per standard values

advanced glycation end products (sRAGE) and the monocyte activation markers may be used as potential biomarkers of a hyperinflammatory state in sepsis. sRAGE is considered as a receptor for danger-associated molecule patterns (DAMPs), and its high level is found to be associated with poor survival in severe sepsis (Bopp et al. 2008; Peters van Ton et al. 2018) .

Elevations of both PCT and CRP were included in the updated definition of sepsis in 2003. Furthermore, for targeted therapy of severe sepsis and septic shock, elevated lactate levels were used to monitor patients at risk of developing sepsis.

9.8 Role of CRP in Sepsis

CRP is an acute phase protein biomarker produced in response to inflammation and/or infection. It is widely used in clinical laboratories to diagnose and manage sepsis patients. This is synthesized from the liver by stimulation with IL-6. It has been reported to bind directly to various pathogens (by surface sugar and other molecules) to play an important role in complement activation, opsonization, and phagocytosis, activating neutrophils and chemotaxis (Póvoa 2002).

As CRP levels rise significantly during acute inflammation, this biomarker has been used for decades in clinics as inflammatory biomarker to designate the presence of significant infectious or inflammatory disease. In neonatology, it is commonly used to screen for early-onset sepsis (Hofer et al. 2012; Henriquez-Camacho and Losa 2014). It is used as an inflammatory biomarker in pediatrics (Jaye and Waites 1997; Henriquez-Camacho and Losa 2014). CRP has also been investigated in many studies as a potential marker of sepsis. In critically ill patients (Póvoa et al. 2006; Reny et al. 2002), high CRP concentration was used (over 80% sensitivity) for distinguishing infectious from non-infectious pathologies and also to screen suspected sepsis cases also.

In 491 neonates in Germany, only 17 neonates presented clinical signs and symptoms of infection within the first 72 h. IL-6 level significantly varies between the patients and control 17 neonates (no infection within the first 72 h). CRP, neutrophil levels, neutrophil extracellular traps (NETs), surrogate markers like cell-free DNA (cfDNA), neutrophil elastase (NE), and myeloperoxidase (MPO) did not show any significant variance. The inability or delayed ability of neonates to form NETs makes them more prone to severe infections. This may potentially lead to morbidity and mortality in neonates with early-onset sepsis (Stiel et al. 2020).

PCT was used for guidance of antimicrobial therapy, CRP was used to monitor the severity of infection (Table 9.2) with limited discriminative capabilities, and IL-6 was used for its fast response toward infectious stimulus. But, none of these noted biomarkers can be reliably employed in differentiating SIRS from sepsis. So biomarkers along with PCR-based pathogen detection might be helpful to the clinicians in differentiating SIRS from sepsis (Bloos 2015).

CRP level > 100 mg/L acts as a prognostic factor in intensive care admissions (ICU; $n = 851$) associated with short-term mortality (30-day) and increased length of stay (LOS) for survivors in ICU in patients of sepsis (Koozi et al. 2020).

High serum CRP level can be a better predictor, along with serum CRP/albumin (CRP/ALB) ratio and low ALB in adult patients with sepsis staying in ICU for at least 3 days. At 72 h, CRP/ALB ratio was cited as one of the best indicators for the prognosis and assessment of clinical therapy and of patients with sepsis (Sun et al. 2016).

CRP/albumin ratio, CRP, albumin, or lactate are used in hospitals to discriminate septic from non-septic patients. But acute physiology and chronic health evaluation (APACHE) II, SOFA, and Simplified Acute Physiology Score (SAPS 3) are better predictors to predict in-hospital mortality of surgical critically ill patients. APACHE II, APACHE II, death probability (DP), SOFA, and to a lesser extent SAPS 3, SAPS 3, DP, and lactate predict the outcome of surgical patients who have survived sepsis after initial 24-h ICU stay from admission (Basile-Filho et al. 2019).

The expression plasma long noncoding RNA metastasis-associated lung adenocarcinoma transcript 1 was positively correlated with high level of CRP in sepsis patients as compared to healthy controls. High lncRNA MALAT1 and CRP concentration was observed in non-survivors. Elevated levels of serum creatinine (Scr), WBC count, PCT, TNF-α, IL-8, IL-17, APACHE II Score, and SOFA Score were also noted. Thus, lncRNA MALAT1 expression along with CRP was linked with severity, risk, inflammatory level, and prognosis in sepsis (Geng et al. 2019).

In elderly sepsis patients (older than 75 years), the performance of natriuretic peptide (NT-pro-BNP) and CRP act as independent predictors of mortality. The inclusion of NT-pro-BNP or CRP or both with APACHE II Score significantly enhances the ability to predict mortality of sepsis patients admitted in ICU. CRP (as biomarker) itself is quite appropriate in predicting ICU outcomes of elderly sepsis patients having multiple organ failure. Also neutrophil count (NEU count), CRP, NT-pro-BNP, APACHE II Score, lactic acid level, and HCT level independently could predict ICU mortality (Li et al. 2020).

Blood cholesterol levels were controlled by the proprotein convertase subtilisin/kexin type 9 (PCSK9) enzyme by downregulating the expression of the low-density lipoprotein receptor (LDLR). Pathogenic lipids like LPS are cleared from the circulation by an LDLR/PCSK9-dependent mechanism. Thus, PCSK9 inhibitors may be used in the treatment of bacterial infections. Patients with a *Streptococcus pneumoniae* bacteremia and lower respiratory tract infection showed significant positive correlation between PCSK9 levels and high CRP level. The plasma PCSK9 level was equally high in patients with Gram-positive or Gram-negative bacterial infections. Low PCSK9 plasma responses in patients were significantly noted with mortality at days 7, 28, and 90 in admitted patients. Plasma PCSK9 mimics/resembles acute phase proteins; its expression is induced high during an infection, showed strong positive correlation with CRP, and reduced expression in liver disease (Rannikko et al. 2019).

The hs-CRP level was significantly higher in mild sepsis, severe sepsis, and septic shock patients, but no significant differences were found among the groups. In non-sepsis group, the hs-CRP level was low. The APACHE II Score and PCT level showed a progressively increasing trend in the non-sepsis, mild sepsis, severe sepsis, and septic shock group, respectively. Significant differences between all pairs of groups were also observed. Interestingly, hs-CRP and PCT were used to assess the severity of pneumonia newborn patients having sepsis. hs-CRP in combination with PCT facilitates early diagnosis, monitoring, treatment response, and prognosis prediction of pneumonia with sepsis in newborns (Liu et al. 2018).

In patients with sepsis, serum tenascin-C (pro-inflammatory glycoprotein found in tissue remodeling, inflammation, and autoimmune diseases) levels were significantly high and positively correlated with SOFA Scores, high CRP, serum creatinine, IL-6, and TNF-α. Serum tenascin-C levels were also significantly elevated in non-survivor patients with sepsis admitted to the ICU. Elevated serum tenascin-C was associated with severity and poor prognosis in septic patients with high CRP levels (Yuan et al. 2018).

The systematic inflammation biomarkers like high CRP level along with greater number of leukocytes and lymphocytes are positively correlated with PCT and could play a vital role in early evaluation and continuous management of polytrauma-induced sepsis and septic shock. CRP, an indicator of severity in pathology progression, was highly manifested during initial stages, but blood lactate level was normalized after ventilation/oxygen therapy (Kanashvili et al. 2018).

The diagnosis of sepsis in postoperative period (POD) is a challenging medical intervention. CRP levels (>100 mg/L) increased highly in the first week of POD due to major surgery. But if the high level persists even after the first week, then it indicates postoperative infection. In contrast, CRP level decreases after first week of surgery due to non-infection (Santonocito et al. 2014).

From admission to up to12–24 h, neonates with sepsis ("proven early-onsetsepsis" and "clinical early-onset sepsis" group) had a higher WBC count, IL-6, PCT, and hs-CRP levels compared with those neonates without sepsis or with uncertain sepsis (uncertain infectious status). The combination of IL-6, hs-CRP, and PCT could be used as predictor in diagnosis of early-onset neonatal sepsis with a specificity on more than 80% (Abdollahi et al. 2012).

Serum lactate, IL-6, and CRP were simultaneously used to differentiate patients with or without sepsis. In this case, the prognosis of acute abdomen was also assessed with these markers (Ravishankaran et al. 2011).

The kinetics of CRP, IL-6, IL-10, and phospholipase A2-II (PLA2-II) was assessed in severely traumatized patients suffering from injury-induced anergy for potential exploration of early sepsis. Mean values of CRP on the first day to seventh day after the injury might be used as predictor of sepsis development in traumatized patients (Lausević et al. 2010).

High CRP level measured concurrently with the first positive monomicrobial blood culture represents a biological interaction between high levels of CRP to independently predict 30-day mortality in adult bacteremia patients (Gradel et al. 2011).

High correlation was observed between temperature and CRP in patients (Table 9.2) with community-acquired infections with undergoing antibiotic therapy. Persistently, high levels of CRP with normalized body temperature were related to comorbid conditions. Though CRP level is represented in a nonspecific way for diagnosis, its usefulness was highlighted (Roger et al. 2009).

Initial serum CRP level has a diagnostic and prognostic role in patients admitted to internal medicine patients. High CRP level was noted in infectious than non-infectious state. Even bacterial infections had significantly higher CRP level (120 mg/L) than nonbacterial infections (32 mg/L). CRP level >200 mg/L was taken as marker of sepsis. Higher CRP level was significantly correlated with mortality, longer hospital stay, and the need for intubation. Differentiation criterion of CRP is better than platelets count or ESR (Keshet et al. 2009).

The quantitative index of CRP at the first hours of life is a rapid, sensitive diagnostic marker corresponded to the severity of the septicemia (Table 9.2) for identification of early-onset sepsis in preterm infants. In patients without sepsis, normal values of CRP along with other examinations determined the discontinuation of antibiotics in preterm infants. Thus, CRP level in dynamics used to assess severity of the course of the disease (Jincharadze et al. 2006).

Patients with severe neonatal sepsis having non-adequate initial antibiotic therapy had continued high level of CRP even after the treatment started. In neonatal sepsis, with appropriate antibiotic therapy, significant decrease of serum CRP levels was noted. Thus, increased level of CRP during initial antibiotic treatment, mainly during the second day of treatment of neonatal septicemia, should be taken as an indicator for substitution of initial antibiotics, even before sensitivity of microbiological sources to given antibiotics is known (Janković et al. 2001).

Lipopolysaccharide-binding protein (LBP) was a better marker/predictor than IL-6, PCT, and CRP in neonates with SIRS within 48 h with high sensitivity, specificity, and predictive values (Pavcnik-Arnol et al. 2004).

Patients (heterogeneous population) with high serum CRP levels (>10 mg/dL) at ICU admission are more prone to severe organ dysfunction, longer higher mortality rates, and longer ICU stays than control patients in ICU admission CRP levels (<1 mg/dL). Decreased CRP level after 48 h was associated with low mortality rate whereas a continuous increased CRP level was associated with high mortality rate (Lobo et al. 2003).

Plasma CRP levels were significantly higher in patients with septic syndrome symptoms on Day 1 to Day 4 as compared to patients without septic syndrome symptoms. Serum concentration of CRP is correlated with severity of the clinical state of patients but not dependent on sepsis etiological factor (like *Neisseria meningitidis*, *Streptococcus pneumoniae,* or *Staphylococcus aureus*). CRP concentration in patients with bacterial sepsis might have evaluating value in identifying clinical state severity, disease monitoring as well as therapeutic efficacy (Kepa and Oczko-Grzesik 2001).

Extremely high CRP level in cerebrospinal fluid (CSF, above 100 μg/L) was seen in bacterial meningitis patients as compared to viral meningitis patients. High level of serum amyloid A protein (SAA) in CSF (more than 10 μg/L) was observed in bacterial meningitis, mycotic meningitis patients, and in viral meningitis patients. Serious bacterial meningitis was reflected by high serum PCT level but not in CSF. PCT expression was very slightly induced by viral infections. Thus, PCT differentiates bacterial meningitis versus viral meningitis in serum whereas CRP and SAA acts as marker of inflammation both in serum and in CSF (Shimetani et al. 2001).

The levels of PLA2-II and PCT were higher in patients upon admission in bacteremia than in non-bacteremic bacterial or viral infections but CRP level peaked 1 day later in same study group. PLA2-II (cut-off level 150 μg/L) and PCT (cut-off level 2–6 μg/L) showed high sensitivity and specificity for bacterial infection within the first 24 h from admission. But after 24 h, CRP level spikes for bacterial infection. Thus, PLA2-II and PCT were useful biochemical markers for early diagnosis of septicemia and CRP was indicators of the severity of inflammation (Rintala et al. 2001).

Sequential determinations of CRP level were useful in childhood acute lymphoblastic leukemia (ALL) with invasive bacterial infections or septicemia as indicator for diagnosis and monitoring. Some children receiving anti-leukemic chemotherapy also developed septicemia. In both cases, CRP level was very high (15–340 mg/L). CRP level above 100 mg/L indicates 100 mg/L. Children with no septicemia had moderate to low CRP level, but should be monitored as they may develop septicemia (Peltola et al. 1983). Regular monitoring of CRP level was done in children beyond the neonatal period, some of them are immune-compromised, some receiving antimicrobial therapy, or some died during the study period. CRP level remained high in septicemic patients and its value gradually lowers with remission/recovery (Peltola and Jaakkola 1988).

9.9 Management of Sepsis

Patients can be presented to the hospital in different stages of sepsis including septic shock. The main goal after provisional diagnosis is to start the treatment immediately. The first step of treatment includes point-to-point definite diagnosis (differentiate sepsis from other bloodstream infections; or to find the non-infectious source also) to minimize sepsis-related morbidity and mortality. A comprehensive and elaborate daily checklist should be maintained in ICU for the long-stay patients and specially for multiple organ support (Box 9.1) (Chawla and Todi 2012).

Box 9.1: Daily Checklist in Long-Stay Cases Multi-Organ Support

Hug Fast for Continuous Quality Improvement(CQI)

- I = Infection control
- H = Hand hygiene and head of bed elevation
- U = Ulcer prophylaxis
- G = Glycemic control
- F = Feeding
- A = Analgesia
- S = Sedation
- T = Thromboprophylaxix

After observing the initial assessment of heart rate, blood pressure, general parameters, systemic examination, and the initial phase of treatment started (Fig. 9.10). According to severity of sepsis, one wide bore peripheral intravenous line is established and preferably 1–2 L of normal saline (or Ringer lactate as crystalloids) is initially transfused within 1–2 h to maintain central venous pressure (CVP) 8–10 cm of H_2O. Fluid resuscitation should be continued as long it does not improve. Furthermore, for a patient on the ventilator, CVP 12–15 cm of H_2O should be maintained for hemodynamic monitoring. For sepsis patients, in ventilation support or not, the mean arterial pressure of 65 mm Hg, systolic pressure of 90 mmHg, and urine output above 0.5 ml/kg/h (by fluid resuscitation) should always be maintained. If these targeted therapies are not achieved, vasopressor therapy, preferably nor-epinephrine in titrated dose, is administered through central venous line. After that, accurate blood pressure is to be recorded by invasive intra-arterial catheter. In sepsis patients, with myocardial dysfunction as reflected by low cardiac output (low ejection fraction) inotropic therapy with dobutamine should be started. If the patient is still in refractory shock, injection hydrocortisone is to be started in a dose of 50 mg every 6 h or 200 mg infusion over 24 h intravenously. If improvement occurs, hydrocortisone should be continued for 5–7 days with slow tapering dose and gradual discontinuation (Fig. 9.10).

A patient with hypoxemia (low PO_2) and hypercapnia (PCO_2 high) as recorded by arterial blood gas analysis and is not improving by noninvasive method of ventilation, then the patient is to be intubated and put on mechanical ventilator. The tidal volume of the patient is to be determined on his/her weight and lung condition. The final outcome is to maintain the plateau pressure within 30 cm H_2O.

Within 1 h of presentation of the patient in the hospital, specimen (blood, urine, sputum, pus, etc.) should be collected and sent to the laboratory for culture/sensitivity report (for identification of strain and type of pathogen). After collection of the specimen, broad-spectrum antibiotics should be administered according to hospital protocol. If the patient is in severe immunosuppressive condition, then empirical antifungal therapy should also be started. For different biochemical laboratory investigations, blood sample of the patient was sent to the laboratory.

For stress ulcer prophylaxis, patient should be given histamine (H_2) receptor blocker or proton pump inhibitor.

Packed blood transfusion is recommended to keep blood hemoglobin level 9 g/dL or above. Care should be taken to maintain coagulation profile specifically before invasive method by transfusing fresh frozen plasma. Nutrition should be maintained preferably by enteral route. If not possible, parenteral nutrition should be started to fight against catabolic phase of sepsis.

In case of severe metabolic acidosis, with high base deficits in ABG (Arterial Blood Gas) and poor urine output, dialysis is to be planned for better outcome. Choosing the method of dialysis is dependent on blood pressure, ultra-filtration, and other factors.

Capillary blood glucose (CBG) of the sepsis patient should be maintained between 120 mg/dL to 180 mg/dL by injecting regular insulin and checking the capillary blood glucose level frequently.

The patient should be covered for thromboprophylaxis by pharmacological drug and non-pharmacologically device (sequential compression device) in lower limbs.

Assessment in time to time and point of care from different clinical condition of the patient should be maintained in liaison with microbiologist, pathologist, and other related personal to play a pivotal role in management of sepsis for final outcome (Fig. 9.10, Box 9.1).

9.10 Discussion

Sepsis is life-threatening organ dysfunction caused by dysregulated host response to infection. It is an important global health problem. It is one of the most common causes of in-hospital deaths and cost billions of monetary loss annually. Infection-prevention efforts are necessarily included to target both healthcare-associated and community-acquired infections to reduce sepsis incidence. Timely implementation of targeted interventions with proper treatment improves the outcomes in sepsis (Fig. 9.10). The different worldwide clinical organizations has strengthened efforts to identify, document all cases, prevent at first-hand, and treat sepsis. Accurate quantification/documentation of sepsis incidence and related mortality is erstwhile important for all public health leaders, researchers, and different funding agencies. It remains a formidable challenge worldwide (Rudd et al. 2020).

The diagnosis of sepsis in clinical medicine is a challenge. Clinical and regular standard laboratory examinations/tests are not very specific because most of the critically ill patients develop some inflammatory response (s), whether they have sepsis or not. Even microbiological culture assessment is not fruitful because many culture samples do not yield microbes in these study population. However, with the ushering of many biomarkers, has not created a pinpoint strategic evaluation in the

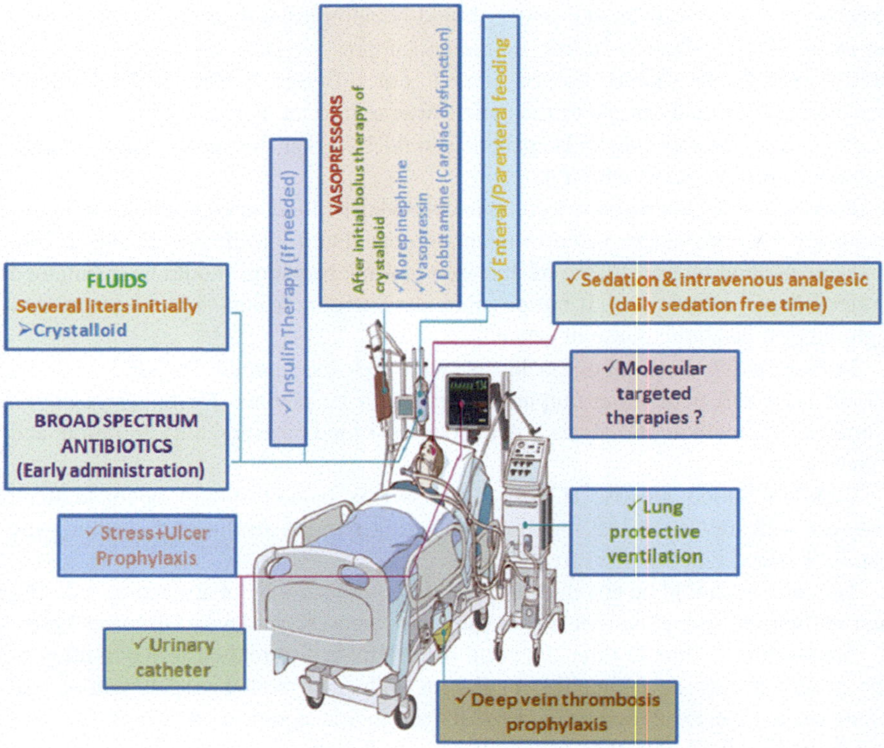

Fig. 9.10 Management of sepsis. Sepsis management of patients in ICU needs lung ventilation, vasopressors, antibiotics administration, crystalloid fluid, urinary catheter and many others. Molecule targeted therapy (?) may be applied if and when needed as per clinical recommendations. The daily checklist for the patient of long-stay in case of multi-organ dysfunction is mentioned in Box 9.1

diagnosis of sepsis. Only a few biomarkers have been assessed as diagnostic markers. No clear-cut biomarker has been being able to differentiate sepsis from other susceptible infections.

At present, no single biomarker or biomolecular assay accessible on hand which alone allows a fast and reliable discrimination between sepsis and SIRS without infection. Indeed, the currently available biomarkers appear to mainly identify different invasive bacterial infections but could not diagnose viruses, fungi, and parasites which may also evoke sepsis. Differentiation between infection and colonization is often challenging. This may cause erroneous result while using biomarkers or other tools of sepsis diagnosis. Thus, fast diagnosis and quick initiation of therapy remains a critical clinical decision by assessing the patient's history, possible visible symptoms of infection, and progression of acute organ dysfunction. Taking into account all the shortcomings of usage of biomarkers, it should be mentioned that biomarkers can aid and shorten the decision-making process in diagnosis.

PCT is currently the widely investigated biomarker for sepsis therapeutics and the only biomarker which has been incorporated into treatment modalities. CRP and IL-6 are also used apart from PCT for the diagnosis of sepsis in most of the studies. Similarly, PCR-based microbial detection may cut down the time course to prescription of a suitable antimicrobial therapy but cannot over rule the presence of infection when negative. It may be too critical to pinpoint that one single measurement of a biomarker can discriminate the complex response to infection from a non-infectious activation. Using a panel of biomarkers revealed some better performances (Table 9.2) in many promising studies. But selection of an optimal set of biomarkers would be too patchy. The clinical algorithms of sepsis must focus on incorporating biomarkers to investigate their usefulness in affective clinical course of the patient (Bloos and Reinhart 2014).

References

Abdollahi A, Shoar S, Nayyeri F, Shariat M (2012) Diagnostic value of simultaneous measurement of procalcitonin, Interleukin-6 and hs-CRP in prediction of early-onset neonatal sepsis. Mediterr J Hematol Infect Dis 4(1):e2012028. https://doi.org/10.4084/MJHID.2012.028

American College of Chest Physicians/Society of Critical Care Medicine consensus conference (1992) Definitions for sepsis and organ failure and guidelines for the use of innovative therapies in sepsis. Crit Care Med 20(6):864–874

Angus DC, Linde-Zwirble WT, Lidicker J, Clermont G, Carcillo J, Pinsky MR (2001) Epidemiology of severe sepsis in the United States: analysis of incidence, outcome, and associated costs of care. Crit Care Med 29:1303–1310. https://doi.org/10.1097/00003246-200107000-00002

Assicot M, Gendrel D, Carsin H, Raymond J, Guilbaud J, Bohuon C (1993) High serum procalcitonin concentrations in patients with sepsis and infection. Lancet 341:515–518. https://doi.org/10.1016/0140-6736(93)90277-N

Basile-Filho A, Lago AF, Menegueti MG, Nicolini EA, Rodrigues LAB, Nunes RS, Auxiliadora-Martins M, Ferez MA (2019) The use of APACHE II, SOFA, SAPS 3, C-reactive protein/albumin ratio, and lactate to predict mortality of surgical critically ill patients: a retrospective cohort study. Medicine (Baltimore) 98(26):e16204. https://doi.org/10.1097/MD.0000000000016204

Bates DW, Sands K, Miller E, Lanken PN, Hibberd PL, Graman PS, Schwartz JS, Kahn K, Snydman DR, Parsonnet J, Moore R, Black E, Johnson BL, Jha A, Platt R (1997) Predicting bacteremia in patients with sepsis syndrome. Academic Medical Center Consortium Sepsis Project Working Group. J Infect Dis 176(6):1538–1551. https://doi.org/10.1086/514153

Becker KL, Snider R, Nylen ES (2008) Procalcitonin assay in systemic inflammation, infection, and sepsis: clinical utility and limitations. Crit Care Med 36:941–952

Bernard GR, Vincent JL, Laterre PF, LaRosa SP, Dhainaut JF, Lopez-Rodriguez A, Steingrub JS, Garber GE, Helterbrand JD, Ely EW, Fisher CJ Jr, Recombinant human protein C Worldwide Evaluation in Severe Sepsis (PROWESS) Study Group (2001) Efficacy and safety of recombinant human activated protein C for severe sepsis. N Engl J Med 344:699–709. https://doi.org/10.1056/NEJM200103083441001

Bloos F (2015) Clinical diagnosis of sepsis and the combined use of biomarkers and culture- and non-culture-based assays. Methods Mol Biol 1237:247–260. https://doi.org/10.1007/978-1-4939-1776-1_19

Bloos F, Reinhart K (2014 Jan 1) Rapid diagnosis of sepsis. Virulence 5(1):154–160

Bone RC, Balk RA, Cerra FB, Dellinger RP, Fein AM, Knaus WA, Schein RM, Sibbald WJ (1992) Definitions for sepsis and organ failure and guidelines for the use of innovative therapies in sepsis. The ACCP/SCCM Consensus Conference Committee. American College of Chest Physicians/Society of Critical Care Medicine. Chest 101(6):1644–1655

Bopp C, Hofer S, Weitz J, Bierhaus A, Nawroth PP, Martin E et al (2008) sRAGE is elevated in septic patients and associated with patients outcome. J Surg Res 147:79–83. https://doi.org/10.1016/j.jss.2007.07.014

Botero JSH, Pérez MCF (2012) The history of sepsis from ancient Egypt to the XIX century. Intechopen

Bouza C, López-Cuadrado T, Saz-Parkinson Z, Amate-Blanco JM (2014) Epidemiology and recent trends of severe sepsis in Spain: a nationwide population-based analysis (2006–2011). BMC Infect Dis 14:3863. https://doi.org/10.1186/s12879-014-0717-7

Canabal JM, Kramer DJ (2008) Management of sepsis in patients with liver failure. Curr Opin Crit Care 14(2):189–197

Casserly B, Phillips GS, Schorr C, Dellinger RP, Townsend SR, Osborn TM et al (2015) Lactate measurements in sepsis induced tissue hypoperfusion. Crit Care Med 43:567–573

Chawla R, Todi S (2012) ICU protocols, a stepwise approach. Springer

Churpek MM, Zadravecz FJ, Winslow C, Howell MD, Edelson DP (2015) Incidence and prognostic value of the systemic inflammatory response syndrome and organ dysfunctions in ward patients. Am J Respir Crit Care Med 192(8):958–964. https://doi.org/10.1164/rccm.201502-0275OC

Chen YJ, Chen FL, Chen JH, Wu MM, Chen YL, Chien DS, Ko Y (2019) Epidemiology of sepsis in Taiwan. Medicine (Baltimore) 98(20):e15725. https://doi.org/10.1097/MD.0000000000015725

Cheng N, He R, Tian J, Ye PP, Ye RD (2008) Cutting edge: TLR2 is a functional receptor for acute-phase serum amyloid A. J Immunol 181:22–26

Crouser E, Exline M, Knoell D, Wewers MD (2008) Sepsis: links between pathogen sensing and organ damage. Curr Pharm Des 14(19):1840–1852

Delano MJ, Scumpia PO, Weinstein JS, Coco D, Nagaraj S, Kelly-Scumpia KM, O'Malley KA, Wynn JL, Antonenko S, Al-Quran SZ, Swan R, Chung CS, Atkinson MA, Ramphal R, Gabrilovich DI, Reeves WH, Ayala A, Phillips J, Laface D, Heyworth PG, Clare-Salzler M, Moldawer LL (2007) MyD88-dependent expansion of an immature GR-1(+)CD11b(+) population induces T cell suppression and Th2 polarization in sepsis. J Exp Med 204(6):1463–1474

Dellinger RP, Levy MM, Carlet JM, Bion J, Parker MM, Jaeschke R, Reinhart K, Angus DC, Brun-Buisson C, Beale R, Calandra T, Dhainaut J-F, Gerlach H, Harvey M, Marini JJ, Marshall J, Ranieri M, Ramsay G, Sevransky J, Thompson BT, Townsend S, Vender JS, Zimmerman JL, Vincent J-L (2008) Surviving Sepsis Campaign: International guidelines for management of severe sepsis and septic shock. Intensive Care Med 34:783–785. https://www.ncbi.nlm.nih.gov/pmc/articles/PMC4969965/. https://dx.doi.org/10.1007%2Fs00134-008-1040-9\tpmc_ext10.1007/s00134-008-1040-9

Deng M, Scott MJ, Loughran P, Gibson G, Sodhi C, Watkins S, Hackam D, Billiar TR (2013) Lipopolysaccharide clearance, bacterial clearance, and systemic inflammatory responses are regulated by cell type-specific functions of TLR4 during sepsis. J Immunol 190(10):5152–5160

Dhainaut JF, Marin N, Mignon A, Vinsonneau C (2001) Hepatic response to sepsis: interaction between coagulation and inflammatory processes. Crit Care Med 29(7 Suppl):S42–S47

Dombrovskiy VY, Martin AA, Sunderram J et al (2007) Rapid increase in hospitalization and mortality rates for severe sepsis in the United States: a trend analysis from 1993 to 2003. Crit Care Med 35:1244–1250

Ferrer R, Artigas A, Suarez D, Palencia E, Levy MM, Arenzana A, Pérez XL, Sirvent JM (2009) Edusepsis study group effectiveness of treatments for severe sepsis: a prospective, multicenter, observational study. Am J Respir Crit Care Med 180:861–866. https://doi.org/10.1164/rccm.200812-1912OC

Fisher CJ Jr, Agosti JM, Opal SM, Lowry SF, Balk RA, Sadoff EA, Schein RM, Benjamin E, The Soluble TNF Receptor Sepsis Study Group (1996) Treatment of septic shock with the tumor

necrosis factor receptor: Fc fusion protein. N Engl J Med 334:1697–1702. https://doi.org/10.1056/NEJM199606273342603
Fleischmann C, Scherag A, Adhikari NK, Hartog CS, Tsaganos T, Schlattmann P, Angus DC, Reinhart K, International Forum of Acute Care Trialists (2016) Assessment of global incidence and mortality of hospital-treated sepsis. Current estimates and limitations. Am J Respir Crit Care Med 193(3):259–272. https://doi.org/10.1164/rccm.201504-0781OC
Fleischmann-Struzek C, Goldfarb DM, Schlattmann P, Schlapbach LJ, Reinhart K, Kissoon N (2018) The global burden of paediatric and neonatal sepsis: a systematic review. Lancet Respir Med 6(3):223–230
Foreman MG, Mannino DM, Moss M (2003) Cirrhosis as a risk factor for sepsis and death: analysis of the National Hospital Discharge Survey. Chest 124:1016–1020. https://doi.org/10.1378/chest.124.3.1016
Fourrier F (2012) Severe sepsis, coagulation, and fibrinolysis. Crit Care Med 40:2704–2708
Freeman BD, Natanson C (2000) Anti-inflammatory therapies in sepsis and septic shock. Expert Opin Investig Drugs 9:1651–1663
Gabay C, Kushner I (1999) Acute-phase proteins and other systemic responses to inflammation. N Engl J Med 340:448–454. https://doi.org/10.1056/NEJM199902113400607
Gabrilovich DI, Nagaraj S (2009) Myeloid-derived suppressor cells as regulators of the immune system. Nat Rev Immunol 9:162–174. https://doi.org/10.1038/nri2506
Gaieski DF, Edwards JM, Kallan MJ, Carr BG (2013) Benchmarking the incidence and mortality of severe sepsis in the United States*. Crit Care Med 41(5):1167–1174
Garcia-Alvarez M, Marik P, Bellomo R (2014) Sepsis-associated hyperlactatemia. Crit Care 18:503
Geng F, Liu W, Yu L (2019) Potential role of circulating long noncoding RNA MALAT1 in predicting disease risk, severity, and patients' survival in sepsis. J Clin Lab Anal 33(8):e22968. https://doi.org/10.1002/jcla.22968
Gonnert FA, Kunisch E, Gajda M, Lambeck S, Weber M, Claus RA et al (2012) Hepatic fibrosis in a long-term murine model of Sepsis. Shock 37(4):399–407
Gradel KO, Thomsen RW, Lundbye-Christensen S, Nielsen H, Schønheyder HC (2011) Baseline C-reactive protein level as a predictor of mortality in bacteraemia patients: a population-based cohort study. Clin Microbiol Infect 17(4):627–632. https://doi.org/10.1111/j.1469-0691.2010.03284.x
Gul F, Arslantaş MK, Cinel İ, Kumar A (2017) Changing definitions of Sepsis. Turk J Anaesthesiol Reanim 45:129–138
Gül F, Arslantaş MK, Cinel İ, Kumar A (2017) Changing definitions of sepsis. Turk J Anaesthesiol Reanim 45(3):129–138. https://doi.org/10.5152/TJAR.2017.93753
Gyawali B, Ramakrishna K, Dhamoon AS (2019) Sepsis: the evolution in definition, pathophysiology, and management. SAGE Open Med 7:2050312119835043. https://doi.org/10.1177/2050312119835043. eCollection 2019
Henriksen DP, Laursen CB, Jensen TG, Hallas J, Pedersen C, Lassen AT (2015) Incidence rate of community-cquired sepsis among hospitalized acute medical patients-a population-based survey. Crit Care Med 43(1):13–21. https://doi.org/10.1097/CCM.0000000000000611
Henriquez-Camacho C, Losa J (2014) Biomarkers for sepsis. Biomed Res Int 2014:547818. https://doi.org/10.1155/2014/547818
Hofer N, Zacharias E, M¨uller W, Resch B (2012) An update on the use of C-reactive protein in early-onset neonatal sepsis: current insights and new tasks. Neonatology 102(1):25–36
Hotchkiss RS, Karl IE (2003) The pathophysiology and treatment of sepsis. N Engl J Med 348:138–150. https://doi.org/10.1056/NEJMra021333
Hotchkiss RS, Nicholson DW (2006) Apoptosis and caspases regulate death and inflammation in sepsis. Nat Rev Immunol 6(11):813–822
Hotchkiss RS, Monneret G, Payen D (2013) Immunosuppression in sepsis: a novel understanding of the disorder and a new therapeutic approach. Lancet Infect Dis 13(3):260–268
Huerta LE, Rice TW (2019) Pathologic difference between Sepsis and bloodstream infections. J Appl Lab Med 3(4):654–663. https://doi.org/10.1373/jalm.2018.026245

Iwashyna TJ, Cooke CR, Wunsch H, Kahn JM (2012) Population burden of long-term survivorship after severe sepsis in older Americans. J Am Geriatr Soc 60(6):1070–1077
James JH, Luchette FA, McCarter FD, Fischer JE (1999) Lactate is an unreliable indicator of tissue hypoxia in injury or sepsis. Lancet 354:505–508
Janković B, Pasić S, Marković M, Veljković D, Milicić M (2001) (C-reactive protein concentrations during initial (empiric) treatment of neonatal sepsis). Srp Arh Celok Lek 129 (Suppl 1):17–22
Jarrar DWP, Chaudry IH (2001) Hepatocellular dysfunction–basic considerations. In: Holzheimer RGMJ (ed) Surgical treatment: evidence-based and problem-oriented. Zuckschwerdt, Munich
Jaye DL, Waites KB (1997) Clinical applications of C-reactive protein in pediatrics. Pediatr Infect Dis J 16(8):735–747
Jincharadze N, Abelashvili D, McHedlishvili M, Kacharava M (2006) Diagnostic value of C-reactive protein test at early-onset sepsis in preterm infants. Georgian Med News 130:87–91
de Jong HK, van der Poll T, Wiersinga WJ (2010) The systemic pro-inflammatory response in sepsis. J Innate Immun 2:422–430. https://doi.org/10.1159/000316286
Kalil AC, Gilbert DN, Winslow DL, Masur H, Klompas M (2018) Infectious Diseases Society of America (IDSA) position statement: why IDSA did not endorse the surviving sepsis campaign guidelines. Clin Infect Dis 66:1631–1635
Kaukonen K, Bailey M, Suzuki S et al (2014) Mortality related to severe sepsis and septic shock among critically Ill patients in Australia and New Zealand, 2000–2012. JAMA 311 (13):308–1316. https://doi.org/10.1001/jama.2014.2637
Kaukonen K-M, Bailey M, Pilcher D, Cooper DJ, Bellomo R (2015) Systemic inflammatory response syndrome criteria in defining severe sepsis. N Engl J Med 372(17):1629–1638. https://doi.org/10.1056/NEJMoa1415236. Epub 2015 Mar 17
Kanashvili B, Saganelidze K, Ratiani L (2018) The role of procalcitonin and blood lactic acid values in prognosis of sepsis and septic shock in polytrauma patients. Georgian Med News 279:102–107
Kepa L, Oczko-Grzesik B (2001) Usefulness of plasma C-reactive protein (CRP) estimation in patients with bacterial sepsis. Przegl Epidemiol 55(Suppl 3):63–67
Keshet R, Boursi B, Maoz R, Shnell M, Guzner-Gur H (2009) Diagnostic and prognostic significance of serum C-reactive protein levels in patients admitted to the department of medicine. Am J Med Sci 337(4):248–255. https://doi.org/10.1097/MAJ.0b013e31818af6de
Koo DJ, Chaudry IH, Wang P (1999) Kupffer cells are responsible for producing inflammatory cytokines and hepatocellular dysfunction during early sepsis. J Surg Res 83(2):151–157
Koozi H, Lengquist M, Frigyesi A (2020) C-reactive protein as a prognostic factor in intensive care admissions for sepsis: a Swedish multicenter study. J Crit Care 56:73–79. https://doi.org/10.1016/j.jcrc.2019.12.009
Kopf M, Baumann H, Freer G, Freudenberg M, Lamers M, Kishimoto T, Zinkernagel R, Bluethmann H, Köhler G (1994) Impaired immune and acute-phase responses in interleukin-6-deficient mice. Nature 368:339–342. https://doi.org/10.1038/368339a0
Kramer L, Jordan B, Druml W, Bauer P, Metnitz PG (2007) Incidence and prognosis of early hepatic dysfunction in critically ill patients--a prospective multicenter study. Crit Care Med 35 (4):1099–1104
Lagu T, Rothberg MB, Shieh MS, Pekow PS, Steingrub JS, Lindenauer PK (2012) Hospitalizations, costs, and outcomes of severe sepsis in the United States 2003 to 2007. Crit Care Med 40 (3):754–761. https://doi.org/10.1097/CCM.0b013e318232db65
Landry A, Docherty P, Ouellette S, Cartier LJ (2017) Causes and outcomes of markedly elevated C-reactive protein levels. Can Fam Physician 63:e316–e323
Lausević Z, Vuković G, Stojimirović B, Trbojević-Stanković J, Resanović V, Lausevic M (2010) Kinetics of C-reactive protein, interleukin-6 and -10, and phospholipase A2-II in severely traumatized septic patients. Vojnosanit Pregl 67(11):893–897
Lee SH, Hsu TC, Lee MG, Chao CC, Lee WC, Lai CC, Lee CC, National Taiwan University Health Economics and Outcome Research Group (2018) Nationwide trend of sepsis: a comparison

among octogenarians, elderly, and young adults. Crit Care Med 46(6):926–934. https://doi.org/10.1097/CCM.0000000000003085

Lelubre C, Anselin S, Boudjeltia KZ, Biston P, Piagnerelli M (2013) Interpretation of C-Reactive protein concentrations in critically ill patients. Biomed Res Int 2013:124021. https://www.ncbi.nlm.nih.gov/pmc/articles/PMC3826426/. https://dx.doi.org/10.1155%2F2013%2F124021 \tpmc_ext10.1155/2013/124021

Levy MM, Fink MP, Marshall JC, Abraham E, Angus D, Cook D, Cohen J, Opal SM, Vincent JL, Ramsay G (2003) SCCM/ESICM/ACCP/ATS/SIS 2001 SCCM/ESICM/ACCP/ATS/SIS International Sepsis Definitions Conference. Crit Care Med 31:1250–1256. https://doi.org/10.1097/01.CCM.0000050454.01978.3B

Li H, Shan-Shan Z, Jian-Qiang K, Ling Y, Fang L (2020) Predictive value of C-reactive protein and NT-pro-BNP levels in sepsis patients older than 75 years: a prospective, observational study. Aging Clin Exp Res 32(3):389–397. https://doi.org/10.1007/s40520-019-0124400

Liu GB, Cui XQ, Wang ZB, Wen L, Duan HL (2018) Detection of serum procalcitonin and hypersensitive C-reactive protein in patients with pneumonia and sepsis. J Biol Regul Homeost Agents 32(5):1165–1169

Lobo SM, Lobo FR, Bota DP, Lopes-Ferreira F, Soliman HM, Melot C, Vincent JL (2003) C-reactive protein levels correlate with mortality and organ failure in critically ill patients. Chest 123:2043–2049. https://doi.org/10.1378/chest.123.6.2043

Luca-Harari B, Ekelund K, van der Linden M, Staum-Kaltoft M, Hammerum AM, Jasir A (2008) Clinical and epidemiological aspects of invasive Streptococcus pyogenes infections in Denmark during 2003 and 2004. J Clin Microbiol 46(1):79–86

MacArthur RD, Miller M, Albertson T, Panacek E, Johnson D, Teoh L, Barchuk W (2004) Adequacy of early empiric antibiotic treatment and survival in severe sepsis: experience from the MONARCS trial. Clin Infect Dis 38:284–288. https://doi.org/10.1086/379825

Marshall JC (2004) Sepsis: current status, future prospects. Curr Opin Crit Care 10(4):250–264

Martin GS, Mannino DM, Eaton S, Moss M (2003) The epidemiology of sepsis in the United States from 1979 through 2000. N Engl J Med 348(16):1546–1554. https://doi.org/10.1056/NEJMoa022139

Medzhitov R (2007) Recognition of microorganisms and activation of the immune response. Nature 449:819–826. https://doi.org/10.1038/nature06246

Meehan M, Murchan S, Gavin PJ, Drew RJ, Cunney R (2018) Epidemiology of an upsurge of invasive group A streptococcal infections in Ireland, 2012-2015. J Infect 77(3):183–190. https://doi.org/10.1016/j.jinf.2018.05.010

Meisner M, Tschaikowsky K, Hutzler A, Schick C, Schuttler J (1998) Postoperative plasma concentrations of procalcitonin after different types of surgery. Intensive Care Med 24:680–684

Mimoz O, Benoist JF, Edouard AR, Assicot M, Bohuon C, Samii K (1998) Procalcitonin and C-reactive protein during the early posttraumatic systemic inflammatory response syndrome. Intensive Care Med 24:185–188

Movahedi K, Guilliams M, Van den Bossche J, Van den Bergh R, Gysemans C, Beschin A, De Baetselier P, Van Ginderachter JA (2008 Apr 15) Identification of discrete tumor-induced myeloid-derived suppressor cell subpopulations with distinct T cell-suppressive activity. Blood 111(8):4233–4244. https://doi.org/10.1182/blood-2007-07-099226

Owen J, Punt J, Stranford S (2013) Kuby immunology, 7th edn. W. H. Freeman, New York

Parrish WR, Gallowitsch-Puerta M, Czura CJ, Tracey KJ (2008) Experimental therapeutic strategies for severe sepsis: mediators and mechanisms. Ann N Y Acad Sci 1144:210–236. https://doi.org/10.1196/annals.1418.011

Pavcnik-Arnol M, Hojker S, Derganc M (2004) Lipopolysaccharide-binding protein in critically ill neonates and children with suspected infection: comparison with procalcitonin, interleukin-6, and C-reactive protein. Intensive Care Med 30(7):1454–1460

Peltola H, Jaakkola M (1988) Serious bacterial infections. C-reactive protein as a serial index of severity. Clin Pediatr (Phila) 27(11):532–537

Peltola H, Saarinen UM, Siimes MA (1983) C-reactive protein in rapid diagnosis and follow-up of bacterial septicemia in children with leukemia. Pediatr Infect Dis 2(5):370–373

Peters van Ton AM, Kox M, Abdo WF, Pickkers P (2018) Precision Immunotherapy for Sepsis. Immunol 9:1926. https://www.ncbi.nlm.nih.gov/pmc/articles/PMC6133985/Front. https://dx.doi.org/10.3389%2Ffimmu.2018.01926\tpmc_ext10.3389/fimmu.2018.01926

Pierrakos C, Vincent JL (2010) Sepsis biomarkers: a review. Crit Care 14(1):R15. https://doi.org/10.1186/cc8872

Pinsky MR, Vincent JL, Deviere J, Alegre M, Kahn RJ, Dupont E (1993) Serum cytokine levels in human septic shock. Relation to multiple-system organ failure and mortality. Chest 103:565–575

Póvoa P (2002) C-reactive protein: a valuable marker of sepsis. Intensive Care Med 28:235–243

Póvoa P, Coelho L, Almeida E, Fernandes A, Mealha R, Moreira P, Sabino H (2006) Early identification of intensive care unit-acquired infections with daily monitoring of C reactive. protein: a prospective observational study. Crit Care 10:R63

Protzer U, Maini MK, Knolle PA (2012) Living in the liver: hepatic infections. Nat Rev Immunol 12(3):201–213

Rannikko J, Jacome Sanz D, Ortutay Z, Seiskari T, Aittoniemi J, Huttunen R, Syrjänen J, Pesu M (2019) Reduced plasma PCSK9 response in patients with bacteraemia is associated with mortality. J Intern Med 286(5):553–561. https://doi.org/10.1111/joim.12946

Ravishankaran P, Shah AM, Bhat R (2011) Correlation of interleukin-6, serum lactate, and C-reactive protein to inflammation, complication, and outcome during the surgical course of patients with acute abdomen. J Interf Cytokine Res 31(9):685–690. https://doi.org/10.1089/jir.2011.0021

Recknagel P, Gonnert FA, Westermann M, Lambeck S, Lupp A, Rudiger A et al (2012) Liver dysfunction and phosphatidylinositol-3-kinase signalling in early sepsis: experimental studies in rodent models of peritonitis. PLoS Med 9(11):e1001338

Reinhart K, Bauer M, Riedemann NC, Hartog CS (2012) New approaches to sepsis: molecular diagnostics and biomarkers. Clin Microbiol Rev 25(4):609–634. https://doi.org/10.1128/CMR.00016-12

Reny J-L, Vuagnat A, Ract C, Benoit M-O, Safar M, Fagon J-Y (2002) Diagnosis and follow-up of infections in intensive care patients: value of C-reactive protein compared with other clinical and biological variables. Crit Care Med 30:529–535

Rhodes A, Evans LE, Alhazzani W, Levy MM, Antonelli M, Ferrer R, Kumar A, Sevransky JE, Sprung CL, Nunnally ME, Rochwerg B, Rubenfeld GD, Angus DC, Annane D, Beale RJ, Bellinghan GJ, Bernard GR, Chiche JD, Coopersmith C, De Backer DP, French CJ, Fujishima S, Gerlach H, Hidalgo JL, Hollenberg SM, Jones AE, Karnad DR, Kleinpell RM, Koh Y, Lisboa TC, Machado FR, Marini JJ, Marshall JC, Mazuski JE, McIntyre LA, McLean AS, Mehta S, Moreno RP, Myburgh J, Navalesi P, Nishida O, Osborn TM, Perner A, Plunkett CM, Ranieri M, Schorr CA, Seckel MA, Seymour CW, Shieh L, Shukri KA, Simpson SQ, Singer M, Thompson BT, Townsend SR, Van der Poll T, Vincent JL, Wiersinga WJ, Zimmerman JL, Dellinger RP (2017) Surviving sepsis campaign: international guidelines for management of sepsis and septic shock: 2016. Crit Care Med 45(3):486–552. https://doi.org/10.1097/CCM.0000000000002255

Riedemann NC, Guo RF, Ward PA (2003) Novel strategies for the treatment of sepsis. Nat Med 9:517–524. https://doi.org/10.1038/nm0503-517

Rintala EM, Aittoniemi J, Laine S, Nevalainen TJ, Nikoskelainen J (2001) Early identification of bacteremia by biochemical markers of systemic inflammation. Scand J Clin Lab Invest 61 (7):523–530

Ritchie DG, Fuller GM (1983) Hepatocyte-stimulating factor: a monocyte-derived acute-phase regulatory protein. Ann N Y Acad Sci 408:490–502. https://doi.org/10.1111/j.1749-6632.1983.tb23268.x

Rittirsch D, Flierl MA, Ward PA (2008) Harmful molecular mechanisms in sepsis. Nat Rev Immunol 8:776–787. https://doi.org/10.1038/nri2402

Roger PM, Hung S, de Salvador F, Allieri-Rosenthal A, Farhad R, Pulcini C, Cua E (2009) Usefulness of C-reactive protein in the therapeutic follow-up of infected patients. Med Mal Infect 39(5):319–324. https://doi.org/10.1016/j.medmal.2009.03.001

Rohde JM, Odden AJ, Bonham C, Kuhn L, Malani PN, Chen LM, Flanders SA, Iwashyna TJ (2013) The epidemiology of acute organ system dysfunction from severe sepsis outside of the intensive care unit. J Hosp Med 8:243–247. https://doi.org/10.1002/jhm.2012

Rudd KE, Johnson SC, Agesa KM, Shackelford KA, Tsoi D, Kievlan DR, Colombara DV, Ikuta KS, Kissoon N, Finfer S, Fleischmann-Struzek C, Machado FR, Reinhart KK, Rowan K, Seymour CW, Watson RS, West TE, Marinho F, Hay SI, Lozano R, Lopez AD, Angus DC, Murray CJL, Naghavi M (2020) Global, regional, and national sepsis incidence and mortality, 1990-2017: analysis for the global burden of disease study. Lancet 395(10219):200–211. https://doi.org/10.1016/S0140-6736(19)32989-7

Sander LE, Sackett SD, Dierssen U, Beraza N, Linke RP, Müller M, Blander JM, Tacke F, Trautwein C (2010) Hepatic acute phase proteins control innate immune responses during infection by promoting myeloid-derived suppressor cell function. J Exp Med 207(7):1453–1464

Santonocito C, De Loecker I, Donadello K, Moussa MD, Markowicz S, Gullo A, Vincent JL (2014) C-reactive protein kinetics after major surgery. Anesth Analg 119(3):624–629. https://doi.org/10.1213/ANE.0000000000000263

Say L, Chou D, Gemmill A, Tunçalp Ö, Moller AB, Daniels J, Gülmezoglu AM, Temmerman M, Alkema L (2014) Global causes of maternal death: a WHO systematic analysis. Lancet Glob Health 2(6):e323–e333. https://doi.org/10.1016/S2214-109X(14)70227-X

Schmit X, Vincent JL (2008) The time course of blood C-reactive protein concentrations in relation to the response to initial antimicrobial therapy in patients with sepsis. Infection 36:213–219

Schottmueller H (1914) Wesen und Behandlung der sepsis. Inn Med 31:257–280

Schulte W, Bernhagen J, Bucala R (2013) Cytokines in sepsis: potent immunoregulators and potential therapeutic targets—an updated view. Mediat Inflamm 2013:1–16

Shah C, Hari-Dass R, Raynes JG (2006) Serum amyloid A is an innate immune opsonin for Gram-negative bacteria. Blood. 108:1751–1757. https://doi.org/10.1182/blood-2005-11-011932

Shiferaw B, Bekele E, Kumar K, Boutin A, Frieri M (2016) The role of procalcitonin as a biomarker in sepsis. J Infect Dis Epidemiol 2(1):006

Shimetani N, Shimetani K, Mori M (2001) Levels of three inflammation markers, C-reactive protein, serum amyloid a protein and procalcitonin, in the serum and cerebrospinal fluid of patients with meningitis. Scand J Clin Lab Invest 61(7):567–574

Singer M, Deutschman CS, Seymour CW, Shankar-Hari M, Annane D, Bauer M, Bellomo R, Bernard GR, Chiche JD, Coopersmith CM, Hotchkiss RS, Levy MM, Marshall JC, Martin GS, Opal SM, Rubenfeld GD, van der Poll T, Vincent JL, Angus DC (2016) The third international consensus definitions for sepsis and septic shock (sepsis-3). JAMA 315(8):801–810. https://doi.org/10.1001/jama.2016.0287

Stiel CU, Ebenebe CU, Trochimiuk M, Pagarols Raluy L, Vincent D, Singer D, Reinshagen K, Boettcher M (2020) Markers of NETosis do not predict neonatal early onset sepsis: a pilot study. Front Pediatr 7:555. https://doi.org/10.3389/fped.2019.00555. eCollection 2019

Sun R, Sun X, Yang H, Liu Q (2016) Retrospective analysis of serum C-reactive protein/albumin ratio for the prognosis of the adult patients with sepsis. Zhonghua Wei Zhong Bing Ji Jiu Yi Xue 28(5):413–417

Vincent J-L, Abraham E (2006) The last 100 years of sepsis. Am J Respir Crit Care Med 173:256–263

Vincent JL, Moreno R, Takala J et al (1996) The SOFA (Sepsis-related Organ Failure Assessment) score to describe organ dysfunction/failure. On behalf of the Working Group on sepsis-Related Problems of the European Society of Intensive Care Medicine. Intensive Care Med 22 (7):707–710

Wacker C, Prkno A, Brunkhorst FM, Schlattmann P (2013) Procalcitonin as a diagnostic marker for sepsis: a systematic review and meta-analysis. Lancet Infect Dis 13:426–435. https://doi.org/10.1016/S1473-3099(12)70323-7

Weiss SJ, Guerrero A, Root-Bowman C, Ernst A, Krumperman K, Femling J, Froman P (2019) Sepsis alerts in EMS and the results of pre-hospital ETCO2. Am J Emerg Med 37 (8):1505–1509. https://doi.org/10.1016/j.ajem.2018.11.009

World Health Organization (2011) WHO Report on the burden of endemic health care-associated infection worldwide. 2017-11-21 15:11:22

Wu Y, Wang F, Fan X, Bao R, Bo L, Li J, Deng X (2012) Accuracy of plasma sTREM-1 for sepsis diagnosis in systemic inflammatory patients: a systematic review and meta-analysis. Crit Care 16(6):R229. https://doi.org/10.1186/cc11884

Yan J, Li S, Li S (2014) The role of the liver in sepsis. Int Rev Immunol 33(6):498–510

Yuan W, Zhang W, Yang X, Zhou L, Hanghua Z, Xu K (2018) Clinical significance and prognosis of serum tenascin-C in patients with sepsis. BMC Anesthesiol 18(1):170. https://doi.org/10.1186/s12871-018-0634-1

Zouki C, Beauchamp M, Baron C, Filep JG (1997) Prevention of in vitro neutrophil adhesion to endothelial cells through shedding of L-selectin by C-reactive protein and peptides derived from C-reactive protein. J Clin Invest 100:522–529. https://doi.org/10.1172/JCI119561

C-Reactive Protein: Is Early Prognostic Marker? 10

S. Yogeshpriya and P. Selvaraj

Contents

10.1 Introduction 293
10.1.1 Acute Phase Response (APR) 293
10.1.2 Acute Phase Proteins (APP) 294
10.1.3 Classification of APP 295
10.2 CRP 297
10.2.1 CRP in Canines 298
10.2.2 Concentration of CRP in Other Infectious Diseases of Dogs 301
10.2.3 CRP in Felines 302
10.2.4 CRP in Ruminants 304
10.2.5 CRP in Equines 306
10.2.6 CRP in Swine 307
10.2.7 CRP in Fish 308
10.2.8 CRP in Captive Asian Elephants 309
10.2.9 CRP in Avian Biology 309
References 310

Abstract

C-reactive protein (CRP) is synthesized by liver and is an acute phase protein (APP). It is upregulated under inflammatory stimulus of infection, injury, and tissue damage. 1970s mark the first identification of human CRP. Acute phase response (APR) includes host inclusive of both animal and human responses to infection, inflammation, injury, stress, and trauma which involves upregulated synthesis of different plasma proteins by the liver which are collectively called as

S. Yogeshpriya (✉)
Department of Veterinary Medicine, Veterinary College and Research Institute, Orathandu, Thanjavur, Tamil Nadu, India

P. Selvaraj
Tamilnadu Veterinary and Animal Sciences University, Chennai, Tamil Nadu, India

W. Ansar, S. Ghosh (eds.), *Clinical Significance of C-reactive Protein*,
https://doi.org/10.1007/978-981-15-6787-2_10

acute phase proteins (APP). APPs are overexpressed, while other plasma proteins reveal downregulation from their basal levels. CRP is an acute inflammatory protein that reveals up to 1000-fold upregulation in expression at infection or inflammation sites. Dramatic increase of CRP levels occurs during inflammatory processes. It increases beyond normal levels of expression within 6 h of trigger and reaches its peak at 48 h post the trigger. It is reported to aid incomplete binding to externally intruding agents, pathogens, and damaged cells. Phagocytic activity of the cells and macrophages are enhanced. Macrophages express a CRP receptor. It is also reported to play a role in early innate immune defenses against infectious agents. A huge number of studies have documented the role of CRP in cardiovascular disease (CVD) and associated disease conditions by readily available commercial automated CRP assays. This chapter highlights the role of CRP in acute phase response, based on clinical observations of CRP in relation to CVD and other diseases in animal species, and possibility of employing CRP as therapeutic target, quantization of CRP in detection, and monitoring systemic inflammatory responses from the view point of scope of research and medical and veterinary practices.

Keywords

Acute phase proteins (APP) · Canine · Aves · Immunity · CRP · Fish · Veterinary science · Inflammation

Abbreviations

A. pleuropneumoniae	*Actinobacillus pleuropneumoniae*
AGP	Acid glycoprotein
AIHA	*Autoimmune hemolytic anemia*
Apo-A1	Apolipoprotein 1
APP	Acute phase protein
APR	Acute phase response
BDV	Border disease virus
BSRV	Bovine respiratory syncytial virus
BTV	Bluetongue virus
CanL	*Canine leishmaniasis*
CIE	Chronic inflammatory enteropathies
Cp	Ceruloplasmin
CRP	C-reactive protein
CSF	Cerebrospinal fluid
CVSM	Cervical vertebral stenotic myelopathy
FMDV	Foot-and-mouth disease virus
Hp	Haptoglobulin
LPS	Lipopolysaccharide
P. multocida	*Pasteurella multocida*
PON1	Paraoxonase 1
PPRV	Peste des petits ruminants virus

S. aureus	*Staphylococcus aureus*
S. typhimurium	*Salmonella typhimurium*
SAA	Serum amyloid A
SAP	Serum amyloid P
SI	Swine influenza
SRMA	Steroid-responsive meningitis-arteritis
T. cruzi	*Trypanosoma cruzi*
WBC	White blood cells

10.1 Introduction

The animal or human body functions are strictly regulated by a variety of homeostatic mechanisms in a controlled internal environment. However, external factors can disrupt the homeostasis leading to imbalances. The host reacts to challenges from altered external conditions like injury, damage infections, trauma, immunologic, neoplastic, or other causes and tries to restore the normal homeostasis, by reducing the tissue damage, and removal of homeostasis disrupting agents. These several immune and metabolic responses of the host to external forces are broadly included as acute phase response (APR) and is considered a part of innate host defenses enhancing chances of host survival during critical early stages of attack.

10.1.1 Acute Phase Response (APR)

As early as in 1930, changes during the pathobiology of inflammatory process has been studied. CRP was discovered by Tillett and Francis and marked the beginning of the understanding of inflammation process and the acute phase response triggered in acute or a chronic inflammatory process. The APR was studied to be regulated by molecules termed as cytokines, synthesized and released by macrophages, triggered by agents including bacterial endotoxin, viruses, free radicals, prostaglandins, or other factors released under various conditions leading to inflammation and release of acute phase proteins (APP).

Synthesis and release of pro-inflammatory cytokines, at the site of injury and tissue injury, promotes different cell types triggering the production of cascade of other cytokines like IL-6-type cytokines, signaling cascades, which leads to synthesis and release of APP from hepatocytes of liver or other tissues (Fig. 10.1). APR enables restoration and maintenance of homeostasis and heals tissue injury. APR is a very fast response, leading to overexpression APP shooting up within a few hours, remaining upregulated till the inflammatory stimulus persists. In the acute phase response, serum/plasma level of some kind of proteins are found to decrease while the levels of some other proteins increase by many folds.

The APR extends for several days post-triggering stimulus; however, the kinetics of APR varies with species and extent of tissue damage. The serum concentration of APP peaks typically within 24–48 h after the initiation of the trigger. As the infection

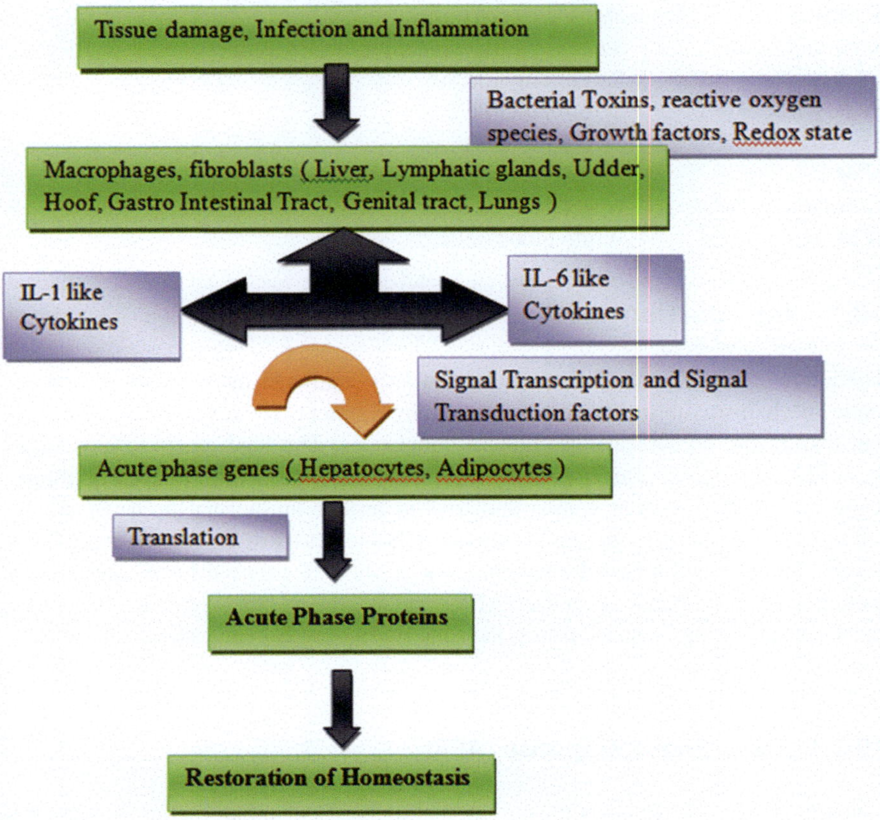

Fig. 10.1 Synthesis of APP

subsides, the APR also subsides with feedback regulations leading to its resolution within 4–7 days after initial trigger and in the absence of any further stimulus. Chronic inflammation in disorders like arthritis is thought to involve series of inflammatory stimuli revealing upregulated expression and release of APP in serum. However, in case of acute inflammation or infection conditions the increase in APP and the APR is higher.

10.1.2 Acute Phase Proteins (APP)

In an APR, the concentration of different APPs varies. The concentration of protein varies in chronic inflammation as compared to acute inflammation. These proteins include protease inhibitors (e.g., α 1 antitrypsin, α 1 antichymotrypsin), coagulation proteins like fibrinogen, prothrombin, complement proteins like C2, C3, C4, C5, etc., transport proteins including haptoglobin (Hp), ceruloplasmin (Cp), hemopexin, and proteins like CRP, serum amyloid A (SAA), serum amyloid P (SAP), acid glycoprotein (AGP), etc.

10.1.3 Classification of APP

Upregulation of APP in response to inflammation, infectious disease, surgical treatment, or trauma (Table 10.1) has been studied in humans and other animals.

APP can be classified according to the upregulation in positive APP or downregulation as in negative APP in their serum during an APR. They have also been classified into major, moderate, or minor, based on their mode of activity and expression.

- **Major proteins:** Include proteins revealing 10- to 100-fold increase markedly within the first 24–48 h post the trigger and often revealing rapid decline due to their very short half-life.
- **Moderate proteins:** Include proteins with an increase in 2- to 10-fold.
- **Minor proteins:** Reveal slight increase.

Moderate and minor proteins may increase at a slow rate and reveal prolonged in duration, based on the status of the trigger. These also reveal overexpression during chronic inflammatory processes. Here, an aberrant continuation of some aspects of APR can lead to underlying tissue damage, increasing the complicacy of disease-like conditions of protein deposition in reactive amyloidosis.

Table 10.1 Biological activities of selected acute phase proteins

APP	Biological activities
Hepatoglobin	• Affects lipid metabolism • Angiogenesis might be stimulated • Can lead to the development of fatty liver • Leads to inhibition of neutrophils respiratory burst activity • Interacts and binds with hemoglobin (Hb) • Reveals bacteriostatic effect • Reveals immunomodulatory effect
CRP	• Activation of complement proteins • Chromatin binding • Cytokine production • Lead to opsonization • Monocytes and macrophage and their modulation • Tissue migration of neutrophils inhibited
SAA	• Calcium (Ca^{2+}) mobilization by monocyte • Cholesterol transport from dying cells to hepatocytes • Inhibition of in vitro immune response • Inhibition of oxidative burst of neutrophilic granulocytes • Leukocytes, monocytes, and T cells chemotaxis • Thrombocyte activation inhibition

Positive Acute Phase Proteins

- An APP is called positive with overexpression by 1.25 fold or more in plasma following trigger. The most common positive APPs are CRP, D-dimer protein, mannose-binding protein (MBP), α-1 antitrypsin, α-1 antichymotrypsin, α-2-macroglobulin, fibrinogen, prothrombin, factor VIII, von Willebrand factor, plasminogen, complement factors, ferritin, serum amyloid P (SAP), SAA, ceruloplasmin (Cp), and haptoglobin (Hp) (Table 10.2). However, the positive APPs like serpins, α-2-macroglobulin reveal different physiological functions like minimization of inflammatory reactions and coagulation factor that affects coagulation through negative feedback mechanism. While others including SAA and Hp act to destroy or inhibit the growth of microorganism. Positive APPs have been categorized into three groups.
- Proteins that reveal an increase of about 1.5-fold like complement factor C3.
- Proteins revealing 2- to 3-fold increase like haptoglobin.
- Proteins rapidly increasing of up to 5–1000 fold including pentraxins like CRP, SAP, and SAA. The APPs can act differently based on the species it is expressed. Like α2-macroglobuin is not considered in humans as an APP while its homologous protein in rat, α2-macrofetoprotein, is considered an APP.
- APP may be a positive or negative based on the species.

Negative APP

Albumin is reported to be negative APP, with a decrease in blood concentrations during APR, with amino acid deviation toward synthesis of positive APP (Cray et al. 2009; Paltrinieri 2007). Other negative APP includes cortisol binding protein, retinol

Table 10.2 Positive acute phase proteins

• Cp
• Complement factors
• CRP
• D-dimer protein
• Ferritin
• Fibrinogen
• Hp
• MBP
• Plasminogen
• Prothrombin, factor VIII
• SAA
• SAP
• von Willebrand factor
• α-1 antichymotrypsin
• α-1 antitrypsin
• α-2-macroglobulin

Table 10.3 Negative acute phase proteins

• Albumin
• Cortisol binding protein
• Proteins involved in vitamin and hormone transport
• Retinol ligand
• Transferrin
• Transthyretin

ligand, transferrin, transthyretin, and proteins with roles in vitamin and hormone transport (Jain et al. 2011). This decrease in serum concentration is associated with their temporarily increased availability of free hormones bound to them (Table 10.3). The negative APP are also termed as acute booster reactants (Ingenbleek and Young 1994).

10.2 CRP

CRP is an APP synthesized predominantly by the hepatocytes in the liver, in response to different inflammatory stimuli such as infection, injury, trauma, inflammation, or tissue damage. It is a non-glycosylated protein pentameric proteins with 5 subunits that bind to polysaccharides from pneumococcus. CRP is one of the APPs and is widely expressed in all vertebrates. Research on CRP in canine (dCRP or dog CRP) finds importance as a model for studies of different human inflammatory diseases.

Detection and quantization of CRP finds importance in monitoring systemic inflammatory activity and finds importance for the practice of veterinary doctors to monitor inflammatory diseases in animals. CRP detection and quantization finds importance in evaluation of treatment efficiency. Performance of detection and quantitative analysis of CRP during and post-treatment indicative of effectively of selected treatment like use of antibiotics.

Serum CRP concentration is indicative in prediction of necrosis, infection, or both, and assess efficiency in protocol of administered treatment in humans with acute pancreatitis. CRP is considered as a major APP in dogs.

Serum CRP concentration can reveal upregulation from 10- to 100-fold in different conditions, including acute pancreatitis, immune-mediated hemolytic anemia, pyometra, infections, and neoplasia. Research in dogs have revealed that downregulation of concentration in serum CRP after appropriate therapy is indicative of clinical improvement. As the stimuli is removed, serum CRP concentration drops sharply and rapidly with a biological half-life of approximately 18–20 h (Ridker 2003).

In the past, CRP has been demonstrated in ruminants, horses, pigs, dogs, and cats (Murata et al. 2004). This review narrows down the roles and uses of CRP in above said species and other wild and avian species.

10.2.1 CRP in Canines

CRP concentrations in normal healthy dogs is reported to be low varying from below 5 mg/L, as compared to those found in dogs with systemic inflammation. Differences in the method or assay in detection of CRP reveal difference in CRP. Normal CRP interval needs to be detected with specific method or analysis. In veterinary medicine, canine serum CRP measurement finds importance as a tool for diagnosis and monitoring of different diseases including acute gastric mucosal injury, chronic valvular disease, bacterial cystitis, idiopathic polyarthritis, immune-mediated hemolytic anemia, and pyometra.

Diseases including allergic bronchitis, atlantoaxial subluxation, brain tumor, bronchitis, diabetes mellitus, epilepsy, hydrocephalus, hyperadrenocorticism, hypoadrenocorticism, hypothyroidism, intervertebral disk protrusion, leiomyosarcoma, urolithiasis, megaesophagus, mitral valve insufficiency, necrotizing meningo-encephalitis, protosystemic shunt, rhinitis, and tracheal collapse is not reported to affect CRP concentrations in dogs.

Acute Abdomen

Initial serum CRP concentrations were not reported to predict outcome in dogs suffering from with acute abdomen syndrome (AAS). Overexpression of serum CRP at presentation of dogs suffering from acute abdomen syndrome and persisting for 48–72 h has been associated with a poor prognosis in dogs. Dogs with serum CRP concentrations reaching at normal levels within 48–72 h of initial presentation could therefore be discharged as served as a marker in disease diagnosis and prognosis. Alternatively, elevated CRP serum levels in spite of appropriate therapy prevents their application as additional diagnostic investigation. This is due to the fact that a secondary/underlying disease process may be contributing to elevated CRP levels. Sequential CRP determination has been reported to provide better insight into tissue injury and together with sequential WBC counts, finds importance as a means of evaluating persistence of disease and efficacy of treatment in dogs with AAS. Persistent elevations carrypoor prognosis; reductions after therapy indicate better recovery.

Osteoarthritis and Suppurative Arthritis

Dogs suffering from with suppurative arthritis revealed expression of elevated serum CRP and serum IL-6 as compared to dogs with osteoarthritis (Christensen et al. 2014; Hillström et al. 2016). Levels of IL-6 and CRP reveal positive correlation, as IL-6 promotes CRP production (Foster et al. 2014) by liver.

Cystitis

In the study carried by Kyoung-won Seo et al. (2012), they reported sharp rise in serum CRP levels in dogs with induced bacterial cystitis infection as compared to those of the tissue damage control group, suggesting that CRP is produced rapidly in response to both tissue damage and bacteria in the urinary bladder. Bacterial infection has been reported to activate different cytokines, including IL-6 which primarily stimulates the synthesis of hepatic CRP, activating the classical complement system needed for initiating opsonization, phagocytosis, and lysis of invading microorganisms (Mold et al. 1982). For infection by Gram-negative bacteria expressing lipopolysaccharides (LPS), CRP by interacting with FcγR leads to production of anti-inflammatory cytokine IL-10 and downregulates IL-12 and finds importance in protecting tissue and limiting inflammatory responses (Mold et al. 1982).

It has been reported in studies encompassing CRP levels and WBC counts in canine polyarthritis that there is rapid decrease in serum CRP in response to treatment whereas the blood WBC counts remained unaltered (Ohno et al. 2006). Thus, while WBC counts do not indicate bacterial cystitis severity or predict responses to antibiotic therapy, serial CRP levels find importance in monitoring primary responses to antibiotic treatment and enable veterinarians to the need for further study or additional treatment.

Autoimmune Hemolytic Anemia (AIHA)

Studies conducted by Mitchell et al. (2009) revealed that the acute hemolytic crisis of canine AIHA is associated with APR, with elevated serum CRP and α1-acid glycoprotein (AAG) and reduced serum albumin. With treatment, APR resolution occurs, with receding of mean serum CRP, AAG, and albumin to within the control range within 7–9 days. However, the studies also revealed that neither APP concentrations at initial evaluation nor the ratios of CRP/AAG, CRP/albumin, AAG/albumin could predict acute mortality or disease severity, duration of hospitalization, and number of blood transfusions. Serum APP concentrations have been reported to positively correlate with the ratio of PCV/Hct and bold WBC count, indicative of the importance of these proteins as valuable, rapid markers of disease remission and therapy. Dogs with AIHA, having survived post-acute illness and achieved clinical remission, have revealed to express serum APP similar to that of control dogs in the experiment.

Pyometra

Pyometra is a common reproductive disorder of female dogs revealing pus accumulation in the uterine lumen for which the conventional mode of treatment is ovariohysterectomy pyometra in England in 1998. It finds application as method of contraception in dogs. Preventing postoperative complications is based on proper monitoring of the health condition of these dogs undergoing the treatment. Serum APP in female dogs is known to indicate the severity of inflammation and enabled the early detection and restoration of abnormal homeostasis even in case of manifestation of any clinical symptoms. Studies by Dabrowski et al. (2007) demonstrated that serum CRP and SAA finds importance in assessing the severity of inflammation in female dogs with pyometra. After ovariohysterectomy, elevations in serum CRP, SAA, and Hp are reported with slow decrease during the uneventful postoperative period. However, these changes reveal variations with time. Decline in postoperative APP was found to be associated with recovery of female dogs. Thus, serum level CRP, SAA, and Hp post-ovariohysterectomy finds importance in indicating inflammatory response to surgical insult and may find importance in monitoring of the postoperative course. It has also been reported that APP finds importance as biomarkers in routine diagnosis in cases of early postoperative complications (Dabrowski et al. 2007).

Chronic Valvular Disease (CVD)

CVD or chronic degenerative valvular disease together with myxomatous atrioventricular valvular degeneration, chronic valvular fibrosis, and endocardiosis is the most common acquired form of heart disease reported in dogs. CRP has been reported to play a role in apoptosis of human coronary vascular smooth muscle cells (Blaschke et al. 2004), and it decreases prostaglandin I_2 release in human aortic endothelial cells (Venugopal et al. 2003). Studies have revealed (Verma et al. 2002a, b) that suggestions that CRP may directly play a role in atherosclerosis lesions, which, was earlier thought not to have an inflammatory component.

Rush et al. (2006) revealed elevated CRP concentration in this population of dogs with CVD, as compared to those of healthy controls. CVD reveals complex etiopathogenesis and is not well studied. Studies have revealed a genetic predisposition for the disease that reveals modification by certain environmental or pathophysiologic mechanisms, such as inflammation. However, more studies are necessitated in these lines to determine the quantitative estimation of CRP and its importance in assessment or management of CVD in dogs.

Steroid-Responsive Meningitis-Arteritis (SRMA)

Studies have revealed that dogs with SRMA reveal elevated level of CRP in both cerebrospinal fluid (CSF) and serum as compared to either healthy dogs or those suffering from other neurological diseases. Serum and CSF concentrations of SAA are also reported to be elevated in dogs with SRMA (Lowrie et al. 2009). Serum CRP level is reported to reduce in parallel with CSF nucleated cell counts in response to prednisolone therapy and in patients with discontinued therapy following recovery.

Surgical Trauma and Post-Surgical Complications

Measured CRP levels in dogs is not reported to indicate the category of surgical trauma. Elevated CRP level prior to surgery, due to trauma, as observed in some orthopedic patients, or infections as observed in patients suffering from pyometra, may lead to in limited, additional elevation during post-surgery period. CRP therefore finds importance as a biomarker of systemic inflammation post-surgery in dogs. However, studies in this area are limited and affected by risks of biases (Kjelgaard-Hansen et al. 2013). Reports from laparoscopic surgery including laparoscopic splenectomy, colopexy, and ovariohysterectomy with open surgery studies reveal lower CRP as compared to open surgery in comparable tissues.

- The serum CRP in dogs may reveal significant deviation with infectious complications post-surgery, as compared to dogs without such complications (Christensen et al. 2015). However, more studies are necessitated to substantiate

these assumptions and to establish criteria for the differentiation of infections from normal post-surgical changes.
- The serum CRP may reflect the degree of operative trauma in different categories of surgery, but comparable studies based on comparable study designs and comparable assays for quantitative CRP analysis are necessitated to understand the role of CRP.

10.2.2 Concentration of CRP in Other Infectious Diseases of Dogs

Parvoviral enteritis

CRP in dogs with severe clinical signs of parvovirus enteritis infection with post-infection death was elevated at admission and at 12 and 24 h time interval after admission when compared with dogs that survived (McClure et al. 2013). A recent study suggested that a higher magnitude increase in CRP values is related to the magnitude of secondary bacterial infections observed in this disease due to sepsis (Kocaturk et al. 2014).

Japanese Encephalitis

Studies of Sheahan et al. (2010) in experimentally inoculated dogs with Japanese encephalitis virus (JEV) revealed that CRP was the only altered APPs parameter varying from 10 to 56 mg/L.

Aspergillosis

Dogs revealing clinical symptoms of nasal disease and positive serology for aspergillosis revealed elevated CRP ranging from 2 to 82 mg/L and Hp ranging from 2 to 19.2 g/L as compared to that of healthy animals.

Infectious Tracheobronchitis

An elevated peak of CRP expression in Beagles 24 h post-infection with *Bordetella bronchiseptica* causing canine infectious tracheobronchitis but revealed rapid decrease afterward (Yamamoto et al. 1994).

Leptospirosis

Post-intraperitoneal inoculation of Leptospira interrogans *serovar canicola* in Beagles, revealed elevated CRP in 2–4 days correlated with a brief or mild pyrexia with increased alkaline phosphatase activity. There was a significant strong correlation observed by Yogeshpriya et al. (2014) between CRP concentration and temperature of the dogs with leptospirosis.

Ehrlichiosis

In canine monocytic ehrlichiosis, occurring naturally, elevations in concentrations in CRP, SAA, and Hp were reported on admission; with no role in prediction of clinical outcome. But elevated CRP and Hp in dogs were reported from chronic myelosuppressive form, finding importance as indicators of the clinical phase of the disease. CRP and AGP revealed elevated in dogs infected with *Ehrlichia canis* during acute stage of infection.

Demodicosis

In demodicosis in dogs, CRP and Hp were reported to be biomarkers indicative of generalized inflammatory skin disease, together with the release of inflammatory cytokines from the follicular epithelium due to tissue damage which is reported to be induced by proliferation of Demodex mites in hair follicles and skin surfaces.

Canine leishmaniasis (CanL)

CanL both naturally occurring and experimentally revealed elevated positive APPs such as CRP, haptoglobin (Hp), and ferritin and reduction in expression of negative APPs such as albumin, paraoxonase 1 (PON1), or apolipoprotein 1 Apo-A1) during different disease stages. Monitoring of APP levels has been reported to play a role in identification of animals in transition from one stage to another, like subclinical infection into clinical disease, or indicative of drug response to therapy. This enables in the decision-making process in the management of CanL.

Heartworm Disease

Cardiac biomarkers including troponin I, myoglobin, and NT-proBNP is indicative of myocardial injury and heart failure, particularly in chronic infections, which may reveal improvement post-treatment by adulticide. An APR in dogs with *Dirofilaria immitis* has been reported to reveal variations of APP, indicative of ongoing inflammatory processes leading to disease progression. Elevated level of CRP has been reported to correlate with the severity of the disease. A strong correlation has been observed between pulmonary hypertension and CRP in dogs.

Chronic Inflammatory Enteropathies (CIE)

Serum CRP finds importance as surrogate marker in assessing disease progression and response to treatment in dogs with CIE characterized by chronic persistent or recurrent signs of gastrointestinal (GI) disease with mucosal inflammation. Biomarkers detecting the disease, at its site of origin, severity of the disease, prognosis, response to treatment, or a combination of all these would be clinically significant in dogs with CIE. Recently, studies have revealed that serum CRP of $\geq$9.1 mg/L concentration distinguished dogs with CIE requiring anti-inflammatory or immunosuppressive treatment from those dogs responding to an elimination diet or antibiotic trial with a sensitivity of 72% with a 100% specificity (Heilmann and Steiner 2018).

10.2.3 CRP in Felines

The APP in cats were identified from clinically normal animals, diseased animals, in experimentally induced inflammation, in postoperative studies by comparative serum measurements (Kajikawa et al. 1999). The SAA is reported to be elevated

Table 10.4 Previous studies of APP in feline diseases

Name of the acute phase proteins	Diseases
CRP	Feline immunodeficiency (FIV) virus cats treated with recombinant feline interferon Induced inflammation and surgery
Haptoglobin	Feline infectious peritonitis (FIP) Induced inflammation and surgery Abscesses, pyothorax, fat necrosis Various diseases (Feline lower urinary tract disease (FLUTD), tumors, diabetes mellitus; renal disease, injury, etc.) Infection by *Hepatozoon felis* and *Babesia vogeli* Feline leukemia virus(FLV), Hemotropic mycoplasms *Dirofilaria immitis*
α1-Acid Glycoprotein (AGP)	Feline immunodeficiency virus Infection by Chlamydophila psittaci; Pancreatitis and pancreatic tumors Feline infectious peritonitis Lymphoma and other tumors Induced inflammation and surgery FIV cats treated with recombinant feline interferon Abscesses, pyothorax, fat necrosis diseases, FLUTD, tumors, diabetes mellitus, renal disease, injury,
Serum Amyloid A (SAA	Pancreatitis, renal failure, FLUTD, tumors, diabetes mellitus; renal disease, injury, etc. Induced inflammation and surgery Sepsis FeLV; *Hemotropic mycoplasms* infections Infection by *Hepatozoon felis* and *Babesia vogeli* *Dirofilaria immitis* FIV cats treated with recombinant feline interferon

initially, with AGP and Hp, with a less significant increase of CRP (Kajikawa et al. 1999). Studies have revealed that CPR behaves similarly to SAA and AGP in feline inflammation (Leal et al. 2014).

Most feline APP studies have revealed highlights on AGP, SAA, and less on Hp—which are reported as major APPs in the cat (Table 10.4). Excepting CRP all these proteins were higher in hospitalized cats as compared to that of in healthy control cats, with 2–4-fold-elevated expression post 24 h of surgery, independent of possible post-surgical complications, or after experimentally induced inflammation (Kajikawa et al. 1999).

Cat CRP is yet to gain the status of an APP. Watanabe et al. (1992) studied the serum CRP levels in kittens as lower as compared to that of adult cats, while kittens revealed quite elevated serum CRP levels.

Of the 12 healthy cats studied with serum CRP varying from 38 to 168 pg/mL, serum CRP levels of the kittens ranged from <20 to 359 pg/mL. Eighty one percent of the sera from the kittens revealed serum CRP levels of <40 pg/mL and therefore CRP does not act as an APR in all species.

10.2.4 CRP in Ruminants

In ruminants, plasma half-life of CRP is reported to be around 19 h and reveals to be constant under health and disease. Concentration of CRP and its synthesis rate reflects the intensity of the pathological process stimulating production of CRP (Burim et al. 2011). With the withdrawal or removal of the stimulus for APR, concentrations of circulating CRP reduces dramatically fast, at almost the rate of plasma CRP clearance. CRP in the serum of healthy dairy cows ranged between 10 and 30 μg/mL (Lee et al. 2003) which may be elevated dramatically after natural or experimental infections. Studies of Lee et al. (2003) revealed a three-fold elevation in CRP level in cattle with mastitis as compared to that of healthy lactating cattle.

Intestinal Obstruction

In intestinal obstruction generating conditions, it is reported that elevations in CRP level ranging from 4.16 ± 0.69 mg/dL may be attributed to inflammatory response due to ischemic necrosis of the obstructed part of intestines. Disease condition have been reported to accelerate the synthesis and release of CRP from liver to the bloodstream (Hussaina et al. 2015).

Viral Diseases

In ruminants, Hp and SAA are considered to be the major APPs. However, they have been mostly studied in case of bacterial infections or multisystemic disorders. In viral infections including bovine viral diarrhea virus, Peste des petits ruminants virus (PPRV), bluetongue virus (BTV), bovine respiratory syncytial virus (BRSV), border disease virus (BDV), and foot-and-mouth disease virus (FMDV), an elevation in SAA expression were reported (Reczyńska et al. 2018). SAA has been known to play a role in lipid binding, and lipids have been reported to play a crucial role in antiviral defense mechanisms.

Lactation Status

The circulating CRP was revealed to show an increase for a few days after parturition in dairy cows together with responses to tissue injury in the uterine epithelium (Morimatsu et al. 1991). Gradual increase in serum CRP has been observed to increase gradually with an increase in milk production during peak lactation in dairy cows. Thus, monitoring of lactation-induced pattern of CRP expression in dairy cows finds importance in monitoring milk production and progression of mastitis (Morimatsu et al. 1991). Recent studies have highlighted a strong correlation between rumen endotoxin and plasma CRP with a suggested potential role for CRP in milk fat depression syndrome in dairy cows.

The serum CRP level has been reported to reveal positive correlation with milk production with the elevated level of milk production, correlating with the elevated serum CRP. Diseases, especially acute infections, induced much elevated levels of CRP expression than stress or lactation.

Pregnancy Status

Studies have reported that serum CRP increased during the first 4 months of gestation to as high as 34.5 μg/mL but gradually reduced to low levels of 4.9 μg/mL during late gestation (Turk et al. 2008). Despite our limited knowledge on the function of CRP during gestation in cattle, research studies in humans indicate that elevated CRP during pregnancy have been associated with complications during pregnancy (Teran et al. 2001). Increased serum CRP at first trimester could be indicative of risk to gestational preeclampsia (Wolf et al. 2001). Studies by Sacks et al. (2004) revealed as early elevation of circulatory CRP at 4 weeks of gestation in non-obese young pregnant women indicating that a maternal low-grade inflammatory response could be associated in earliest phases of ovule implantation.

Endometritis

Uterine infections in ruminants could lead to huge economic loss by causing decreased reproductive performance and milk production, extending the time between parturition-first insemination and parturition-conception. APP are known to play an important role in combating infection, and CRP being one such major APP in ruminants. In a study conducted by Kaya et al. (2016) showed that serum-elevated CRP level were reported from ruminant groups with endometritis. While CRP expression in mild endometritis ranged from 82.86 ± 4.28 μg/mL, endometritis ranged from 98.86 ± 8.2 μg/mL and severe endometritis ranged from 122.34 ± 12.72 μg/mL as compared to the control group that expressed CRP ranging from 48.88 ± 3.92 μg/mL. CRP has been reported to increase significantly in uterus infections in cows as compared to other inflammatory conditions.

Colostral Fed Calves and CRP in Blood

CRP is known as a mediator of innate immunity in ruminants. Proteins are known to be passively transferred from colostrum to blood in calves (Talukder et al. 2002). A study conducted by Schroedl et al. (2003) observed that the CRP could be detected in all blood plasma of calves before the first colostrum intake, and revealed elevated expression in blood at day 1 post-colostrum application, indicating that such increased CRP concentration at 1 day postpartum is contributed by the passive transfer of CRP from colostrum to blood.

Mastitis

CRP concentration revealed significant variation with the causative pathogen of mastitis indicating variable levels of stimulation by the virulence factors of different pathogens. Studies by Thomas et al. (2018) and Kruger and Neumann (1999) revealed that different pathogens could stimulate variable levels of milk CRP in milk from mastitis with agents as E. coli, S. aureus, S. dysgalactiae, and CNS organisms.

APPs therefore have found importance in prediction and study of animal health and welfare and antemortem finds importance in meat inspection. Dairy cows with infectious and metabolic diseases at slaughter revealed around six-fold elevation in

Hp levels in serum as compared to animals with minor lesions. Culled dairy cattle with acute lesions revealed an elevation of 40- and 7-fold in Hp and SAA, respectively, as compared to the healthy beef-type animals (Eckersall and Bell 2010). Thus, APP finds importance as potential biomarkers and diagnostic tools in the study of inflammatory conditions of domestic animals and aids in understanding the pathophysiology of the APR in clinical settings.

10.2.5 CRP in Equines

In horses, expression of SAA, α1-glycoprotein, and CRP have been reported to be with intermediate response times. Serum CRP is reported to be a minor APP in the horse. Elevated expression of CRP has been reported in horses suffering from pneumonia, enteritis, arthritis, experimentally induced laminitis, and in aseptic inflammation induced by intramuscular turpentine injection. CRP reported from normal animals ranges from $\leq$0.1 mg/L while plasma concentrations for inflammatory conditions varied in the range of 10–35 mg/L. Serum CRP is measured by assay similar to detection in cerebrospinal fluid (CSF). However, currently there is dearth of established reference ranges for serum or CSF fluid CRP values in horses.

Age and Physiology

In 5-year-old horse, serum CRP revealed around 7–8 μg/mL. In mares during the perinatal period, serum CRP revealed to be downregulated at 2 months prior to delivery, with moderate changes just before and after delivery, and then elevated to relatively high level by 2 months post delivery (Yamashita et al. 1991). Serum CRP revealed elevation at 24 h after experimental inflammatory stimuli in adult horses, peaking with 3–6 elevation as compared to that of baseline values within 3–5 days. Serum CRP returned to baseline values by 2–3 weeks post treatments. However, CRP is yet to gain the status of APR proteins in horses.

Equine Protozoal Myeloencephalitis (EPM)

Neither serum CRP nor SAA was reported to be elevated significantly in disorders including equine protozoal myeloencephalitis (EPM) or cervical vertebral stenotic myelopathy (CVSM). Thus in horses CRP, as of now does not gain status of diagnostic marker in these diseases (Mittelman et al. 2018). However, it is also not known whether horses do not develop clinical signs of neurologic disease until waning of systemic inflammatory response. This requires further studies and holds scope for future research.

Neosporiasis

Neospora spp. infection in horses has been reported to lead to reproductive disorders including infertility, abortion, and neonatal mortality. It is known that to combat infection by *N. caninum* a cell-mediated immune (CMI) response is mounted by the host (Da Silva et al. 2014). Seropositive horses reveal elevated levels of CRP, an important APP and immune component. CRP in horses is known to play a role in opsonization of the pathogen by activating the classical complement pathway and modulation of monocytes and macrophages activities (Petersen et al. 2004). Therefore, this protein is observed to probably play an important role in controlling infection by *Neospora* in horses and in inhibiting invasion of cardiac cells by *Trypanosoma cruzi (T. cruzi).*

Other Causes

Elevated levels of serum CRP horses have been reported from those suffering from aseptic inflammation induced by intramuscular turpentine injections (Takiguchi et al. 1990), arthritis and after castration, enteritis, and pneumonia (Nunokawa et al. 1993). Elevated plasma level expression of CRP has been observed in carbohydrate-induced laminitis (Fagliari et al. 1998). Expression of SAA has been correlated with expression of CRP in horses with clinical signs of inflammation (Nunokawa et al. 1993), fever, and in those horses revealing clinical signs of respiratory disease.

10.2.6 CRP in Swine

In pigs, a higher CRP serum concentration was observed without any clinical signs of acute inflammation. Other important APP reported in the pigs being CRP and Hp revealed an increase of 8–10 and 2–10 times, respectively, in the turpentine inflammation model. While SAA has been indicated as a major APP, Ceruloplasmin has been observed to be a minor APP in pigs and albumin, fetuin, and transferrin are recorded as negative APPs with decreased expression by 20–40% in pigs. However, clinical studies on swine CRP is still at its infancy. In swine, subcutaneous injections of turpentine have been reported to lead to non-septic inflammation with an increase in 6- to 8-fold CRP concentrations within 48 h (Lampreave et al. 1994). Studies have revealed that swine inoculated with *Actinobacillus pleuropneumoniae* (*A. pleuropneumoniae*) exhibited increased body temperature, acute respiratory inflammation, and a rapid and significant overexpression of CRP levels (Heegaard et al. 1998). Plasma concentrations of CRP has been reported to bear positive correlation with the clinical score and the effect of antibiotic treatment in experimental *A. pleuropneumoniae* infection in pigs.

Swine Influenza

Swine influenza (SI) is a contagious, respiratory disease that affects the health of the pigs and can be a huge economic loss to the countries. The CRP, Hp, and SAA serum level concentrations were reported to be overall higher in pigs with clinical course of influenza. The significant overexpression of these APPs were observed for prolonged time period, as compared to subclinically infected pigs. The rise in rectal temperature revealed correlation with alterations in CRP, Hp, and SAA concentrations. Maximum serum SAA concentration revealed correlation with changes in the lungs regarding both clinical and subclinical course of influenza.

The rise in rectal temperature coincided with changes in CRP, Hp, and SAA. All these parameters reflected the onset of systemic acute phase response to infection. Post-influenza infection, without any observation of relevant clinical signs APP response was revealed to be weaker than that observed during clinical influenza. The increase in concentration of CRP and Hp during subclinical infection was reported for a shorter period of time as revealed by studies conducted by Pomorska-Mol et al. (2014).

10.2.7 CRP in Fish

The mammalian APP bear homology to the APP proteins identified from fish with predicted similarity in functions (Christiansen et al. 2015). However, the APP synthesis and the mechanism involved in acute phase plasma cation changes in fish are yet to be studied. With the growing number of reports that some major acute phase cytokines and their receptors are highly conserved throughout evolution (Jang et al. 1995); it is predicted that the basic signaling mechanisms of APR in fish bear similarity to those reported in mammals.

The protein revealed in mammals as APP was the same APP discovered in fish in 1973. CRP belongs to the ancient multifunctional pentraxin family of protein and is pentameric in nature. The microbial cells and macromolecules on the disrupted surfaces of eukaryotic cells like phosphorylcholine (PC) and pneumococcal C-Polysaccharide acted as ligand for fish CRP. CRP in fish reveals similar functions as mammalian CRP in activating the complement system, mediating bacterial fungi and parasitic optimization, binding with free chromatin, agglutinating particles, and the release of cytokine. CRP concentration in the plasma of healthy fish CRP is reported very trace amounts ranging from <1 mg mL^{-1}. However, post-acute infection, injury or inflammatory stimuli the CRP production is overexpressed in plasma concentration, thus confirming to its role as an APP in fish. With the removal of the stimuli for triggering APR, the plasma concentration of CRP reduces drastically returning to normal range.

Research of Roy et al. (2017) revealed that proteins similar to CRP has been isolated and are characterized in invertebrates such as Limulus polyphemus commonly called as horseshoe crab, land snail Achatina fulica, and marine bivalve mollusc. CRP-like molecules have been reported from several fish species, i.e., eggs of Cyclopterus lumpus commonly called as lumpsucker, sera of O. mossambicus commonly called as Tilapia, I. punctatus or channel catfish, O. *m*ykiss commonly called as rainbow trout, S. salar or Atlantic salmon, P. platessa or plaice, A. japonica commonly called as Japanese eel and C. catla common carp Catla, elasmobranch, and Mustelus canis (dogfish).

Further in cods, a CRP-like protein was identified for the first time by Lund and Olafsen in the year 1998. Studies by Gisladottir et al. (2009) revealed two CRP-like proteins, CRP-PI and CRP-PII in cods. CRP-PI was found to be similar to the CRP-like protein described by Lund and Olafsen with the amino acid (aa) sequence bearing more homology to SAP in species including Anarhichas lupus, Hippoglossus *hippoglossus*, *Salmo salar*, and rainbow trout than to CRP (Gisladottir et al. 2009). It was also reported to be phosphorylcholine (PC) specific, which is a typical characteristic of the CRP type pentraxins. The CRP-PII protein, Gisladottir et al. (2009) was a new identity, with amino acid sequence homology with CRP from other species and a striking unusual degree of heterogeneity in subunit size and density between individual cod.

During performing diagnostic procedures on selected cases, the quantitative analysis of CRP finds importance in distinguishing infections caused by virus and bacteria or different bacterial infections. Elevated CRP levels (>100 mg/L) in serum have been reported to indicate bacterial infection. Measurements of CRP find importance in its accuracy when done serially. The overexpression of CRP finds importance in monitoring the effect of treatment, and early diagnosis of infections and inflammatory reactions. Characterization of many definitive and defensive roles of acute phase proteins defensive role in fish are yet to be done. Thus, several APPs have importance in diagnosis of disease, correlating with clinical symptoms of disease.

10.2.8 CRP in Captive Asian Elephants

Isaza et al. (2014) reported the low expression of CRP in Asian elephant serum which is typical of several other species. However, expression of CRP differs with the species. As in human beings, CRP is a major APP and biomarker for cardiovascular disease (CVD). In dogs, CRP is a major APP with normal reference intervals of 0–20 mg/L, revealed to be higher than that of expressed in Asian elephants (Isaza et al. 2014) where the reference interval for CRP is reported to be (1.3–12.8 mg/L) that reveals to be similar to CRP intervals reported in harbor seals (Phoca vitulina, P. vitulina) and bottlenose dolphins (Tursiops truncat*us*, T. truncat*us*). However, there is every scope of further studies to address the reactivity of the CRP reagents with proteins in elephant and to consider the application of either elephant-specific reagents or non-antibody-based assays.

10.2.9 CRP in Avian Biology

Several subclinical conditions of diseases in birds are difficult to be detected by postmortem analysis. These conditions can however be diagnosed by the overexpression of APPs. Plasma concentration of APPs of domestic fowl finds importance in monitoring various aspects of health of the bird. The APR in chickens has been reported to be associated with depression in growth with reduced productivity. Secondly, there is a possibility of diagnosing deep-seated inflammatory processes on inspection of poultry meat. Antemortem inspection protocols of chickens do not include taking rectal temperatures with very little individual bird assessment or clinical examinations unlike the protocol in mammals. Thirdly, there is a possibility to detect early inflammatory changes occurring before antibodies can be detected. Finally, quantization of plasma APPs could be used to complement the routine necropsy results which can be used to determine stages and severity of lesions, thus finds importance in planning of treatment protocols or for the monitoring the disease prognosis.

Parasitic Diseases

High levels overexpression of CRP were reported post-natural infections by Coccidia and Histomonas.

Bacterial Diseases

Patterson and Mora (1964) purified CRP from serum of chicken and developed an assay to detect their presence in CRP in chicken serum. Their studies revealed that CRP is overexpressed in birds with infections by pathogens including E. coli, Staphylococcus aureus (S. aureus*) and* Pasteurella multocida (P. multocida*),* Histomoniasis and adjuvant. CRP-positive birds were also reported in a clinically normal population which revealed lesions consistent with chronic respiratory disease, in postmortem studies, emphasizing the importance of CRP as a potential biomarker for identification of a preclinical forms of a disease.

Viral Diseases

Recent reports revealed an increase in the CRP-positive birds due to experimental infections with IBV, *S. typhimurium* LPS and Fumonisin toxicity during heat stress. Basal CRP concentration of 1.3 ± 0.014 mg/L has been reported to increase significantly to 1.65 ± 0.011 mg/L 2 days post-inoculation with IBV.

References

Blaschke F, Bruemmer D, Yin F et al (2004) C-reactive protein induces apoptosis in human coronary vascular smooth muscle cells. Circulation 110:579–587

Burim N. Ametaj, Afshin Hosseini, John F. Odhiambo, Summera Iqbal, Sumeet Sharma, Qilan Deng, Tran H. Lam, Umar Farooq, Qendrim Zebeli, Suzanna M. Dunn (2011) Application of acute phase proteins for monitoring inflammatory states in cattle, acute phase proteins as early non-specific biomarkers of human and veterinary diseases. In: Francisco Veas (Ed.), INTECH Open Access Publisher. ISBN: 978–953–307-873-1

Christensen MB, Langhorn R, Goddard A, Andreasen EB, Moldal E, Tvarijonaviciute A et al (2014) Comparison of serum amyloid A and C-reactive protein as diagnostic markers of systemic inflammation in dogs. Can Vet J 55(2):161–168

Christensen MB, Eriksen T, Kjelgaard-Hansen M (2015) C-reactive protein: quantitative marker of surgical trauma and post-surgical complications in dogs: a systematic review. Acta Vet Scand 57:71

Christiansen EF, Cray C, Lewbart GA, Harms CA (2015) Plasma protein electrophoresis and acute phase proteins in Koi Carp (Cyprinus carpio) following exploratory coeliotomy. J Exotic Pet Med 24(1):76–83

Cray C, Zaias J, Altman NH (2009) Acute phase response in animals: a review. Comp Med 59 (6):517–526

Da Silva AS, Radavelli WM, Moura AB, Duarte T, Duarte MM, Stefani LM (2014) Horses seropositive for Neospora spp.: immunoglobulins, cytokines, and C-reactive protein levels. J Equine Vet Sci 34:1240–1243

Dabrowski R, Wawron W, Kostro K (2007) Changes in CRP, SAA and haptoglobin produced in response to ovariohysterectomy in healthy bitches and those with pyometra. Theriogenology 67:321–327

Eckersall PD, Bell R (2010) Acute phase proteins: biomarkers of infection and inflammation in veterinary medicine. Vet J 185:23–27

England GCW (1998) Rozro'd i połoz˙ nictwo pso'w według Allena (Reproduction and gynecol-ogy in dogs acording to Allen). Warszawa:221–225

Fagliari JJ, McClenahan D, Evanson OA, Weiss DJ (1998) Changes in plasma protein concentrations in ponies with experimentally induced alimentary laminitis. Am J Vet Res 59:1234–1237

Foster JD, Sample S, Kohler R, Watson K, Muir P, Trepanier LA (2014) Serum biomarkers of clinical and cytologic response in dogs with idiopathic immune-mediated polyarthropathy. J Vet Intern Med 28(3):905–911

Gisladottir B, Gudmundsdottir S, Brown L, Jonsson ZO, Magnadottir B (2009) Isolation of two C-reactive protein homologues from cod (Gadus morhua L.) serum. Fish Shellfish Immunol 26 (2):210–219

Heegaard PM, Klausen J, Nielsen JP et al (1998) The porcine acute phase response to infection with Actinobacillus pleuropneumoniae. Haptoglobin, C-reactive protein, major acute phase protein and serum amyloid A protein are sensitive indicators of infection. Comp Biochem Physiol B Biochem Mol Biol 119:365–373

Heilmann RM, Steiner JM (2018) Clinical utility of currently available biomarkers in inflammatory enteropathies of dogs. J Vet Intern Med 32:1495–1508

Hillström A, Bylin J, Hagman R, Björhall K, Tvedten H, Königsson K, Fall T, Kjelgaard-Hansen M (2016) Measurement of serum C-reactive protein concentration for discriminating between suppurative arthritis and osteoarthritis in dogs. BMC Vet Res 12:240

Hussaina SA, Uppala SK, Randhawaa CS, Sood NK (2015) Bovine intestinal obstruction: blood gas analysis, serum C-reactive protein and clinical, haematological and biochemical alterations. J Appl Anim Res 43(2):224–230. https://doi.org/10.1080/09712119.2014.963093

Ingenbleek M, Young V (1994) Transthyretin (prealbumin) in health and disease: nutritional implications. Ann Rev Nutr 14:495–533

Isaza R, Wiedner E, Hiser S, Cray C (2014) Reference intervals for acute phase protein and serum protein electrophoresis values in captive Asian elephants (Elephas maximus). J Vet Diagn Investig 26(5):616–621. https://doi.org/10.1177/1040638714543923. Epub 2014 Jul 23. 25057161

Jang SI, Hardie LJ, Secombes CJ (1995) Elevation of rainbow trout Oncorhynchus mykiss macrophage respiratory burst activity with macrophage-derived supernatants. J Leukoc Biol 57:943–947

Jain S, Gautam V, Naseem S (2011) Acute-phase proteins: as diagnostic tool. J Pharm Bioallied Sci 3(1):118–127

Kajikawa T, Furuta A, Onishi T (1999) Changes in concentrations of serum amyloid A protein, alpha-1-acid glycoprotein, haptoglobin, and C-reactive protein in feline sera due to induced inflammation and surgery. Vet Immunol Immunopathol 68:91–98

Kaya S, Merhan O, Kacar C, Colak A, Bozukluhan K (2016) Determination of ceruloplasmin, some other acute phase proteins, and biochemical parameters in cows with endometritis. Vet World 9 (10):1056–1062. https://doi.org/10.14202/vetworld.2016.1056-1062

Kjelgaard-Hansen M, Strom H, Mikkelsen LF, Eriksen T, Jensen AL, Luntang-Jensen M (2013) Canine serum C-reactive protein as a quantitative marker of the inflammatory stimulus of aseptic elective soft tissue surgery. Vet Clin Path 42:342–345

Kocaturk M, Tvarijonaviciute A, Martínez-Subiela S, Tecles F, Eralp O, Yilmaz Z, Cerón JJ (2014) Inflammatory and oxidative biomarkers of disease severity in dogs with parvoviral enteritis. J Small Anim Pract 56:1–6

Kruger M, Neumann A (1999) Investigations to dynamics of C-reactive protein, somatic cell counts, lactose content, electrical conductivity in initial quarter milk samples of subclinically diseased quarters of the udders in relationship to bacteriological results. Tierarztl Prax Ausgabe Grobtiere Nutztiere 27(3):164–167

Lampreave F, González-Ramón N, Martínez-Ayensa S et al (1994) Characterization of the acute phase serum protein response in pigs. Electrophoresis 15:672–676

Leal RO, Gil S, Sepúlveda N, McGahie D, Duarte A, Niza MMRE, Tavares L (2014) Monitoring acute phase proteins in retrovirus infected cats undergoing feline interferon-ω therapy. J Small Anim Pract 55(1):39–45

Lee WC, Hsiao HC, Wu YL, Lin JH, Lee YP, Fung HP, Chen HH, Chen YH, Chu RM (2003) Serum C-reactive protein in dairy herds. Can J Vet Res 67(2):102–107. ISSN 0830-9000

Lowrie M, Penderis J, Mclaughlin M, Eckersall PD, Anderson TJ (2009) Steroidresponsive meningitis-ateritis: a prospective study of potential disease markers,prednisolone treatment and long-term outcome (2006–2008). J Vet Intern Med 23:862–870

Lund V, Olafsen JA (1998) A comparative study of pentraxin-like proteins in different fish species. Dev Comp Immunol 22:185–194

McClure V, van Schoor M, Thompson PN, Kjelgaard-Hansen M, Goddard A (2013) Evaluation of the use of serum C-reactive protein concentration to predict outcome in puppies infected with canine parvovirus. JAVMA 243:361–336

Mitchell KD, Kruth SA, Wood RD, Jefferson B (2009) Serum acute phase protein concentrations in dogs with autoimmune hemolytic anemia. J Vet Intern Med 23:585–591

Mittelman NS, Stefanovski D, Johnson AL (2018) Utility of C-reactive protein and serum amyloid A in the diagnosis of equine protozoal myeloencephalitis. J Vet Intern Med:1–5

Mold C, Du Clos TW, Nakayama S, Edwards KM, Gewurz H (1982) C-reactive protein reactivity with complement and effects on phagocytosis. Ann N Y Acad Sci 389:251–262

Morimatsu M, Watanabe A, Yoshimatsu K, Fujinaga T, Okubo M, Naiki M (1991) Elevation of bovine serum C-reactive protein and serum amyloid P component levels by lactation. J Dairy Res 58(3):257–261. ISSN 0022-0299

Murata H, Shimada N, Yoshioka M (2004) Current research on acute phase proteins in veterinary diagnosis: an overview. Vet J 168(2004):28–40

Nunokawa Y, Fujinaga T, Taira T, Okumura M, Yamashita K, Tsunoda N, Hagio M (1993) Evaluation of serum amyloid A protein as an acute-phase reactive protein in horses. J Vet Med Sci 55:1011–1016

Ohno K, Yokoyama Y, Nakashima K, Setoguchi A, Fujino Y, Tsujimoto H (2006) C-reactive protein concentration in canine idiopathic polyarthritis. J Vet Med Sci 68:1275–1279

Paltrinieri S (2007) Early biomarkers of inflammation in dogs and cats: the acute phase proteins. Vet Res Commun 31(Suppl. 1):125–129

Patterson LT, Mora EC (1964) Occurance of a substance analogous to C-reactive protein in the blood of the domestic fowl. Tex Rep Biol Med 22:716–721

Petersen HH, Nielsen JP, Heegaard PMH (2004) Application of acute phase protein measurements in veterinary clinical chemistry. Vet Res 35:163–187

Pomorska-Mol M, Krzysztof K, Pejsak Z, Markowska-Daniel I (2014) Analysis of the acute-phase protein response in pigs to clinical and subclinical infection with H3N2 swine influenza virus. Influenza Other Respiratory Viruses 8(2):228–234

Reczyńska, D., Zalewska, M., Czopowicz, M., Kaba, J., Zwierzchowski, L., & Bagnicka, E. (2018). Acute phase protein levels as an auxiliary tool in diagnosing viral diseases in ruminants—a review. Viruses

Ridker PM (2003) Clinical application of C-reactive protein for cardiovascular disease detection and prevention. Circulation 107:363–369

Roy S, Kumar V, Behera BK (2017) Acute phase proteins and their potential role as an indicator for fish health and in diagnosis of fish diseases. Protein Peptide Lett 24:1–13

Rush JE, Lee ND, Freeman LM, Brewer B (2006) C-Reactive Protein in concentration in dogs with chronic valvular disease. J Vet Intern Med 20:635–639

Sacks GP, Seyani L, Lavery S, Trew G (2004) Maternal C-reactive protein levels are raised at 4 weeks gestation. Hum Reprod 19(4):1025–1030. ISSN 0268-1161

Schroedl W, Jaekel L, Krueger M (2003) C-Reactive protein and antibacterial activity in blood plasma of colostrum-fed calves and the effect of lactulose. J. Dairy Sci 86:3313–3320

Seo K, Lee J, Ahn J, Lee H, Hwang C, Youn H, Lee C (2012) C-reactive protein as an indicator of inflammatory responses to experimentally induced cystitis in dogs. J Vet Sci 13(2):179–185

Sheahan D, Bell R, Mellanby RJ, Gow AG, Friend E, Heller J, Bence LM, Eckersall PD (2010) Acute phase protein concentrations in dogs with nasal disease. Vet Rec 167:895–899

Takiguchi M, Fujinaga T, Naiki M, Mizuno S, Otomo K (1990) Isolation, characterization and quantitative analysis of C-reactive protein from horses. Am J Vet Res 51:1215–1220

Talukder MJ, Takeuchi T, Harada E (2002) Transport of colostral macromolecules into the cerebrospinal fluid via plasma in newborn calves. J Dairy Sci 85:514–524

Teran E, Escudero C, Moya W, Flores M, Vallance P, Lopez-Jaramillo P (2001) Elevated C-reactive protein and proinflammatory cytokines in andean women with pre-eclampsia. Int J Gynaecol Obstet 75(3):243–249. ISSN 0020-7292

Thomas FC, Geraghty T, Simões PBA, Mshelbwalac FM, Hainingb H, Eckersall PD (2018) A pilot study of acute phase proteins as indicators of bovine mastitis caused by different pathogens. Res Vet Sci 119:176–181

Turk R, Juretić D, Geres D, Bacić G, Milesević M, Flegar-Mestrić Z, Turk N, Svetina A (2008) Bovine Platelet-activating Factor Acetylhydrolase (PAF-AH) activity related to fertility. Anim Reprod Sci 105(3–4):344–353. ISSN 0378-4320

Venugopal SK, Devaraj S, Jialal I (2003) C-Reactive protein decreases prostacyclin release from human aortic endothelial cells. Circulation 108:1676–1678

Verma S, Li SH, Badiwala MV et al (2002a) Endothelin antagonism and interleukin-6 inhibition attenuate the proatherogenic effects of C-reactive protein. Circulation 105:1890–1896

Verma S, Wang CH, Li SH et al (2002b) A self-fulfilling prophecy: C-reactive protein attenuates nitric oxide production and inhibits angiogenesis. Circulation 106:913–919

Watanabe A, Morimatsu M, Yoshimatsu K, Yamamoto S, Terao A, Tsukazaki K, Saito M, Naiki M (1992) Isolation of C-reactive protein from cat serum. J Small Anim Pract 33:71–77. https://doi.org/10.1111/j.1748-5827.1992.tb01078.x

Wolf M, Kettyle E, Sandler L, Ecker JL, Roberts J, Thadhani R (2001) Obesity and preeclampsia: the potential role of inflammation. Obstet Gynecol 98(5):757–762. ISSN 0029-7844

Yamamoto S, Shida T, Honda M, Ashida Y, Rikihisa Y, Odakura M, Hayashi S, Nomura M, Isayama Y (1994) Serum C-reactive protein and imune responses in dogs inoculated with Bordetella bronchiseptica (phase I cells). Vet Res Commun 18:347–357

Yamashita K, Fujinaga T, Okumura M, Takiguchi M, Tsunoda N, Mizuno S (1991) Serum C-reactive protein (CRP) in horses: the effect of aging, sex, delivery and inflammations on its concentration. J Vet Med Sci 53(6):1019–1024

Yogeshpriya S, Pillai UN, Ajithkumar S, Madhavan Unny N (2014) Haemato-biochemical changes and serum C-reactive protein. concentration in dogs with acute renal failure. Shanlax Int J Vet Sci 4(1):29–31

The manufacturer's authorised representative in the EU is Springer Nature Customer Service Centre GmbH, Europaplatz 3, 69115 Heidelberg, Germany. If you have any concerns regarding our products, please contact ProductSafety@springernature.com

Printed and bound by CPI Group (UK) Ltd, Croydon, CR0 4YY
30/09/2026
02225328-0004